D1637531

Volume 73

Advances in Genetics

Advances in Genetics, Volume 73

Serial Editors

Theodore Friedmann
University of California at San Diego, School of Medicine, USA
Jay C. Dunlap
Dartmouth Medical School, Hanover, NH, USA
Stephen F. Goodwin
University of Oxford, Oxford, UK

Volume 73

Advances in Genetics

Edited by

Theodore Friedmann

Department of Pediatrics
University of California at San Diego
School of Medicine, CA, USA

Jay C. Dunlap

Department of Genetics
Dartmouth Medical School
Hanover, NH, USA

Stephen F. Goodwin

Department of Physiology, Anatomy and Genetics
University of Oxford
Oxford, United Kingdom

AMSTERDAM • BOSTON • HEIDELBERG • LONDON
NEW YORK • OXFORD • PARIS • SAN DIEGO
SAN FRANCISCO • SINGAPORE • SYDNEY • TOKYO
Academic Press is an imprint of Elsevier

Academic Press is an imprint of Elsevier

525 B Street, Suite 1900, San Diego, CA 92101-4495, USA
30 Corporate Drive, Suite 400, Burlington, MA 01803, USA
32 Jamestown Road, London, NW1 7BY, UK
Radarweg 29, POBox 211, 1000 AE Amsterdam, The Netherlands

First edition 2011

ISBN: 978-0-12-380860-8
ISSN: 0065-2660

For information on all Academic Press publications
visit our website at elsevierdirect.com

Printed and bound in USA

11   12   13   10 9 8 7 6 5 4 3 2 1

Contents

6   **Restless Genomes: Humans as a Model Organism for Understanding Host-Retrotransposable Element Dynamics   219**
Dale J. Hedges and Victoria P. Belancio

Contributors

Numbers in parentheses indicate the pages on which the authors' contributions begin.

Victoria P. Belancio (219) Department of Structural and Cellular Biology, Tulane School of Medicine, Tulane Cancer Center, Tulane Center for Aging, New Orleans, Louisiana, USA

Michael Coleman (185) The Babraham Institute, Babraham, Cambridge, United Kingdom

François-Loic Cosset (121) Université de Lyon, UCB-Lyon1, IFR128, Lyon, France, and INSERM, U758, Lyon, France; and Ecole Normale Supérieure de Lyon, Lyon, France

Dale J. Hedges (219) Hussman Institute for Human Genomics, Dr. John T. Macdonald Foundation Department of Human Genetics, Miller School of Medicine, University of Miami, Miami, Florida, USA

Marion Hogg (87) MRC Human Genetics Unit, Institute of Genetics and Molecular Medicine, Western General Hospital, Edinburgh, United Kingdom

Liam P. Keegan (87) MRC Human Genetics Unit, Institute of Genetics and Molecular Medicine, Western General Hospital, Edinburgh, United Kingdom

Dimitri Lavillette (121) Université de Lyon, UCB-Lyon1, IFR128, Lyon, France, and INSERM, U758, Lyon, France; and Ecole Normale Supérieure de Lyon, Lyon, France

Christopher J. McInerny (51) Division of Molecular and Cellular Biology, Faculty of Biomedical and Life Sciences, Davidson Building, University of Glasgow, Glasgow, United Kingdom

Mary A. O'Connell (87) MRC Human Genetics Unit, Institute of Genetics and Molecular Medicine, Western General Hospital, Edinburgh, United Kingdom

Simona Paro (87) MRC Human Genetics Unit, Institute of Genetics and Molecular Medicine, Western General Hospital, Edinburgh, United Kingdom

Alexander J. Whitworth (1) MRC Centre for Developmental and Biomedical Genetics, Department of Biomedical Sciences, University of Sheffield, Sheffield, United Kingdom

1

Drosophila Models of Parkinson's Disease

Alexander J. Whitworth

MRC Centre for Developmental and Biomedical Genetics, Department of Biomedical Sciences, University of Sheffield, Sheffield, United Kingdom

0065-2660/11 $35.00
DOI: 10.1016/B978-0-12-380860-8.00001-X

ABSTRACT

Parkinson's disease (PD) is the second most prevalent neurodegenerative disorder principally affecting the dopaminergic neurons of the substantia nigra. The pathogenic mechanisms are unknown and there are currently no cure or disease-modifying therapies. Recent genetic linkage studies have begun to identify single-gene mutations responsible for rare heritable forms of PD and define genetic risk factors contributing to disease prevalence in sporadic cases. These findings provide an opportunity to gain insight into the molecular mechanisms of this disorder through the creation and analysis of appropriate genetic models. One model system that has proven surprisingly tractable for these studies is the fruit fly, *Drosophila melanogaster*. Analysis of a number of Drosophila models of PD has revealed some profound and sometimes surprising insights into PD pathogenesis. Moreover, these models can be used to investigate potential therapeutic strategies that may be effective *in vivo*, and tests have highlighted the efficacy of a number of neuroprotective compounds. Here, I review the methodologies employed in developing the various Drosophila models, and the recent advances that these models in particular have contributed to our understanding of the mechanisms that underlie PD pathogenesis and possible treatment strategies. © 2011, Elsevier Inc.

I. INTRODUCTION

Neurodegenerative diseases such as Parkinson's, Alzheimer's, and motor neuron diseases will be a considerable medical challenge in the twenty first century, due to their extremely debilitating effects and because of a profound lack of effective therapies that replace dead or dying neurons. Another plausible therapeutic approach is to develop disease-modifying treatments that halt or reverse neuronal deterioration before they are lost. For this method to be realistic, another major challenge must be overcome, that of identifying early stage or presymptomatic patients who still retain some intact neuronal architecture. While considerable advances are being made in the identification of prodrome biomarkers, effective therapies to halt neurodegeneration remain to be developed. This is itself held back by our lack of understanding into the causes of neurodegenerative diseases. Hence, a more complete insight into the pathogenic mechanisms represents the first link in the chain of developments toward effective therapeutics.

　　　Parkinson's disease (PD) is the most common neurodegenerative movement disorder, affecting approximately 1% of the retirement-age population. There are currently no cures and no effective disease-modifying therapies. It is not only typically recognized by the classic motor system disturbances such as resting tremor, difficulty initiating movements, and postural instability but it is also becoming recognized that nonmotor symptoms, including autonomic and cognitive disturbances, are an integral part of the disease. Although there are signs of pathology in many brain regions, the motor symptoms are largely attributable to the selective degeneration of dopaminergic (DA) neurons in the substantia nigra pars compacta (Forno, 1996). The mainstay of therapeutic intervention, DA replacement treatment by levodopa (L-DOPA) can, for a time, substitute for the lost striatal DA innervation and effectively ameliorate the motor symptoms. However, the nonmotor symptoms are likely to have a different pathological basis and are largely refractory to DA drugs (Poewe, 2009). "Pure" PD is also characterized pathologically by the presence of intraneuronal inclusions called Lewy bodies and Lewy neurites.

The majority of cases occur sporadically; hence, it is generally considered that a combination of both genetic susceptibility and environmental factors plays a role in the pathogenic cause. Identification of those causative factors is, of course, paramount in the pursuit of a disease-modifying therapy. While the precise pathologic mechanisms remain unclear, epidemiological and genetic studies have begun to yield insights into putative pathogenic causes. Currently, the commonly favored mechanisms include aberrant protein degradation, mitochondrial dysfunction, calcium imbalance, and inflammation (Gupta *et al.*, 2008). In addition, oxidative stress is a prominent and common feature in all forms of PD and likely represents a convergent toxic event leading to neuronal cell death. It is still unclear whether experimentally induced defects, observed in either genetic or toxin models, represent nodal junctures suitable for therapeutic targeting. Indeed, it is also unclear whether these observed defects are common across the spectrum of Parkinsonism disorder. However, the pursuit of mechanism-based studies in experimentally tractable model systems still likely presents the best route toward a tailored disease-modifying therapy.

Definitive evidence for environmental factors in sporadic cases has remained relatively elusive; however, the evidence for the influence of genetic contribution has been accelerating (Gasser, 2009). In a little over a decade, linkage studies have begun to identify single-gene mutations responsible for rare heritable forms of PD which in many cases is clinically indistinguishable from the more common sporadic cases, with the exception of early onset. While this approach is technically advantageous, to be applicable to all forms of PD, it assumes a shared pathogenic mechanism between inherited and sporadic cases. To identify genetic susceptibility factors for sporadic PD, the incredible technical advances in robust genome-wide methodologies have allowed interrogation of thousands of disease and control samples to potentially correlate disease with commonly occurring polymorphisms. Genome-wide association studies (GWAS) have revealed susceptibility risk for a relatively small set of genes (reviewed by Gandhi and Wood, 2010). Encouragingly, these included genes are already known as causative agents in neurodegeneration. The contributions to the field from GWAS are still materializing; hence, the majority of insights into pathogenesis have so far come from analysis of genes linked to rare Mendelian forms of PD. While the identification of these genes has tremendous potential to provide insight into the mechanisms of PD, we do not currently fully understand their biological functions and how their mutation results in neuronal death. Generating animal models that recapitulate the genetic lesions leading to disease is an essential tool on the path to fully elucidating the disease biology. One promising approach to this problem involves the use of classical genetic analysis in the fruit fly *Drosophila melanogaster* to identify the genetic pathways leading to pathology in fly models of this disease. This review describes the features that make Drosophila useful in studies of the genes implicated in heritable forms of PD and how these studies have begun to contribute to our understanding of the pathogenesis of this disorder and the identification of potential treatment strategies for PD.

II. DROSOPHILA AS A MODEL ORGANISM

D. melanogaster, commonly referred to as fruit flies reflecting their natural habitat on rotting fruit, belong to the family Drosophilidae, which derives its name from the Greek (*drosos* = dew and *philos* = loving). Over the past century, studies of Drosophila genetic and molecular pathways have provided pivotal advances in our understanding of fundamental biological processes, such as chromosome structure and segregation, regulation of gene expression, mechanisms of development, and genetic control of behavior (Ashburner and Novitski, 1976). These studies, together with more recent genome sequencing efforts, have made it clear that gene sequence and gene function are highly conserved between flies and humans (Rubin *et al.*, 2000). This conservation has contributed enormously to our current understanding of human biology and the pathogenesis of particular human diseases. For example, the Drosophila counterparts to numerous tumor suppressor and proto-oncogenes have provided a wealth of understanding about the pathways involved in cancer. While much of the insight into human disease mechanisms derived from genetic studies of Drosophila has been a byproduct of investigations of more fundamental biological questions, the increasing appreciation of the usefulness of Drosophila to address questions of human pathogenesis has led more recently to overt efforts to model specific human diseases (Bier, 2005). Over the past several years, significant effort has been invested in the development of Drosophila models of neurological diseases, such as polyglutamine diseases, Alzheimer's disease, PD, and others (Lessing and Bonini, 2009; Lu and Vogel, 2009). The extensive degree of conservation of neuronal function and development at the cellular level between Drosophila and higher vertebrates, coupled with the versatility and rapidity of Drosophila techniques, makes them an ideal system for advancement of our understanding of neurological disease mechanisms. Such progress may ultimately yield preventative treatments for these disorders.

The completion and annotation of the genome sequence of *D. melanogaster* (Adams *et al.*, 2000) and the availability of internet accessible homology search algorithms make the identification of candidate Drosophila disease homologs relatively trivial (e.g., see http://superfly.ucsd.edu/homophila/). Use of these search algorithms indicates that the Drosophila genome encodes excellent homologs for many of the currently identified PD-related genes (Table 1.1). In particular, the Drosophila genome encodes genes highly homologous to *Dardarin/LRRK2*, *parkin*, *PINK1*, *Omi/HtrA2*, *DJ-1*, *UCH-L1*, *GIGYF2*, *PLA2G6*, and *GBA* (Table 1.1). Notably, there is a clear lack of Drosophila

Table 1.1. Putative Function of Human Genes Linked to PD and Their Fly Homologs

PD locus	Gene/protein	Mode of inheritance	Fly homolog CG#	Fly homolog(s) identity, similarity	Putative function
PARK1 PARK4	SNCA/ α-synuclein	AD	*no homolog*	no homolog	Synaptic plasticity?
PARK2	Parkin	AR	*parkin/* CG10523	42%, 59%	E3 ubiquitin-protein ligase
PARK5	UCH-L1	AD(?)	*Uch/* CG4265	45%, 66%	Ubiquitin hydrolase/ ligase
PARK6	PINK1	AR	CG4523	32%, 50%	Mitochondrial kinase
PARK7	DJ-1	AR	*DJ-1α /* CG6646	56%, 70%	Oxidative stress sensor, chaperone?
			DJ-1β/ CG1349	52%, 69%	
PARK8	*Dardarin/* LRRK2	AD	CG5483	26%, 43%	Kinase, GTPase
PARK9	ATP13A2	AR	*no homolog*	no homolog	Lysosome
PARK11	GIGYF2	AD	CG11148	50%, 71%	Insulin signaling
PARK13	Omi/HtrA2	AD(?)	*HtrA2/* CG8464	51%, 70%	Serine protease
PARK14	PLA2G6	AR	CG6718/ *iPLA2- VIA*	50%, 67%	Phospholipase
PARK15	FBXO7	AR	*no homolog*	no homolog	Protein complex scaffolding
	GBA	AD(?)	*CG31414* *CG31148*	31%, 49% 32%, 50%	Glucocerebrosidase

counterparts to α-synuclein, ATP13A2, and FBXO7. Nevertheless, while it is early days for ATP13A2 and FBXO7, the dominant, toxic gain-of-function mechanism by which α-synuclein is thought to act in DA neuron death supports the use of a transgenic approach using the human α-synuclein gene to study the α-synuclein pathogenesis in Drosophila (see below). The existence of Drosophila homologs of most of the genes implicated in heritable forms of PD implies that the pathways regulated by these genes are also likely to be conserved in Drosophila.

A. Advantages and limitations of Drosophila to model PD

The widespread use and tractability of Drosophila have spurred the development and refinement of a versatile and powerful array of molecular and genetic tools. Both mutational and transgenic approaches have been used to create and study the Drosophila models of neurological disease. The mutational approach involves the generation of mutations in Drosophila counterparts of human disease genes, whereas the transgenic approach involves the introduction and expression of a human disease gene in Drosophila. The former approach is typically used when the corresponding human disease results from a loss-of-function mutation in a particular gene, whereas the latter approach is more often used when the corresponding human disease appears to result from a dominant toxic gain-of-function mechanism. However, a combination of both techniques is typically employed for a full and thorough genetic analysis of the functional role of an endogenous gene of interest. This section outlines some of the typical methodologies involved in the modeling approaches discussed throughout.

1. Mutagenesis and related loss-of-function techniques

There are a number of different techniques that can be employed to generate mutations in a particular Drosophila gene of interest. Many of the classical methods for recovering mutations in defined genes in Drosophila require a predictable phenotype that can be readily identified, such as developmental defects or preadult lethality. However, there are now a number of methods available that allow the recovery of mutations in defined genes irrespective of the mutant phenotype. One particularly powerful approach involves the use of transposable elements. A number of specifically engineered transposons have been designed that can be used for insertional mutagenesis in Drosophila. These transposons are being utilized by a large consortium, for example, the Berkeley Drosophila Genome Project, in an effort to generate mutations in most Drosophila genes as a service to the Drosophila research community. Primarily as a result of these efforts, it is currently estimated that more than 50% of the predicted

Drosophila genes have associated transposon insertions (Bellen *et al.*, 2004; Hiesinger and Bellen, 2004). Depending on the type of transposon, secondary manipulations can be employed to induce additional localized lesions, for example, to induce local hops or small deletions (Ryder and Russell, 2003).

However, this still leaves a significant proportion of genes that do not have readily available genetic lesions. For Drosophila, homologous recombination methodologies similar to that used in making mouse knockouts (Gong and Golic, 2003; Rong and Golic, 2000) are still in their infancy but are now gaining recognition as a primary route rather than a last resort. Briefly, this method requires the generation of a transgenic line bearing a targeting construct that will be used to inactivate the gene of interest. With the use of flp recombinase sites, the targeting construct is subsequently excised to form a linear extrachromosomal fragment. This then recombines with the homologous chromosomal locus, possibly through a double-strand break repair mechanism, and the endogenous locus is replaced by the engineered construct. The recent integration of site-specific recombination sites makes feasible the ability to replace the endogenous locus with an engineered sequence of choice, such as a fluorescent fusion or a mutationally altered version, but importantly still expressed under the control of endogenous regulatory elements (reviewed by Venken and Bellen, 2005).

Another approach that is being used with increasing frequency in place of traditional genetic analysis in Drosophila is RNA interference (RNAi) (Kalidas and Smith, 2002; Lee and Carthew, 2003). This approach capitalizes on the finding that double-stranded RNA (dsRNA) molecules corresponding in sequence to endogenous transcripts can trigger the degradation of the endogenous transcript. This approach requires the construction of a transgenic construct bearing an inverted repeat sequence corresponding to the target transcript. Through the use of the GAL4 system (see below), targeted expression of the dsRNA in (almost) any desired tissue can be achieved in one generation simply by conducting the appropriate cross. One of the reasons this approach has become popular is the speed with which a transgenic line bearing an RNAi construct can be created relative to more traditional genetic approaches. However, it is clear from the many studies involving RNAi approaches that this method effects only partial gene inactivation. Thus, a phenotype resulting from RNAi is typically equivalent to a hypomorphic loss-of-function mutation (e.g., compare Andrews *et al.*, 2002 and Murphy *et al.*, 2003). This feature of RNAi can be an advantage or a limitation depending on the context of the experiment and the goals of the experimenter. Another potential concern with RNAi technology is the increasing evidence that RNAi can sometimes affect the abundance of unintended mRNA targets (Sarov and Stewart, 2005). Because it can be difficult to control for such "off-target" effects of RNAi, the phenotypes resulting from RNAi must be interpreted with caution until a comparison with a traditional mutant allele can be made.

2. Transgenic misexpression

One of the most powerful and versatile tools available in Drosophila is the ability to generate transgenic constructs that can be used to drive the expression of a chosen gene in a tissue-specific manner by exploiting a yeast transcriptional activation mechanism. The GAL4/UAS system uses the yeast transcriptional activator GAL4 expressed under the temporal and spatial control of endogenous enhancer/promoter elements, to drive the expression of a specific transgene (Brand and Perrimon, 1993). The transgene is cloned in a vector containing a minimal promoter coupled with upstream activator sequences (UASs) that are specifically recognized by the GAL4 transcription factor. Upon binding of GAL4 to the UAS sites, expression of the transgene is induced in a tissue-specific manner dependent on the endogenous enhancer/promoter elements controlling GAL4 expression. There are now many different GAL4 lines available that drive the expression in a wide variety of tissues (e.g., nervous system, muscle tissue, or ubiquitously). This technique has been used with great effect to determine the *in vivo* consequence of misexpression or over-expression of fly genes and has been adapted to study the ectopic expression of human genes and their disease causing aberrant forms. An additional advantage to modeling late-onset, age-dependent processes, relevant for most neurodegenerative disorders, is the facility to bypass confounding developmental defects and study the effects of transgenes only in the adult stage by the use of steroid-induced transgene expression (Roman *et al.*, 2001).

3. Genetic screening

Probably, the single greatest advantage of using Drosophila for any area of biological interest is the ability to conduct relatively unbiased genetic screens for mutations in other genes that suppress or enhance the phenotypes associated with the disease model. This approach has the potential to identify cellular factors that act in the same or parallel pathways to the disease gene and does not require any *a priori* knowledge of the function of the disease gene. The power of such screening approaches cannot be overstated; human counterparts corresponding to suppressors identified from screens using Drosophila define potential targets for therapeutic intervention. It is primarily the feasibility of conducting such high-throughput screens that sets Drosophila apart from verte-brate models of disease.

There are a large number of approaches available for conducting genetic modifier screens in Drosophila (St. Johnston, 2002). One of the most commonly used approaches involves crossing a collection of enhancer *P* (*EP*) element insertions (Rorth, 1996) into a disease model background and investigating the effects of these insertions on the disease model phenotypes. The *EP* transposons have been engineered to drive overexpression of sequences flanking the transposon in a GAL4-dependent fashion. Thus, when used in conjunction with a particular GAL4 line, these transposons might suppress or enhance the disease model phenotype as a result of insertional inactivation of the genes they reside in or as a result of overexpression of flanking genes. These potential effects can be easily distinguished in subsequent studies. This modifier screening approach is compatible with a variety of phenotypes and could be used to identify modifiers of behavioral, morphological, or recessive lethal phenotypes. A particularly useful attribute of this screening method is that the insertion location of all of the EP lines has been determined and thus modifier genes are readily identified. Use of the EP collection in a modifier screening context has proven extremely valuable in studies to identify modifiers of polyglutamine, PD, and tau pathology in Drosophila (Fernandez-Funez et al., 2000; Shulman and Feany, 2003).

Finally, a number of Drosophila cell lines are available for cell biological studies of neurodegeneration. While there are also many vertebrate cell lines available for analysis, an advantage of many Drosophila cell lines is that they are highly amenable to RNAi manipulation. Highly efficient target transcript degradation can be achieved with many Drosophila cell lines by simply adding microgram amounts of dsRNA molecules of 200–400 bp directly to cell culture media (Boutros et al., 2004; Steinbrink and Boutros, 2008). RNAi could potentially be used to create a Drosophila cell culture model of neurodegeneration, or to conduct a whole genome screen for dsRNA molecules that modify a cell culture phenotype. An RNAi screening center, consisting of all of the necessary resources for conducting genome-wide screens in Drosophila cell lines, was recently established at Harvard and another recent established in Sheffield which are currently available for such screens (see http://teratogen.med. harvard.edu/ and http://www.rnai.group.shef.ac.uk/). To bring this technology full circle, the same dsRNAs that make up a "second generation" RNAi library have been used to generate a publically accessible inducible transgenic RNAi collection (Vienna Drosophila Resource Centre; Dietzl et al., 2007). This facility provides the unprecedented ability to move from in vitro screen to in vivo analysis all through community-based initiatives.

B. The neuroanatomy and function of DA neurons in Drosophila

Drosophila has a complex nervous system consisting of approximately 100,000 neurons including a subset of approximately 200 neurons that secrete the neurotransmitter DA (Fig. 1.1). The anatomical locations of all of the Drosophila DA producing neurons have been identified, and their development has been traced throughout the life cycle. Thus, genetic perturbations affecting the number, morphology, or locations of DA neurons in Drosophila can be readily identified. Although the anatomy of the fly brain and the distribution of DA neurons in the Drosophila central nervous system differ vastly from the vertebrate brain,

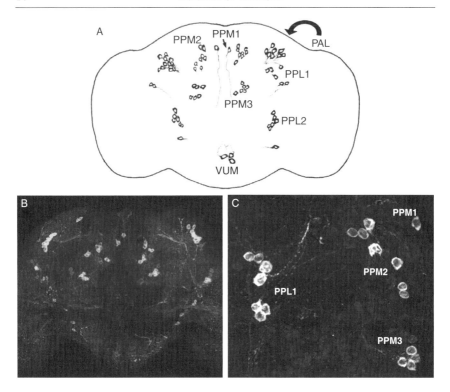

Figure 1.1. (A) Schematic representation of the distribution of dopaminergic neurons in the Drosophila adult brain. Dopaminergic neurons are grouped in small clusters arranged with bilateral symmetry. PPM, protocerebral posterior medial; PPL, protocerebral posterior lateral; PAL, protocerebral anterior lateral; VUM, ventral unpaired medial. (B) Confocal micrograph revealing the dopaminergic neurons in the adult drosophila brain. Expression of a GFP reporter (green) is induced by the endogenous tyrosine hydroxylase promoter and counter-stained with anti-tyrosine hydroxylase antiserum (red) demonstrating significant but not complete overlap. (C) Detail of a projected Z-series of posterior clusters. Axons are revealed as punctate staining with anti-tyrosine hydroxylase antiserum. (See Color Insert.)

previous work indicates that many fundamental cellular and molecular biological features of neuronal development and function are conserved between verte-brates and invertebrates (Ashburner and Novitski, 1976). Recent work indicates that this conservation makes Drosophila a powerful system for cell biological studies of neuronal dysfunction.

During embryonic and larval brain development, DA is found to be expressed in approximately 80 cells in the central nervous system (Budnik and White, 1988; Lundell and Hirsh, 1994). Many of these cells are grouped together into three bilaterally symmetrical clusters in the two lobes of the brain, with the remaining DA neurons distributed singly along the length of the ventral ganglion. These DA neurons are retained in the central nervous system of adult flies and are primarily grouped together into six major clusters (Nassel and Elekes, 1992). These six clusters are arranged symmetrically about the midline with the neuronal cell bodies residing at the periphery of the brain and their axons projecting toward the center (Fig. 1.1). There are four additional clusters of DA neurons on the posterior side of the brain: two medial clusters, designated the protocerebral posterior medial (PPM) 2 and 3, and two lateral clusters, named the protocerebral posterior lateral (PPL) 1 and 2. On the anterior side of the brain, there is a small cluster of approximately five DA neurons, designated the protocerebral anterior lateral (PAL), and a larger cluster of approximately 60 DA neurons with characteristically small cell bodies, designated the protocerebral anterior medial (PAM). In addition to these clusters, there are also a small number of DA neurons that are separated from the clusters, such as the PPM1, the deutocerebral 1 (D1), and the ventral unpaired medial (VUM) neurons.

To date, most studies of DA neuron integrity in Drosophila models of PD have used antisera against tyrosine hydroxylase (TH), an enzyme required for DA biosynthesis, to image DA neurons. However, several different immunocytochemical methods have been utilized to conduct these imaging studies. Early studies used thick sections of paraffin-embedded CNS samples in conjunction with light microscopy to analyze DA neuron integrity, but the sensitivity of this technique has been questioned (Auluck et al., 2005; Pesah et al., 2005). Confocal microscopy of whole-mount brain samples is now more commonly used. Another potential advantage of confocal microscopy relative to the methodology employing paraffin sections is that this technique allows a better visualization of the three-dimensional arrangement of the neurons within the intact brain. Thus, this methodology might facilitate studies of more subtle aspects of DA neuron dysfunction, such as axonal projection and synaptic defects (e.g., see Fig. 1.1C). In contrast, some studies have assessed the presence of DA neurons by monitoring the expression of a reporter transgene such as green fluorescent protein (GFP) under the influence of GAL4 expressed from the endogenous TH or ddc promoters. The reliability of this method is questionable however, since it is clear that TH-GAL4- or ddc-GAL4-induced reporters do not exactly mirror the pattern of TH protein abundance detected by anti-TH immunostaining (see Fig. 1.1B; Pesah et al., 2005). In general, anti-TH immunostaining analyzed by whole-mount confocal microscopy is considered the preferential technique.

In addition to the spatial distribution of DA neurons in Drosophila, the functional effects of DA depletion and DA neuron perturbation have also been studied. Genetic or pharmacologic depletion of DA in Drosophila results in a variety of characteristic phenotypes. Mutations affecting the dopa decarboxylase gene result in decreased learning ability, while mutations in the TH encoding gene *pale* cause a dose-dependent loss of general locomotor ability (Pendleton *et al.*, 2002; Tempel *et al.*, 1984). The systemic administration of chemical inhibitors of TH synthesis, such as 3-iodo-tyrosine, resulted in developmental delay, decreased fertility and inhibition of a simple learning paradigm (Neckameyer, 1996, 1998a,b). However, it is unclear from these studies whether these phenotypes result from loss of DA signaling in the nervous system or from a nonneuronal requirement for dopamine. The Drosophila *pale* gene encodes two alternatively spliced isoforms of TH: one isoform is neuronally expressed, while the other is expressed in the developing mesoderm and required for cuticle hardening and pigmentation (Birman *et al.*, 1994; Neckameyer and White, 1993).

Several recent studies have utilized the GAL4/UAS system to address the behavioral effects of perturbing DA neuron signaling in the Drosophila nervous system. In one report, investigators expressed tetanus toxin in TH-expressing neurons to block DA neuron signaling. Drosophila with most of their DA neurons silenced are viable, display normal locomotion, and have a wild-type (WT) appearance; however, these flies exhibit a hyperexcitable startle response (Friggi-Grelin *et al.*, 2003). Another study made use of a transgene encoding an ATP-gated calcium channel in conjunction with a photolabile caged ATP. This technique allows exquisite control over depolarization, and thereby activation, of neuron signaling in an intact animal, an emerging field called optogenetics. Unexpectedly, photostimulation of DA neurons produced quite differing effects dependent on the locomotor state of the flies preceding photostimulation. One population of flies with low locomotor activity prior to photostimulation exhibited an increased frequency of locomotion and an alter-ation in the routes traversed following photostimulation. Photostimulation of a second population of flies with high locomotor activity prior to photostimulation was found to result in a transient locomotor arrest (Lima and Miesenbock, 2005). Together, these results demonstrate that a function of DA signaling in the Drosophila nervous system is to regulate locomotor behavior in a complex manner.

III. DOMINANT TRAITS

A. α-Synuclein models

The first gene linked to a heritable form of PD, α-synuclein, was also the first of the PD-related genes to be studied in Drosophila. α-Synuclein is a small (140 amino acids) but abundant neuronal protein that is particularly enriched in pre-synaptic terminals. Although the precise physiological function of α-synuclein remains to be elucidated, several studies have suggested a regulatory role for the protein in membrane and vesicular dynamics (Tofaris and Spillantini, 2007). Although mutations of the α-synuclein gene (A53T, A30P, E46K, duplication, and triplication; see Gasser, 2009) appear to be an extremely rare cause of PD, the perceived importance of α-synuclein in the pathobiology of PD was significantly increased with the discovery that α-synuclein is a component of the Lewy body inclusions in sporadic PD patients and has a propensity to self-aggregate (Conway et al., 1998; Spillantini et al., 1997). Thus, investigations into the mechanism by which α-synuclein induces neuronal loss could lead to the development of treatment strategies impacting on multiple forms of PD.

While most of the human genes that have been implicated in PD have Drosophila counterparts (Table 1.1), there appears to be no Drosophila ortholog of the α-synuclein gene. However, because both missense mutations and increased dosage of the α-synuclein gene confer dominant forms of Parkinsonism (Bennett, 2005; Eriksen et al., 2005), transgenic flies expressing WT or mutated human α-synuclein under the transcriptional control of a GAL4-responsive UAS element were generated to assess the contribution of α-synuclein to PD pathogenesis. The main salient features that one might anticipate, and certainly that warrant investigation, are selective DA neuron degeneration, the presence of proteinaceous Lewy body-like inclusions, locomotor deficits, and perhaps early death. In addition, to capitalize on the potential to exploit a Drosophila model for genetic screening, it would be advantageous to observe an easily scorable phenotype such as that described for polyglutamine disease models; progressive degeneration of eye tissue (Fernandez-Funez et al., 2000; Warrick et al., 1998).

To date, a number of α-synuclein-expressing transgenic lines have been independently generated and characterized. To investigate the effects of α-synuclein on disease-relevant features, these lines have been crossed to a variety of tissue-specific GAL4 lines and the effects of α-synuclein expression have been assessed on DA neurons, the entire nervous system, and eye tissue. Remarkably, initial reports indicated that α-synuclein-expressing flies recapitulate all the major features of PD stimulating intense interest in this model. Initial studies reported that pan-neuronal expression of α-synuclein led to a near complete and relatively selective degeneration of DA neurons (the PPM1/2 clusters were primarily studied), a concomitant decline in climbing ability, the accumulation of proteinaceous inclusions, and progressive degeneration of retinal tissue (Feany and Bender, 2000). Intriguingly, similar effects were observed with the expression of either WT or the A30P and A53T mutated α-synuclein, suggesting that simply mis-regulated α-synuclein abundance may drive toxicity.

However, while these initial observations have largely been corroborated by recent reports from the same group (Chen and Feany, 2005; Chen et al., 2009), there have been considerable inconsistent effects reported by other groups, raising some questions about the "genuine" effects of transgenic α-synuclein expression. Other groups using independently derived equivalent lines or the same transgenic lines observed variable effects on all phenotypes, in particular with degeneration of DA neurons and climbing deficits. For example, a couple of groups have reported no quantifiable defects in locomotion upon expression of WT or mutant α-synuclein (Auluck et al., 2002; Pesah et al., 2005). This may not be entirely surprising given that behavioral assays are well known to be sensitive to subtle environmental fluctuations. Coupled with findings that DA neuron dysfunction results in a hyperexcitable or startle phenotype upon vigorous mechanical disturbance of flies raises the possibility that over-vigorous handling of flies may impact the outcome of a climbing assay (Friggi-Grelin et al., 2003). Furthermore, as described above, recent work has shown that apparently homogeneous fly populations consist of subpopulations with varying "high" or "low" locomotor activity (Lima and Miesenbock, 2005). While still unresolved, such technical influences still deserve consideration.

Moreover, several groups have reported significantly less pronounced DA neuron degeneration, seeing only a partial ($\sim$50%) decrease (Auluck and Bonini, 2002; Auluck et al., 2002; Bayersdorfer et al., 2010). Again, reports vary as to the differential effects of WT and mutated α-synuclein; for example, Auluck et al. concur with no difference observed between WT and A30P or A53T α-synuclein (Auluck and Bonini, 2002; Auluck et al., 2002), while Botella et al. (2008) describe only significant effects with A30P. Perhaps, most concerning is a report describing total lack of effect for all phenotypes with multiple lines of WT, A30P, or A53T α-synuclein (Pesah et al., 2005). Although singular in their formal publication, these observations are somewhat corroborated by the very modest effects reported by Du et al. (2010) and others (T. Riemensperger and S. Birman, personal communication). Consequently, other investigators have taken to significantly increase the expression level of α-synuclein by expressing multiple copies of a modified translation-efficient transgene, and still effecting only modest loss of DA neurons (this time in the PPL1 cluster; Trinh et al., 2008).

While these conflicts are not easily reconciled, several possible explanations can be offered. The construction of equivalent lines is essentially identical; however, there may exist subtle differences, such as genetic background or insertion site effects on expression levels. These aspects are hard to control with traditional methodologies, and may have a profound effect on the toxicity of an exogenous gene, but may be circumvented with the use of recent advances in transgenesis such as site-specific genomic integration (Groth et al., 2004). However, a recent study by Auluck et al. (2005) strongly suggests that the conflicting results of DA neuron analysis in α-synuclein transgenic flies are likely explained by differences in the methodology used to analyze DA neurons. While many of the studies that report profound neuronal loss in α-synuclein-expressing flies utilized paraffin-embedded sectioning and light microscopy techniques to visualize TH-positive neurons (Auluck and Bonini, 2002; Auluck et al., 2002; Chen and Feany, 2005; Chen et al., 2009; Feany and Bender, 2000), more recent work observing less dramatic effects involved the use of confocal microscopy of whole-mount brains to detect TH-positive neurons. Thus, one possible explanation for these conflicting results is that the different methods employed to identify TH-positive neurons may differ in their sensitivity of detection. Auluck *et al.* directly addressed this possibility and demonstrated that the subpopulation of DA neurons reported to degenerate in α-synuclein-expressing flies could still be detected by confocal analysis of whole-mount brains from flies expressing α-synuclein. Furthermore, they demonstrated that this same cell population generally stains less intensely than those in clusters reportedly unaffected by α-synuclein expression (Auluck *et al.*, 2005, supplementary data). These findings indicate that the loss of TH staining previously reported in α-synuclein transgenic flies reflects reduced TH levels rather than overt cell loss. While the reduced TH expression conferred by α-synuclein expression suggests that these cells are dysfunctional, whether this phenotype reflects an early stage leading to the death of these neurons, as proposed by Auluck *et al.* (2005), remains an open question. Irrespective of the final outcomes between individual laboratory environments, these findings underscore the paramount requirement for scientific rigor and the absolute necessity to conduct DA neurons analysis in a blinded manner. The standardization of reliable and efficient staining protocols, which has recently been attempted (Drobysheva *et al.*, 2008), would also help. If these standards are upheld, one can be confident to interpret the results.

1. Mechanistic insights from the α-synuclein models

The coincidence of neuronal loss and the formation of α-synuclein-containing Lewy body aggregates in PD prompted speculation of the possible involvement of protein aggregation in promoting DA neuron loss. This hypothesis was addressed directly using the Drosophila α-synuclein models. Overexpression of the chaperone, HSP70, abrogated the α-synuclein-induced loss of TH-positive neurons without detectably influencing the appearance of Lewy body-like aggregates (Auluck *et al.*, 2002). Furthermore, feeding α-synuclein transgenic flies the chaperone-inducing compound geldanamycin was found to phenocopy the protective effect of HSP70 expression (Auluck and Bonini, 2002). These findings, together with other work from the field, suggest that the formation of large α-synuclein-containing aggregates is not the primary toxic species and that HSP70 induction is protective from the harmful effects of α-synuclein. This model is supported by observations that upregulation of the K48-linked ubiquitin–proteasome-dependent degradation can also ameliorate the toxic effects of α-synuclein expression (Lee *et al.*, 2009), again arguing that the toxic species are mono- or oligomeric rather than aggregates. This is further supported by analysis of synthetically mutated α-synuclein with an increased propensity to exist as soluble oligomers that demonstrate a higher neurotoxicity (Karpinar *et al.*, 2009).

To explore further the relationship of α-synuclein modification and aggregate formation to DA neuron loss, investigators analyzed the effects of posttranslational modifications of α-synuclein on aggregate formation and neuronal viability. Previous work showed that α-synuclein is extensively phosphorylated in the brains of PD individuals, particularly at serine residue 129, and that the phosphorylation status of α-synuclein influences its propensity to form aggregates *in vitro* (Fujiwara *et al.*, 2002; Okochi *et al.*, 2000). Interestingly, the phosphorylation of α-synuclein at Ser129 appears to be conserved in Drosophila (Takahashi *et al.*, 2003). To explore the consequence of Ser129 phosphorylation on aggregate formation and neuronal integrity, Chen and Feany generated mutationally altered α-synuclein transgenic constructs; a Ser129 to Ala (S129A) mutation to prevent phosphorylation and a Ser129 to Asp (S129D) mutation to mimic the phosphorylated state (Chen and Feany, 2005). The extent and timing of pathology of these transgenic constructs was compared with WT α-synuclein. Expression of the S129D construct accelerated the onset of DA neuron loss, increased retinal degeneration, and reduced the amount of α-synuclein in aggregates relative to WT α-synuclein. By contrast, expression of the S129A construct resulted in reduced DA neuron loss and increased the total load of aggregated α-synuclein protein. These results suggest that phosphorylation of Ser129 could maintain α-synuclein in a nonaggregated but more toxic conformation. In contrast, recent analysis of a putative phosphorylation at Y125 suggests that this modification led to a decrease in soluble α-synuclein oligomers and was less toxic than WT or S129 phosphorylation (Chen *et al.*, 2009). Together these findings suggest that α-synuclein's toxicity, and likely its normal function, may be modulated by phosphorylation and imply the presence of specific kinases.

Overall, these findings suggest that an increase in α-synuclein aggregates correlates with decreased cellular toxicity, implying that Lewy body formation is a neuronal detoxification response and has significant implications for potential therapeutic approaches. The finding that phosphorylation enhances α-synuclein toxicity suggests that the kinases responsible for phosphorylation might represent therapeutic targets for small-molecule inhibitors. It will be important to identify these kinases and characterize their activity to know if their inhibition may be a viable therapy. More crucially, the evidence that inclusion formation is protective challenges current therapeutic strategies to prevent inclusion formation which may even augment disease progression. While the precise mechanisms of toxicity remain unclear, this evidence lends support to the early findings that administration of the HSP70-inducing geldanamycin is able to suppress α-synuclein toxicity, suggests this avenue as a possible treatment strategy for PD (Auluck and Bonini, 2002).

Another application of Drosophila models of PD is to use these models in genomic studies to identify pathways involved in pathogenesis. A recent genomic study of the α-synuclein transgenic fly model identified 51 Drosophila transcripts that displayed an altered abundance relative to control flies (Scherzer et al., 2003). Importantly, these transcripts were unaffected in transgenic flies expressing the neurodegeneration-inducing tau protein, indicating that the abundance of these 51 transcripts is specifically altered in response to α-synuclein expression. The 51 transcripts that display altered abundance in the α-synuclein transgenic flies encode proteins involved in lipid metabolism, energy production, and membrane transport, potentially implicating these pathways in α-synuclein pathogenesis. Interestingly, a recent study of α-synuclein pathogenesis in yeast identified lipid metabolism and vesicle transport components as a major category of genetic modifiers of α-synuclein toxicity (Willingham et al., 2003). Together, these two studies suggest an evolutionarily conserved pathway of PD pathogenesis involving lipid metabolic and vesicle trafficking defects. Additional work will be required to validate the functional significance of these pathways to DA neuron degeneration.

B. LRRK2

The relatively high prevalence of LRRK2 mutations in familial as well as in sporadic cases has provoked significant interest in developing various model systems to investigate the function and dysfunction of LRRK2, including in Drosophila. Although mutations in LRRK2 are inherited in a dominant manner which can be replicated by transgenic expression, it is relevant and informative to gain insight into the normal function by studying loss-of-function or knockout scenarios. Consequently, a number of groups have reported on a variety of models characterizing the effects of manipulating LRRK2 and its Drosophila homolog.

LRRK2 encodes a large, complex, multidomain protein, comprising 2527 amino acids protein ($\sim$280 kDa) which contains leucine-rich repeat (LRR), Ras of complex (ROC), C terminal of ROC (COR), putative serine/threonine kinase and GTPase domains, and several WD40 protein–protein interactions domains. More than 20 different mutations in LRRK2 have been linked to autosomal dominant Parkinsonism and the G2019S mutation has been shown a high prevalence in sporadic cases particularly in certain genetically isolated communities. A number of groups have shown that G2019S may confer an increase in native kinase activity, and an apparent increase in toxicity. However, not all mutations act likewise; I2020T causes decrease in kinase activity, while the other mutations cause a propensity to aggregate. Thus, mutation of LRRK2 has the potential to cause a multitude of pathogenic effects. Furthermore, the normal physiological function(s) of LRRK2 currently remain poorly understood.

1. Transgenic models

Consistent with the significant interest in LRRK2 pathobiology in the field, a number of groups have independently generated and characterized transgenic lines expressing WT and mutated forms of human LRRK2. However, so far the reported observations do not show a clear or consistent pattern of observations despite a variety of mutated forms being investigated, including G2019S, I2020T, I1915T, Y1699C, R1441C, Y1383C, I1122V, and G2385R (see Table 1.2).

For instance, Ng *et al.* observed late stage degeneration of DA neurons (PPM1/2 and PPM3 only) with G2019S, Y1699C, and G2385R, and essentially no retinal degeneration, with no effects from WT LRRK2 (Ng *et al.*, 2009); however, Liu *et al.* reported that expression of both WT and G2019S caused retinal degeneration and progressive loss of DA neurons in all clusters (Liu *et al.*, 2008). While it may be argued that the toxic effects of WT LRRK2 observed by Liu *et al.* suggest that deregulation of LRRK2 can trigger pathogenesis, it is hard to be confident at this stage in light of negative results of essentially the same experiment, especially since a lack of effect by WT LRRK2 was also reported by Imai *et al.* (2008). In this study, they also observed late stage selective DA neuron loss caused by I1915T and Y1385C but this time in the PPM1/2 and PPL1 clusters. However, Venderova *et al.* report a rather confusing pattern of sensitivity under different conditions. They saw significant but rather variable effects of WT, I1122V, Y1699C and I2020T on DA neurons (PPM1/2 and PPL1 only) under basal conditions but all clusters seemed to be sensitive upon rotenone exposure (Venderova *et al.*, 2009). While they observed modest effects on locomotion and survival, dramatic disruption of eye tissue was reported; however, these experiments were conducted at 29 °C which as noted below can confound analysis of eye morphology.

The discrepancies are hard to reconcile; however, experimental variations, especially the use of different GAL4 drivers, are commonly invoked. In these situations, neuronal expression has been differentially achieved by the use of *elav-*, *ddc-*, and *TH-GAL4* lines which, coupled with the effect of each unique insertion site genomic structure, quite likely induce expression at subtly different levels. In contrast, the studies appear to commonly use GMR-GAL4 to drive eye-specific expression, but it is well appreciated in the field that different GMR-GAL4 lines exist which induce expression at varying levels and high GAL4 levels can themselves be toxic (Kramer and Staveley, 2003).

At present, all of these observations await independent replication, which is undoubtedly facilitated by the sharing of unique reagents including transgenic lines for which the Drosophila community is well known. With the expectation of more independently derived lines to be forthcoming, a clearer picture of the cellular effects of WT and mutated LRRK2 can be anticipated. Indeed, it would be interesting to compare the effects of all transgenic lines under

Table 1.2. Summary of the Neurodegenerative Phenotypes Associated with the LRRK2-Related Drosophila Models

Study	LRRK2 form	Eye phenotype	DA neuron loss
Liu *et al.*	WT	+	All clusters
	G2019S	+	All clusters
Imai *et al.*	WT	N/A	0
	I2020T	N/A	N/A
	I1915T	N/A	PPM1/2, PPL1
	Y1385C	N/A	PPM1/2, PPL1
	Y1699C	N/A	N/A
	dLRRK−/−	N/A	0
Ng *et al.*	WT	0	0
	G2019S	Versus rare (<1%)	PPM1/2, PPM3
	Y1699C	0	PPM1/2, PPM3
	G2385R	0	PPM1/2, PPM3
Venderova *et al.*	WT	++	All clusters affected variably
	Y1699C	++	under different conditions
	I2020T	++	
	I1122V	++	
Lee *et al.*	dLRRK	N/A	0
	dLRRK [=R1441C]	N/A	0
	dLRRK−/−	0	0
Wang *et al.*	dLRRK−/−	0	0

the same controlled conditions in one laboratory, such as was done for α-synuclein. For future consideration, emerging technologies such as site-specific transgene integration will help circumvent some of the common technical concerns.

2. Loss-of-function models

To complement the analysis of mutated forms of human LRRK2, mirroring the dominant inheritance pattern, and to gain an insight into the normal function of LRRK2 homologs, a valuable approach is to analyze and perturb expression of the endogenous gene. The Drosophila genome appears to contain a single gene that shares significant homology to human *LRRK2*, though phylogenetic studies have suggested that *dLRRK* may be somewhat distant homolog of human *LRRK2* and appears to lack certain N-terminal repeat sequences (Marin, 2006, 2008). Nevertheless, the current state of knowledge of LRRK2 biology is very limited including the relative contributions of the various protein domains to pathobiology; hence, valuable understanding may still be gained from studying this ancestral gene.

Similar to vertebrate *LRRK2*, *dLRRK* appears to be broadly expressed (Lee *et al.*, 2007) which does not reveal any secrets as to its function but does suggest a common cellular role. A couple of mutant alleles have so far been described but the most reliably informative is an insertion of a piggyBac transposon element which resides between exons 5 and 6, *dLRRK*[e03680]. This insertion causes exon 6 to be mis-spliced (Wang *et al.*, 2008); however, it is still conceivable that a small proportion is correctly spliced and so homozygotes may not represent a complete genetic null.

Homozygous *dLRRK* mutants are viable to adult stage with a normal lifespan and overall appear grossly normal if a little smaller than WT animals. Several studies are in agreement that the loss of *dLRRK* has no effect on survival of DA neurons, but mild locomotor deficits have been measured (Lee *et al.*, 2007; Tain *et al.*, 2009b). However, variable effects have been noted upon exposure to oxidative stressors; Wang *et al.* observed a slight sensitivity to hydrogen peroxide, while Imai *et al.* reported a noticeable resistance to the same toxin (Imai *et al.*, 2008; Wang *et al.*, 2008). Precise experimental conditions and genetic background easily vary under these tests, which would warrant careful controlling before meaningful interpretation.

One intriguing and unexpected phenotype was observed; while mutant males are fully fertile, females have severely reduced fecundity. While the exact basis of this defect is currently unknown, Drosophila oogenesis is an intensely studied system, thus, this phenotype offers an attractive opportunity to tap this knowledge and investigate the function of dLRRK. Furthermore, it may also provide a model system for experimental manipulation, for example, to test genetic interaction of putative functional partners.

3. Mechanistic insights from LRRK2 models

To date, relatively few studies on the Drosophila models have extended beyond initial characterization, but significantly the most intriguing is the report of a novel putative target of LRRK2 kinase activity. Following a speculative previous link between the PI3K/Akt/TOR signaling pathway and other fly models, Imai *et al.* (2008) demonstrated genetic interactions between dLRRK and several components of this pathway. Candidate testing revealed that dLRRK showed some potential kinase activity toward 4E-BP, a negative regulator of the eIF4E protein translation machinery. 4E-BP is known to be negatively regulated by phosphorylation via the TOR signaling pathway, and its activity is important for survival under a wide variety of stresses including starvation, oxidative stress, unfolded protein stress, and immune challenge (Bernal and Kimbrell, 2000; Kapahi *et al.*, 2004; Teleman *et al.*, 2005; Tettweiler *et al.*, 2005). Consistent with this, genetic analyses showed that loss of 4E-BP exacerbated toxicity of

transgenic dLRRK, while its overexpression, in particular expression of a constitutively active form, was sufficient to prevent DA neuron loss (Imai *et al.*, 2008).
These findings suggest that LRRK2 may normally regulate the activity of 4E-BP,
in a similar manner to TOR but in response to an unknown signal, and that
mutated LRRK2 may aberrantly inactivate 4E-BP. This would prevent 4E-BP
from regulating protein translation and inhibit its effects in promoting cellular
survival.

Presently, it is unclear how specific or robust the direct phosphorylation
of 4E-BP by dLRRK is, as data for other candidate proteins were not shown.
Others have even suggested that human LRRK2 may have a poor affinity for 4E-
BP1 (Kumar *et al.*, 2010), so further work on this is warranted. Nevertheless, the
potential importance of 4E-BP and regulated protein translation has attracted
much attention recently due to a number of supportive findings. First, a genetic
screen independently identified 4E-BP as a genetic interactor in an unrelated fly
model of PD, *parkin* mutants (see below; Greene *et al.*, 2005), and subsequently
my group demonstrated that genetic or pharmacologic activation of 4E-BP was
sufficient to prevent neurodegeneration (Tain *et al.*, 2009b). Second, additional
links to aberrant protein translation are emerging from reports that indicate
mutations in *EIF4G1*, encoding part of the translation machinery, have been
identified as a new *PARK* locus causing dominant inherited PD. The inability of
a cell to make an appropriate response to environmental changes by altering its
proteomic profile would render it susceptible to increased toxic insults. Furthermore, recent work has demonstrated that 4E-BP mediates its protective effects, at
least in part, by the preferential increase in production of numerous mitochondrial proteins (Zid *et al.*, 2009). The growing emphasis on mitochondrial dysfunction and oxidative stress as a key contributor to PD makes this an attractive
therapeutic mechanism. In addition, the modulation of mutant *dLRRK* phenotypes by 4E-BP suggests that dLRRK, and perhaps LRRK2, toxicity may be partly
due to mitochondrial dysfunction.

A very recent study has provided some potentially very interesting
novel insights into the LRRK2-mediated pathogenic mechanism, which again
points toward dysregulation of protein production. A number of tentative links
had previously been made between PD and the microRNA pathway—small
noncoding RNAs that bind to transcript 3′-untranslated regions to regulate
translation. Gehrke and colleagues identified several intriguing genetic interactions between transgenic dLRRK I1915T or hLRRK2 G2019S and components
of the RNA-induced silencing complex (RISC), *ago1* and *dicer1*, and also two
miRNAs, let-7 and miR-184*, such that antagonizing these factors enhanced the
pathogenicity of mutant LRRK2 (Gehrke *et al.*, 2010). In turn, these miRs were
shown to regulate the translation of two transcriptional regulators, *e2f1* and *dp*,
respectively. Consistently, these transcripts were regulated by LRRK2 activity,
overexpressed by pathogenic LRRK2, and downregulated in *LRRK* mutants, but

importantly their genetic reduction was also sufficient to suppress DA neuron loss by pathogenic LRRK2. Interestingly, biochemical analyses revealed that hLRRK2/dLRRK physically interacted with hAgo2/dAgo1 which also bound 4E-BP. The strength of this interaction appeared to increase with mutant LRRK2 which correlated with 4E-BP hyperphosphorylation, but this mechanism remains to be proven. It was hypothesized that this increased interaction would aberrantly relieve the miRNA translational repression. Consistent with this nonphosphorylatable 4E-BP antagonized the inhibition of let-7 by mutant LRRK2. Since 4E-BP itself plays an important role in regulating protein translation, it will be interesting to determine whether the downstream effects of pathogenic LRRK2 on protein translation are mediated principally via miRNA regulation or through the direct action of 4E-BP on the protein translation machinery. Nevertheless, together, these findings have highlighted regulated protein translation as a therapeutic avenue to explore further.

IV. RECESSIVE TRAITS

In contrast to the dominant toxic gain-of-function of α-synuclein and LRRK2 mutations, at least three of the genes associated with heritable forms of PD clearly involve loss-of-function mutations (Table 1.1). A precise understanding of the normal biological functions of the corresponding genes and the pathways regulated by these genes will be invaluable in uncovering the mechanisms by which these mutations cause PD. One of the most powerful approaches to address these issues involves the use of classical genetic analysis in a simple model organism such as Drosophila to explore the biological functions of evolutionarily conserved homologs of these genes. This approach has recently been used to analyze the biological functions of Drosophila parkin, DJ-1, and PINK1 homologs.

A. Parkin

To explore the biological role of parkin, we and others generated a series of mutations in the Drosophila ortholog of parkin, including deletion, nonsense, and missense mutations. Flies lacking the parkin gene are semi-viable and display reduced longevity, DA neuron degeneration, motor deficits, and male sterility (Greene et al., 2003; Pesah et al., 2004; Whitworth et al., 2005). Although initial studies of Drosophila parkin mutants failed to detect DA neuron loss, subsequent analysis using whole-mount confocal analysis indicates that a subset of DA neurons restricted to the PPL1 cluster do indeed degenerate in parkin mutants, while serotonergic neurons remain intact (Whitworth et al., 2005). In addition to the degeneration of neurons in the PPL1 cluster, parkin mutants also manifest reduced TH staining in the PPM1/2 cluster and significantly reduced DA content in the brain

(Cha *et al.*, 2005; Greene *et al.*, 2003). The inability of previous work to detect neuron loss in *parkin* mutants is likely explained by technical limitations of the approaches used, as discussed above, and the simple fact that most of these studies focused exclusively on the PPM1/2 cluster (Cha *et al.*, 2005; Pesah *et al.*, 2004; Yang *et al.*, 2003) owing to the reported effects of α-synuclein expression on this cluster.

The motor deficit of *parkin* mutants is associated with a dramatic and widespread apoptotic degeneration of muscle tissue, and the male sterility derives from a late defect in spermatid formation. Ultrastructural studies revealed that profound loss of mitochondrial integrity was the earliest manifestation during muscle degeneration and also apparent in spermatids, suggesting a role for parkin in mitochondrial integrity (Greene *et al.*, 2003). While humans and mice with *parkin* mutations do not appear to manifest similar muscle and germline pheno-types, mitochondrial defects are a common characteristic of sporadic PD and a conserved feature in all organisms with *parkin* mutations, including humans (Flinn *et al.*, 2009; Mortiboys *et al.*, 2008; Muftuoglu *et al.*, 2004; Palacino *et al.*, 2004; Ved *et al.*, 2005). These observations provide compelling evidence that parkin may act to regulate mitochondrial integrity and that mitochondrial dysfunction is an important contributing factor to DA neuron death in patients with mutated parkin function. Moreover, our finding that Drosophila *parkin* mutants exhibit DA neuron loss further suggests that the pathogenic mechan-isms responsible for neurodegeneration under these circumstances are conserved and hence tractable by analysis of the Drosophila models.

B. DJ-1

Drosophila has also proven useful in dissecting the role of the *DJ-1* gene. Similar to *parkin*, loss-of-function mutations in *DJ-1*, a small protein of unclear function with homology to proteases, kinases, and small heat shock proteins, result in Parkinsonism in humans (Bonifati *et al.*, 2003). The Drosophila genome encodes two homologs of the human *DJ-1* gene, designated *DJ-1α* and *DJ-1β* (Table 1.1). The *DJ-1β* gene appears to be ubiquitously expressed, whereas *DJ-1α* expression is largely, or exclusively, restricted to testes (Menzies *et al.*, 2005; Meulener *et al.*, 2005; Park *et al.*, 2005). To explore the biological roles of *DJ-1α* and *DJ-1β*, we and others have used traditional genetics and RNAi to perturb the functions of these genes (Lavara-Culebras and Paricio, 2007; Menzies *et al.*, 2005; Meulener *et al.*, 2005; Park *et al.*, 2005; Yang *et al.*, 2005).

The results of studies involving traditional mutant alleles of the *DJ-1* genes indicate that flies lacking one or both of the *DJ-1* genes are fully viable, fertile, and display no evidence of DA neuron loss. However, further studies of these mutants revealed that *DJ-1β* single mutants and *DJ-1α;DJ-1β* double mutants display a striking sensitivity to particular stress-inducing agents, includ-ing paraquat and rotenone—chemicals which have been epidemiologically linked to PD (Lavara-Culebras and Paricio, 2007; Meulener *et al.*, 2005; Park *et al.*, 2005). This sensitivity is unique to chemical agents that induce oxidative stress and is not manifest by chemical agents that induce other types of cellular stress implicated in PD, including ER stress and proteasome inhibition (Meulener *et al.*, 2005). Moreover, treatment of WT flies with these oxidative stress-inducing chemical agents results in a modification of DJ-1β protein electrophoretic mobility that can be detected by Western blot analysis or isoelectric focusing, suggesting that DJ-1β is a direct target of an oxidative modification as has been seen with vertebrate DJ-1. *DJ-1α* mutants manifest no apparent sensitivity to these same oxidative stress-inducing agents, although transgenic expression of either DJ-1α or DJ-1β is able to rescue the chemical sensitivity of *DJ-1β* single mutants and *DJ-1α;DJ-1β* double mutants.

   In contrast to these findings, work by Menzies *et al.*, using independently generated alleles of the *DJ-1β* gene, found that these mutants exhibit delayed age-dependent reduction of TH staining in the central nervous system and were less sensitive to paraquat exposure than to WT flies (Menzies *et al.*, 2005). The authors ascribe these two phenotypes to a compensatory induction of *DJ-1α* expression in response to loss of *DJ-1β* function, and provide expression and transgenic data in support of this model. Even more surprisingly, two studies of the *DJ-1α* gene (Lavara-Culebras and Paricio, 2007; Yang *et al.*, 2005) found that strong ubiquitous knockdown of the *DJ-1α* transcript with RNAi resulted in larval lethality (Yang *et al.*, 2005), while partial knockdown dramatically reduced lifespan (Lavara-Culebras and Paricio, 2007). These results clearly contrast with previous finding that *DJ-1α* null mutants are fully viable and exhibit no apparent phenotype, and should be interpreted with caution. Moreover, these investigators found that targeted knockdown of *DJ-1α* expression in the compound eye results in photoreceptor cell loss, and that knockdown of *DJ-1α* expression in the nervous system results in the progressive loss of TH-positive neurons in the PPM1/2 cluster (documented using the paraffin sectioning/light microscopy methodology) and reduced DA content in fly heads.

   While overall the studies of the Drosophila *DJ-1* gene family reached similar conclusions, a disconcerting feature of this work involves the stark phenotypic differences reported in the individual studies. Although further work will be required to explain these discrepancies, there are several likely sources. One potential source of the discordant results is methodological. In particular, while some studies (Lavara-Culebras and Paricio, 2007; Meulener *et al.*, 2005; Park *et al.*, 2005) made use of definitive null alleles of the Drosophila *DJ-1* gene family, the alleles used by Menzies *et al.* may well be hypomorphic in nature, and other work involved RNAi techniques (Menzies *et al.*, 2005; Yang *et al.*, 2005). The more severe phenotypes resulting from RNAi-mediated knockdown of *DJ-1α* activity relative to null alleles of *DJ-1α* cannot be readily explained by a hypomorphic effect of RNAi. These conflicting results are better explained by a possible developmental compensation to loss of *DJ-1α* in null animals relative to the acute loss of *DJ-1α* activity in RNAi treated animals later in development, or to an unexpected off-target effect of the *DJ-1α* RNAi constructs. While off-target effect of RNAi is a well-recognized phenomenon, it is tricky to predict but relatively trivial to exclude. The availability of genomic tools for other Drosophila species, as described above, should allow definitive phenotypic rescue by expression of an RNAi-resistant ortholog. Nevertheless, it is difficult to envision how perturbation of a gene that is predominantly or exclusively expressed in the male germline could result in loss of viability and neuronal degeneration.

Regardless of the explanations for the discordant results of studies of the Drosophila *DJ-1* gene family, all of these studies support the increasing body of evidence linking DJ-1 to the oxidative stress response pathway. A number of studies have demonstrated that human DJ-1 is modified upon oxidative stress by sulfonification of cysteine residues. Importantly, mutational analysis of conserved cysteine residues in Drosophila DJ-1β revealed that modification at C104 (analogous to C106 in human DJ-1) is critical for the antioxidant protective function of DJ-1 *in vivo* (Meulener *et al.*, 2006). Interestingly, this modification increases with age in Drosophila and human tissues (Meulener *et al.*, 2006); nevertheless, DJ-1 double mutants display an age-dependent decline in mitochondrial function (Hao *et al.*, 2010). An important challenge of future work will be to discern between the myriad biological functions ascribed to this protein family to define the mechanism by which loss of DJ-1 function protects against neuron loss, but it seems most likely that the major contribution of DJ-1 is in the response to oxidative stress to maintain mitochondrial integrity. In support of this, an interesting recent study found that feeding *DJ-1β* mutant flies various dietary antioxidants, including vitamin C, melatonin, and vitamin E, was able to significantly extend lifespan of these mutants (Lavara-Culebras *et al.*, 2010), hinting at a potential therapeutic benefit of antioxidants.

C. PINK1

Mutations in *PINK1*, which encodes a putative protein kinase, are a rare cause of recessive Parkinsonism, but has provided some significant insights into possible pathogenic mechanisms. The presence of a putative mitochondrial targeting sequence in PINK1 coupled with its predominant localization to mitochondria (Silvestri *et al.*, 2005; Valente *et al.*, 2004) strongly suggested that loss-of-function mutations in *PINK1* somehow compromise mitochondrial integrity. This prediction is born out in three studies that simultaneously described the effects of perturbation of a Drosophila homolog of *PINK1* (Clark *et al.*, 2006; Park *et al.*, 2006; Yang *et al.*, 2006). Two papers show that null mutations of the Drosophila *PINK1* ortholog result in degeneration of indirect flight muscles and defective spermatid formation (Clark *et al.*, 2006; Park *et al.*, 2006).

Using a combination of ultrastructural and biochemical assays, these papers further report that mitochondrial defects accompany both of these phenotypes. The third paper reports equivalent findings using an RNAi approach to target the *PINK1* gene, although phenotypes are only documented in the indirect flight muscle, presumably because the GAL4 line chosen for this study is not expressed in the male germline (Yang *et al.*, 2006). Again, there is some inconsistency in the reported effect on DA neuron integrity. Two of these studies reported DA neuron loss in the PPM1/2 and PPL1 clusters (Park *et al.*, 2006; Yang *et al.*, 2006), while one found no difference (Clark *et al.*, 2006). Currently, it is unclear why the third study failed to detect neuron loss although, in contrast, in a later study they do observe a subtle decrease in PPL1 (Yun *et al.*, 2008). Interestingly, one study also reports mitochondrial swelling in DA neurons of *PINK1* mutants, suggesting that the pathogenic mechanisms responsible for muscle degeneration, spermatid defects and DA neuron loss all involve mitochondrial dysfunction (Park *et al.*, 2006). It should also be noted that an independent concurrent study analyzing *PINK1*-RNAi knockdown reported multiple pathogenic defects including profound loss of DA neurons in all clusters even at 10 days old and degeneration of eye tissue (Wang *et al.*, 2006). However, these findings must be considered distinctly unreliable since ubiquitous knockdown using this construct caused early stage larval lethality, a phenomenon clearly discordant with the genetic null mutants, strongly implicating confounding off-target effects.

While the finding that Drosophila *PINK1* mutants display mitochondrial defects advances our knowledge, the real surprise of these studies came from genetic analyses with *parkin*. The striking similarity in phenotypes between *PINK1* and *parkin* mutant animals raised the hypothesis that they may act in a common pathway. In support of this idea, it was found that *PINK1*;*parkin* double mutants are phenotypically indistinguishable from the respective single mutants (Clark *et al.*, 2006; Park *et al.*, 2006), consistent with the two mutations affecting one pathway rather than two separate or independent pathways. Furthermore, genetic epistasis experiments showed that overexpression of either Drosophila or human *parkin* is able to rescue the *PINK1* phenotypes, but conversely *PINK1* overexpression does not detectably influence the *parkin* phenotypes (Clark *et al.*, 2006; Park *et al.*, 2006; Yang *et al.*, 2006). Although a caveat of this finding is that *parkin* overexpression can be generally protective, Park *et al.* (2006) showed that *parkin* overexpression is unable to suppress the toxicity of several other cellular insults, demonstrating at least some specificity in this interaction. Together, these findings provide compelling evidence that parkin acts downstream from PINK1 in a common pathway.

These findings provoke the most obvious hypothesis that PINK1 may phosphorylate parkin and stimulate its activity. While there is some evidence that supports this in Drosophila (Kim *et al.*, 2008b), it has been questioned by negative findings in mammalian systems (Vives-Bauza et al., 2010). Although the molecular mechanism of their relationship remains to be elucidated, the epistatic hierarchy established in Drosophila has since been validated in mammalian cells showing human parkin can suppress mitochondrial phenotypes caused by loss of PINK1 (Dagda et al., 2009; Exner et al., 2007). Clinical reports that PINK1 and parkin mutations confer similar symptoms in humans adds support that these cases may be caused by a common pathogenic mechanism (Zadikoff et al., 2006). Surprisingly however, a distinctive feature of parkin-mediated Parkinsonism, the absence of Lewy body inclusions, is not shared in common with PINK1 affected patients, as a recent report describes the presence of Lewy bodies on the first analyzed postmortem sample (Samaranch et al., 2010). Although the current consensus supports PINK1 and parkin acting in a common pathway to maintain mitochondrial integrity (discussed more below), the precise mechanisms by which their dysfunction leads to neuronal death are unclear and may indeed cause different pleiotropic effects at the end stage of pathology.

D. Omi/HtrA2

High-temperature requirement A2 (HtrA2/Omi) is a mitochondrial protease that exhibits proapoptotic and cell protective properties and has been linked to PD. Two mutant alleles of HtrA2 (A141S and G399S) have been found in PD patients, leading to the classification of HtrA2 as PARK13 by OMIM (Strauss et al., 2005). Although one of these genetic variants was later found in non-PD controls (Ross et al., 2008; Simon-Sanchez and Singleton, 2008), Bogaerts et al. (2008) identified a new mutation (Arg404) in a large cohort of Belgian PD patients, confirming a role for HtrA2 in PD susceptibility. Importantly, recent studies have shown that HtrA2 forms a complex with PINK1 (Plun-Favreau et al., 2007). Moreover, HtrA2 is phosphorylated in a PINK1-dependent manner in response to p38 SAPK (stress-activated protein kinase) pathway activation, suggesting that PINK1 can modulate HtrA2 activity as part of mitochondrial stress response.

To date, only two studies have so far described the characterization of independently derived mutant alleles of Drosophila HtrA2/Omi (Tain et al., 2009a; Yun et al., 2008). One study isolated a small deletion removing HtrA2 (Tain et al., 2009a), while the other identified two chemically induced mutations, a nonsense mutation causing and early truncation and a substitution of conserved V110E (Yun et al., 2008). Phenotypic analysis of these mutants revealed that HtrA2 is required for male fertility, normal lifespan, and stress resistance but not for maintenance of DA neuron. In addition, only very mild mitochondrial defects were quantified in one study (Tain et al., 2009a) but not qualitatively noted in the other V110E allele (Yun et al., 2008). Interestingly, Drosophila HtrA2 has also been reported to be released from the mitochondria upon cellular insults such as UV irradiation, and to cleave DIAP1, the principal

Drosophila inhibitor of apoptosis (IAP), and activate apoptosis (Challa *et al.*, 2007; Igaki *et al.*, 2007; Khan *et al.*, 2008). Tain *et al.* (2009a) addressed the contribution of Drosophila HtrA2 to apoptosis induced by a variety of conditions, including γ-ray or UV irradiation, staurosporine exposure and developmentally regulated cell death but surprisingly found no requirement for HtrA2 in these mechanisms.

Overall, the reports indicate that *HtrA2* mutants have a markedly weaker phenotype than *PINK1* and suggest HtrA2 does not play a central role in the major function(s) of PINK1 that are responsible for conferring *PINK1* phenotypes. However, genetic interaction studies have the potential to reveal less obvious functional relationships. A previous study had shown using over-expression paradigms that *PINK1* and *HtrA2* may be functionally related (Whitworth *et al.*, 2008). First, targeted overexpression of either *PINK1* or *HtrA2* under standard conditions (25 °C) caused mild defects in eye morphology or pigmentation; however, co-overexpression of both genes synergistically en-hanced the phenotype, suggesting a cooperative function. This result was corro-borated by Yun *et al.* (2008). Second, *PINK1* overexpression phenotype is partially suppressed by loss of *HtrA2*, while conversely an *HtrA2* overexpression phenotype is not suppressed by *PINK1* mutations (Whitworth *et al.*, 2008), suggesting that HtrA2 functions downstream of PINK1. In contrast, Yun *et al.* were not able to detect suppression of the eye phenotypes in a similar manner; however, these experiments were conducted at unusually high temperatures (29 °C) which is well known to confound analysis of eye morphology (Kramer and Staveley, 2003), so perhaps the subtle effects reported by Whitworth *et al.* were missed. Complementary studies on loss-of-function phenotypes revealed that *PINK1;HtrA2* double mutants do not show enhancement of *PINK1* mutants alone, consistent with them acting in a common pathway (Tain *et al.*, 2009a; Yun *et al.*, 2008). Furthermore, overexpression of *HtrA2* can partially substitute for loss of *PINK1* (Tain *et al.*, 2009a), similar to the suppression of *PINK1* by *parkin* overexpression described above. Together, these findings are consistent with HtrA2 acting in a common pathway downstream of PINK1, which would support the previously reported molecular interactions (Plun-Favreau *et al.*, 2007).

But what about HtrA2's relationship to parkin? Interestingly, no genet-ic interaction was detected between *HtrA2* and *parkin* using overexpression paradigms, suggesting they may do not share a common function (Whitworth *et al.*, 2008). Tain *et al.* also reported that *parkin:HtrA2* double mutants display a stronger locomotor phenotype than either mutant alone (Tain *et al.*, 2009a), again supporting a view that HtrA2 acts in parallel pathway to parkin. Finally, while loss of either *HtrA2* or *parkin* can partially suppress a PINK1 overexpres-sion phenotype, attenuating both genes together is sufficient to completely suppress the overexpression of PINK1 (Whitworth *et al.*, 2008). Taken together these findings support the view that HtrA2 and parkin are both acting as downstream effectors of PINK1 but in parallel pathways suggesting the PINK1 pathway bifurcates.

The contribution of HtrA2 to PINK1 pathology remains to be determined but it will be interesting to understand how this pathway is triggered and how its activation leads to protection from mitochondrial stress. The p38 SAPK pathway is known to be triggered by reactive oxygen species (ROS), which are produced when mitochondrial electron transport is perturbed (Bradham and McClay, 2006). It is possible that the PDZ domain of HtrA2 acts as a sensor for unfolded proteins, and once activated by p38, HtrA2 may cleave unfolded proteins and/or elicit a response to clear the damaged proteins, analogous to bacterial DegS (Vaux and Silke, 2003). Loss of such a mechanism would have important implications in conditions of mitochondrial dysfunction whether by *PINK1* mutation or by another insult.

V. FUNCTION OF THE PINK1/PARKIN PATHWAY

A recent wave of reports has begun to reveal new insights into the process in which PINK1 and parkin normally function, and how their dysfunction may lead to neuronal death (reviewed by Whitworth and Pallanck, 2009). These findings indicate an important function of the PINK1/parkin pathway is to regulate the interaction of mitochondria, help segregate damaged or dysfunctional units, and promote their degradation by autophagy. This process likely acts as part of a quality control mechanism to recognize terminally damaged mitochondria and safely degrade them to prevent increased ROS production when senescent, and potentially catastrophic rupture and release of proapoptotic factors.

Perhaps, the most significant clue into the mechanism by which PINK1 and parkin influence mitochondrial integrity came from analysis of the Drosophila models which showed that mutations in *PINK1* or *parkin* grossly affect mitochondrial morphology. Although mitochondria are often depicted as static kidney bean shaped organelles, mitochondria are highly motile and interact with one another to form interconnected tubular networks. These dynamic networks undergo continual cycles of fission and fusion controlled by evolutionarily conserved fission- and fusion-promoting factors (Chen and Chan, 2009). Among the known mitochondrial fission- and fusion-promoting factors are the large dynamin-related GTPases dynamin-related protein 1 (Drp1), optic atrophy 1 (Opa1), and mitofusin (Mfn) (Chen and Chan, 2009). Drp1 is a cytosolic factor that assembles with mitochondria to promote mitochondrial fission, whereas Opa1 and Mfn reside in the inner and outer mitochondrial membranes, respectively, where they act to promote mitochondrial fusion (Chen and Chan, 2009). Given that mutations in *PINK1* and *parkin* exhibit aberrant mitochondrial morphology, several laboratories tested the hypothesis that PINK1 and parkin regulate mito-chondrial morphology. These studies revealed that removing a single copy of the fission-promoting factor *Drp1* in *PINK1* or *parkin* mutants dramatically reduces their viability. In contrast, *PINK1* and *parkin* mutant phenotypes are suppressed by overexpressing *Drp1* to enhance mitochondrial fission, or by introducing loss-of-function mutations in genes encoding the fusion-promoting factors *Opa1* and *Mfn* (Deng *et al.*, 2008; Park *et al.*, 2009; Poole *et al.*, 2008; Yang *et al.*, 2008). Together, these findings suggested that the PINK1/parkin pathway promotes mitochondrial fission and/or inhibits mitochondrial fusion. Indeed, these same conclusions were drawn from a largely overlooked work by Riparbelli and Callaini (2007) which described detailed analysis of the spermatogenesis defect in *parkin* mutants. This study found that the failure of spermatids to individualize is preceded by a defect in the remodeling of the specialized mitochondrial derivative, the Nebenkern, which fails to divide into the normal major and minor derivatives, thus appearing like excess fusion or aberrant fission. Admit-tedly, because of the specialized nature of spermatogenesis it cannot be assumed *a priori* that the same mechanism would be shared in the neuromuscular tissues; however, the subsequent studies validate these early observations.

Various conditions, however, indicate that PINK1 and parkin are not obligatory components of the mitochondrial morphogenesis machinery. For ex-ample, mutations in core components of the fission/fusion machinery such as *Drp1*, *Opa1*, and *Mfn1/2* in various organisms have much more severe phenotypes than that exhibited by loss of *PINK1* or *parkin* (Chen *et al.*, 2007; Frezza *et al.*, 2006; McQuibban *et al.*, 2006; Verstreken *et al.*, 2005). More likely, these factors regulate the mitochondrial morphogenesis machinery only in a specific biological context. Some recent findings have suggested possible mechanisms by which derangements in mitochondrial fission could impact tissue viability. Genetic studies of Drosophila *Drp1* showed that loss-of-function mutations result in a failure to efficiently traffic mitochondria to presynaptic terminals in neurons, which in turn impairs calcium buffering and synaptic transmission (Verstreken *et al.*, 2005). While these findings are consistent with some of the phenotypes documented in *PINK1* and *parkin*-deficient flies and mice, the distribution of mitochondria in motor neurons appears to be unaffected in *PINK1*-deficient flies (Morais *et al.*, 2009), and a mitochondrial trafficking defect does not readily account for the flight muscle and male germline defects of *PINK1* and *parkin*-deficient flies, or the selective vulnerability of DA neurons to loss of PINK1 and parkin activity. However, another recent study has provided evidence that PINK1 may interact with the mitochondrial trafficking machinery, Miro and Milton (Weihofen *et al.*, 2008), so further investigation into this mechanism is warranted.

Another mechanism by which defective mitochondrial fission could impact tissue viability derives from a recent study involving live-cell imaging of mitochondrial dynamics in cultured vertebrate cells (Twig *et al.*, 2008a,b).

This study showed that while many products of mitochondrial fission rapidly fuse again with the mitochondrial network, a proportion of the fission products exhibit a decreased membrane potential and a decreased probability of fusion. These defective fission products are frequently targeted to the lysosome for degradation through a process termed mitophagy (Twig et al., 2008a). Autophagic turnover of organelles is becoming recognized as another of the myriad cellular functions of ubiquitin modification, and has been linked to the degradation of peroxisomes (Kim et al., 2008a) and paternally delivered mitochondria (Sutovsky et al., 1999). These findings raise the possibility that derangements in PINK1 and parkin could impair the selective turnover of damaged and dysfunctional mitochondria.

Subsequent work has provided substantial support for this hypothesis by showing that parkin is selectively recruited to damaged mitochondria upon treatment of cultured cells with mitochondrial damaging agents, and that parkin promotes the turnover of these damaged mitochondria (Narendra et al., 2008). Moreover, several studies have also shown that PINK1 is required for the translocation and mitophagy-promoting activity of parkin, consistent with their known genetic hierarchy, ultimately leading to mitochondrial ubiquitination and recruitment of the autophagy machinery (Geisler et al., 2010; Kawajiri et al., 2010; Matsuda et al., 2010; Narendra et al., 2010; Vives-Bauza et al., 2010). Importantly, this mechanism is conserved in Drosophila further supporting it as a major and important function of the pathway (Ziviani et al., 2010).

A couple of key questions arise from these studies; how is PINK1 regulated to stimulate the recruitment of parkin, and how does parkin-mediated ubiquitination of mitochondria help segregate damaged mitochondria and stimulate mitophagy? For the first question, surprisingly, little is still known about the regulation of PINK1. It is generally recognized that PINK1 is rapidly imported into the mitochondria where it has been found in several locations with some part in the intermembrane space and some part on the outer surface (Silvestri et al., 2005; Zhou et al., 2008). It is also recognized that PINK1 exists in at least two major forms: a full-length form and a processed form resulting from an N-terminal cleavage (Silvestri et al., 2005). The cleaved form has been proposed to be exported to the cytoplasm where it is sufficient to exert a neuroprotective function against mitochondrial toxins (Haque et al., 2008). In another key advance from the Drosophila models, the putative processing protease has been proposed to be a member of the rhomboid intramembrane proteases.

Rhomboid proteases, named after the founding member the Drosophila rhomboid gene which is involved in EGF signaling, are an unusual family of proteases that have a specialized function to catalyze a hydrolytic cleavage within the hydrophobic environment of transmembrane domains (Freeman, 2008, 2009). Although not very numerous, the function of rhomboid proteases have been linked to a wide variety of biological processes, including mitochondrial membrane remodeling (Herlan et al., 2003; McQuibban et al., 2003). Coincident with the description of

Drosophila *PINK1* mutants was a characterization of Drosophila *rhomboid-7* mutants (McQuibban *et al.*, 2006), which shared remarkable phenotypic similarity with *PINK1/parkin* mutants, including muscle degeneration, locomotor deficits, male sterility, and mitochondrial pathology. Intriguingly, rhomboid-7 and its homologs are the only known mitochondrially targeted rhomboids, provoking the hypothesis that rhomboid-7 may cleave PINK1 in the mitochondrion. Hence, it was exciting to see that in Drosophila *rhomboid-7* mutants PINK1 is found almost exclusively in the full-length form, suggesting that rhomboid-7 may well be the protease responsible for PINK1 cleavage (Whitworth *et al.*, 2008). Additional experiments revealed that *rhomboid-7* genetically interacted with *PINK1* and *parkin*, in a pattern consistent with it acting upstream of *PINK1* (Whitworth *et al.*, 2008). While coexpression of *rhomboid-7* with *PINK1* or *parkin* caused a synergistically enhanced rough eye phenotype, the *rhomboid-7* overexpression rough eye could be significantly suppressed by removal of *PINK1* but not vice versa (Whitworth *et al.*, 2008). Interestingly, it was also shown in this study that *rhomboid-7* genetically interacts with *HtrA2* and contributes to the generation of one of the cleaved forms of HtrA2.

The cleavage of PINK1 provides a potential mechanism to release it from mitochondria and activate the recruitment of parkin. As mentioned above, it was suggested that PINK1 may directly phosphorylate parkin to stimulate its translocation (Kim *et al.*, 2008b) but this has been disputed by others (Vives-Bauza *et al.*, 2010). More recently, others have described that the processed form of PINK1 is rapidly degraded by the proteasome, and furthermore suggested the cleavage may be a mechanism to constitutively remove PINK1 from healthy mitochondria inactivating the signal and preventing parkin recruitment (Narendra *et al.*, 2010). This inference comes principally from compelling evidence that full-length PINK1 is stabilized upon mitochondrial toxification, a condition which stimulates parkin recruitment. In contrast, this study suggests that the mammalian homolog of rhomboid-7, PARL, does not contribute to the cleavage of mammalian PINK1; however, the data for this are rather weak and inconclusive and are disputed by others (Deas, Plun-Favreau, and Wood, personal communication). Nevertheless, the model that stabilization of full-length PINK1 is the active form that recruits parkin is in complete contradiction to the genetic interactions of *rhomboid-7* with *PINK1* and *parkin*. While the mechanism that regulates PINK1 activity remains to be resolved, it seems certain that some as yet known signal activates PINK1 to stimulate the recruitment of parkin to dysfunctional mitochondria.

Recent work has also advanced our understanding of how parkin-mediated ubiquitination may affect mitophagy. Ubiquitination of the outer mitochondrial surface is thought to promote mitophagy by acting as a recruitment signal of the adaptor molecule p62/SQSTM1 which directly interacts with both ubiquitin on the degradation substrate and LC3 on the nascent autophagic membrane (Kirkin *et al.*, 2009). Hence, elucidating the identity of parkin substrates has come under intense scrutiny. Two proteins localized to the mitochondrial outer surface have recently been reported as putative parkin substrates, mammalian VDAC1 and Drosophila Mfn (Geisler *et al.*, 2010; Poole *et al.*, 2010; Ziviani *et al.*, 2010). While the modification of the abundant protein VDAC1 would provide a robust signal for autophagosome recruitment, the ubiquitination of Mfn is a particularly exciting target due to its role in mitochondrial fusion. As discussed above, dysfunctional mitochondria destined for destruction appear to be sequestered from the rest of the healthy network prior to engulfment (Twig *et al.*, 2008a), but it is unclear how such a mechanism works.

Interestingly, instead of observing a "smear" of polyubiquitin adducts Mfn was modified as two distinct ubiquitinated forms which likely correspond to mono- and multiubiquitinated Mfn. A number of hypotheses have been proposed as to how this modification may be crucial to regulated mitophagy (Ziviani and Whitworth, 2010). The first proposes that ubiquitinated Mfn may be degraded by the proteasome, thus removing the profusion factor from the outer surface in a process similar to the removal of misfolded proteins from the endoplasmic reticulum (a process known as Endoplasmic Reticulum-Associated Protein Degradation [ERAD]). A similar mechanism could remove ubiquitinated Mfn specifically from damaged mitochondria and reduce their refusion capacity. However, monoubiquitination does not typically lead to degradation. The ubiquitinated protein may develop new characteristics that affect its activity, subcellular localization, or its interaction with other proteins. Indeed, mono-ubiquitination is becoming recognized as an important regulatory mechanism in controlling protein activity and signaling. A second hypothesis proposes that Mfn ubiquitination might physically interfere with the formation of *trans* Mfn dimers and prevent mitochondria tethering. This process is required for fusion, and its abrogation would thus preclude refusion of damaged mitochondria.

Regardless of the mechanism, these findings provide a molecular explanation for the previously reported genetic interactions; briefly, genetic perturbations that promote mitochondrial fragmentation suppress *PINK1/parkin* mutant phenotypes. Consistent with this, loss of *parkin* or *PINK1* leads to an increased abundance of Mfn and hyperfused mitochondria (Poole *et al.*, 2010). While it remains to be shown that mammalian parkin acts in an analogous fashion to regulate mitophagy, some reports indicate this may be likely (Tanaka and Youle, personal communication). As Drosophila Mfn is the sole ubiquitously expressed Mitofusin homolog and likely performs functions of both mammalian Mfn1 and Mfn2, it will be interesting to determine whether this mechanism may be preferentially mediated by Mfn1 or Mfn2.

In consideration of all the current data, the following mechanism by which regulated fusion and fission of mitochondria may contribute to a quality control process that recognizes and removes damaged mitochondria is proposed (Fig. 1.2). Following a fission event, a terminally damaged daughter

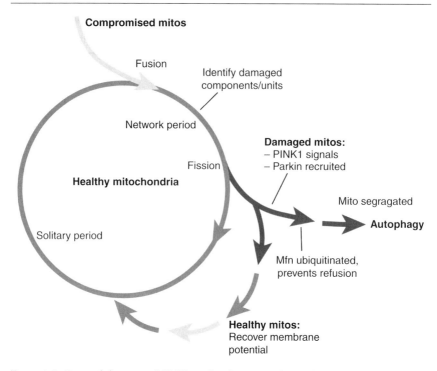

Figure 1.2. Proposed function of PINK1 and parkin in regulating the segregation of damaged mitochondria. Mitochondrial fusion provides a mechanism for partially damaged (compromised) mitochondria to mix their contents with healthy mitochondria and allow complementation of dysfunctional components (proteins, lipids, and mtDNA). Subsequent fission can generate a daughter mitochondrion that has accumulated a significant degree of damage such that complementation would be insufficient, and must be degraded to avoid potentially catastrophic rupture. An unknown mechanism stimulates PINK1 to signal to the recruitment of parkin to that mitochondrion. Parkin ubiquitinates Mfn, which prevents refusion and segregates that mitochondrion for subsequent autophagy. (See Color Insert.)

mitochondrion is detected, perhaps by an unidentified internal mechanism, and stimulates PINK1 to promote the recruitment of parkin. parkin translocates to that mitochondrion where it ubiquitinates Mfn. The initial effect of parkin monoubiquitination of Mfn may be to interfere with mitochondria tethering and exclude damaged mitochondria from rejoining the mitochondria network. Subsequently, p62/SQSTM1 may recognize multiubiquitinated Mfn or VDAC1 on these mitochondria and recruit autophagosomes. In this way, mitochondria that have acquired an unsustainable amount of damage, for example, through prolonged exposure to oxidative stress, are sequestered and safely destroyed, and a healthy mitochondrial network is maintained.

Overall, such a mechanism is attractive for several reasons: first, it offers an explanation for the variety of mitochondrial defects that have been documented in PINK1 and parkin-deficient cell lines, including decreased membrane potential, deficits in the electron transport chain complexes, reduced ATP synthesis, decreased mitochondrial DNA synthesis, and aberrant mitochondrial calcium efflux (Gandhi et al., 2009; Gegg et al., 2009; Morais et al., 2009) by suggesting that these pleiotropic phenotypes derive from the accumulation of damaged mitochondria in the absence of a functional mitochondrial quality control system. Second, these findings would explain the protective effects of PINK1 and parkin overexpression from exposure to mitochondrial toxins (Haque et al., 2008; Paterna et al., 2007; Rosen et al., 2006); and finally, the abundant mitochondrial DNA mutational load of substantia nigra DA neurons (Bender et al., 2006; Kraytsberg et al., 2006) would neatly account for the selective vulnerability of this population of cells to the loss of a mitochondrial quality control system. While many of the details of the mechanism still remain to be resolved, it is clear that significant advances have been made in our understanding of how the PINK1/parkin pathway promotes neuronal longevity.

VI. CONVERGENT THERAPEUTIC APPROACHES

A vast amount of evidence links many neurodegenerative diseases with some contribution, direct or secondary, from oxidative stress (Sayre et al., 2008). This situation is not entirely surprising since a natural and unavoidable consequence of oxygen metabolism is the production of potentially damaging ROS. These molecules can chemically modify, and presumably damage, all manner of cellular components including protein, lipid, DNA, and so on. A host of complex and normally robust mechanisms exists to combat or repair ROS-mediated oxidative damage; however, the effect of compromised antioxidant defenses is most obviously threatening to highly energy-demanding, postmitotic tissues such as the nervous system. Here again, Drosophila models offer a unique advantage to modeling this process in vivo since adult flies are composed almost entirely of postmitotic tissues, with the exception of germ cells and some intestinal tissues, thus avoiding the technical limitations of needing to focus solely on the nervous system. As a result, much effort has been invested in studying the role of oxidative stress in the pathogenesis of PD and methods to combat it using the Drosophila models.

All of the models to date have exhibited, to a greater or lesser degree, some sensitivity to oxidative stress, typically induced via either exposure to high levels or oxygen or oxidative chemicals such as hydrogen peroxide or paraquat. This has primed the field to rapidly address the relative efficacy of transgenic or pharmacologic antioxidant mechanisms. One of the earliest leads came from a genetic screen for modifiers of *parkin* phenotypes which identified a number of factors that influence the partial lethality of *parkin* mutants including genes involved in the oxidative stress pathway (Greene *et al.*, 2005). The most potent enhancer of the *parkin* partial lethal phenotype was a loss-of-function allele of the *glutathione S-transferase S1* (*GstS1*) gene. Furthermore, subsequent work found that transgenic overexpression of GstS1 was able to significantly suppress the loss of DA neurons in *parkin* mutants (Whitworth *et al.*, 2005).

The GST family of polypeptides is thought to act in cellular detoxification pathways and by regulating the cellular redox balance in a number of ways, such as covalently coupling GSH to a variety of products of oxidative damage, including 4-hydroxynonenol, or by maintaining the correct redox state of protein thiol groups (Hayes *et al.*, 2005). The interest in GSTs as a potential therapeutic is significantly elevated with the recognition that there are several dietary components and drugs that are able to induce the expression of glutathione S-transferase and GSH production in vertebrates (Nguyen *et al.*, 2003). In an interesting development to this finding, the protective effects of GSTS1 have also been demonstrated against α-synuclein toxicity (Trinh *et al.*, 2008). Furthermore, the therapeutic potential of chemical inducers of GSTs was demonstrated with the administration of allyl disulfide and sulforaphane, which are known to induce phase II detoxifying enzymes, were able to potently suppress α-synuclein-induced DA neurodegeneration (Trinh *et al.*, 2008).

In another exciting development, Pallanck's group has also begun to reveal the mechanisms behind the previously recognized association of coffee and tobacco use with decreased risk of PD. Epidemiological studies have provided some remarkable evidence that drinking coffee or smoking tobacco can reduce the incidence of PD but the reasons have been unclear. One hypothesis has proposed that caffeine and nicotine may confer symptomatic relief; alternatively, coffee and tobacco may contain neuroprotective agents (Quik *et al.*, 2008; Schwarzschild *et al.*, 2006). To address these hypotheses directly, Trinh *et al.* used two models of PD, the transgenic α-synuclein and *parkin* mutant lines, and revealed that feeding these flies extracts of tobacco and coffee was able to potently suppress age-dependent DA neuron loss (Trinh *et al.*, 2010), mirroring the epidemiological findings. Furthermore, they showed that caffeine and nicotine did not contribute to the neuroprotective effects, suggesting the "regular" coffee and tobacco contains some chemical(s) that actively promotes neuroprotective mechanisms. Interestingly, coffee and tobacco also appear to provide protection in diverse neurodegenerative models including Alzheimer's and polyglutamine disease, implicating the induction of a general neuroprotective program. To understand the underlying mechanism, the Keap–Nrf2 pathway was assessed as a likely candidate. Nrf2 is a transcription factor that under normal conditions is sequestered in the cytoplasm by Keap1, however, upon oxidative stress Keap1 releases Nrf2 which can drive transcription of cytoprotective proteins. Using a genetic approach, Trinh and colleagues found that attenuation of the Drosophila Nrf2 homolog, Cnc, ablated the protective effects of coffee and tobacco. These findings support the need to investigate further the potential for manipulation of Nrf2 as a therapeutic approach, especially through the action of bioactive compounds such as those found in coffee and tobacco.

The general role of oxidative stress and potential of antioxidant approaches is evident from multiple additional studies using transgenic lines, expressing Cu/Zn superoxide dismutase or methionine sulfoxide reductase, and small-molecule antioxidants such as glutathione, S-methyl L-cysteine, N-acet-ylcysteine, ascorbic acid, and various other antioxidants, all having positive effects in α-synuclein, parkin, and DJ-1 models (Botella et al., 2008; Faust et al., 2009; Park et al., 2007; Saini et al., 2010; Wassef et al., 2007). The therapeutic potential is further extended to sporadic models by the positive effects of various polyphenol antioxidants against paraquat-induced neurotoxicity (Jimenez-Del-Rio et al., 2010). One potential source of oxidative stress, also suggested to be a contributing to environmental risk factor, is the chemical reaction catalyzed by heavy metals (iron, copper, and manganese) known as the Fenton chemistry. This reaction results in the conversion of relatively labile hydrogen peroxide to the much more potent hydroxyl radicals. Unsurprisingly, parkin mutants are highly sensitive to Fe and Cu exposure; conversely, the administration of metal chelating agents significantly improved lifespan of parkin mutants (Saini et al., 2010). In contrast, it was also reported that uptake of dietary Zn markedly improved parkin phenotypes (Saini and Schaffner, 2010). The reasons for this are currently unclear but are thought to be independent from the Fenton chemistry and may be due to antioxidant properties of Zn.

While all these findings add further weight to the magnitude of evidence that oxidative stress has a contributing role to play in the pathogenesis of PD, the general lack of evidence that antioxidant therapies so far attempted provide any benefit is disappointing. These previous approaches may be ham-pered by technical difficulties, not only least in delivering small-molecule thera-peutics across the blood–brain barrier but they also fail to shed light on a key area of consideration for PD and almost all neurodegenerative disorders, that of cell type specificity—what makes DA neurons particularly susceptible? While it should be immediately cautioned that nigral DA neurons are prominently affected in PD, it is certainly true that other neuronal types eventually succumb to disease. So the problem is not simply a matter of DA specificity, although this is an obvious culprit to probe. Indeed, an unavoidable linkage to our old adversary, oxidative stress, is provoked by the well-recognized fact that the normal metabolism of DA produces ROS and dopa–quinone conjugates (Hattori et al., 2009). Hence, a number of groups have utilized the tractability of the Drosophila models to address the contribution of DA to selective neuronal vulnerability.

DA is synthesized in the cytosol and packaged into synaptic vesicles by the conserved vesicular monoamine transporter (VMAT). A consensus appears to be emerging that increased production of cytoplasmic DA, by transgenic expression of the rate limiting enzyme TH (encoded by the *pale* gene in Drosophila) or by downregulation of *VMAT*, enhances toxicity of α-*synuclein* or mutant human *parkin* models (Park *et al.*, 2007; Sang *et al.*, 2007). In contrast, depletion of cytosolic DA by RNAi of *pale* or overexpression of *VMAT* to accelerate DA packaging was protective in these models. Furthermore, parallel studies have revealed potential significance for DA metabolism in sporadic PD since loss of Drosophila *VMAT* function confers sensitivity to paraquat and the mitochondrial toxin rotenone (Lawal *et al.*, 2010), whereas overexpression of *VMAT* or downregulation of *pale* is protective against rotenone toxicity (Bayersdorfer *et al.*, 2010; Lawal *et al.*, 2010). Interestingly, VMAT expression does not protect against paraquat toxicity suggesting the principal insult from paraquat and rotenone may be different (Lawal *et al.*, 2010). The overall findings from these studies suggest that promoting the packaging of cytoplasmic DA into vesicles reduces toxicity; however, contradictory findings results have also been reported. One study characterizing of the effects of various genetic and pharmacologic manipulation of DA upon a paraquat and hydrogen peroxide-induced neurotoxicity described a complex effect of increasing and decreasing the production of DA and its cofactor tetrahydrobiopterin on Drosophila survival and DA neurodegeneration (Chaudhuri *et al.*, 2007). Again, some differences were observed between effects on paraquat versus hydrogen peroxide treatment, which may reflect different toxic mechanisms. While some of these details await verification, the data are consistent with the notion that the presence of DA may impact on the burden of toxicity contributing to cell type specificity.

A final link to novel therapeutic strategies in which Drosophila models have contributed involves members of the histone deacetylase (HDAC) protein family. First, numerous studies of yeast and metazoan Sir2 homologs indicate that they play a role in multiple cellular processes including cell survival and promotion of organismal life span (Blander and Guarente, 2004). A number of chemical inhibitors were identified that selectively inhibit yeast Sir2 activity (Outeiro *et al.*, 2007), and were found to decrease the number but increase the size of α-synuclein aggregates. Administration of the Sir2 chemical inhibitors to α-synuclein-expressing flies was shown to prevent the toxicity of α-synuclein in a dose-dependent manner (Outeiro *et al.*, 2007). These findings are particularly compelling since other HDAC inhibitors have been shown to ameliorate expanded polyglutamine-induced toxicity (Steffan *et al.*, 2001). The mechanisms by which inhibiting HDAC activity relieves toxicity in these models of neurodegeneration remain unclear. However, universal HDAC inhibition may not be a sensible strategy. In contrast to these results, genetic ablation of Drosophila HDAC6 was found to exacerbate α-synuclein neurotoxicity (Du *et al.*, 2010).

Conversely, HDAC6 overexpression ameliorated α-synuclein retinal disruption, but suppression of DA neurodegeneration was not reported in this study. Interestingly, HDAC6 overexpression did cause a decrease in the amount of soluble α-synuclein oligomers correlating with neuroprotection, consistent with previously discussed mechanisms of chaperone-mediated protection. Finally, another valuable link should be noted here since HDAC6 has been implicated in the autophagic removal of protein aggregates and dysfunctional organelles. Hence, this mechanism may also have relevance to the PINK1/parkin-induced mitochondrial dysfunction.

In summary, as we learn more about the pathogenic mechanism(s) underlying the disease more putative therapeutic interventions will come to light. Without doubt the biggest advantage that the Drosophila models can bring to the field is genetic analyses that reveal new insights into the pathogenic mechanism and potentially protective pathways. Our current understanding would indicate that there is no single pathogenic insult decreasing the likelihood of a single "magic bullet" disease-modifying therapy. More likely, such an intervention will need to consider multiple toxic events. However, the findings discussed here have already highlighted a few key regulators of important protective mechanisms. One could envisage a combined approach to activate the production of protective factors such as chaperone, antioxidant, and phase II detoxifying enzymes, perhaps through Nrf2- or 4E-BP-mediated mechanisms, alongside damage removal processes by stimulation of autophagy. In any case, the availability and validity of Drosophila models for PD provide an excellent opportunity for early stage drug testing to reduce the cost and risk associated with developing such a therapeutic, prior to further testing in mammalian preclinical models.

VII. CONCLUSIONS

The fruit fly *D. melanogaster* has proven itself to be an invaluable model system in basic studies of genetics and biology for over 100 years. Given the remarkable degree of genetic, molecular, and cell biological conservation between flies and mammals that has been revealed over the decades, Drosophila remains a valid model system in which to address novel biological questions in the future, including those relevant to human health. Indeed, work aimed at an understanding of the genes involved in heritable forms of PD has already made significant contributions to our understanding of the causes of this debilitating disease. Moreover, these studies have begun to define small-molecule compounds that could potentially impinge on the pathways implicated in PD. Drosophila models also lend themselves directly to unbiased high-throughput screening of small-molecule libraries in the search for better compounds that are able to combat pathogenesis.

While the Drosophila system has enormous potential for studies aimed at a mechanistic understanding of PD, it is also important to point out that PD modeling in Drosophila is still relatively new and that the reagents and methodologies for conducting these studies are evolving. In particular, important priorities of future work should be to define the most appropriate methods for analyzing DA neuron integrity and to resolve conflicts in studies of some of the current Drosophila models of PD. It is also imperative that we have realistic expectations from these models. For example, at present the phenotypes of the loss-of-function Drosophila models of PD do not precisely match the phenotypes of humans with mutations in the corresponding genes. While these findings may be of concern, phenotypic differences resulting from mutations in orthologous genes in different species often belies significant underlying molecular pathway conservation. Indeed, the weight of evidence suggests that the phenotypes associated with the Drosophila *parkin*, *PINK1*, and *DJ-1* models, which include sensitivity to oxidative stress agents and mitochondrial dysfunction, are directly relevant to the mechanisms implicated in PD. Nevertheless, Drosophila are still a relatively simple model organism, far less complex than humans, hence, it is understandable that some aspects that may be relevant or even crucial to our understanding of a particular human disease may not be evident in Drosophila. For example, certain triplet repeat sequences underlying numerous neurological diseases are known to expand during germline transmission, causing a more severe syndrome in subsequent generations, but appear to be refractory to expansion in Drosophila (Jackson *et al.*, 2005). However, to date we have only just begun to tap the insight that modeling PD in Drosophila can potentially provide. I believe that these advantages will be best realized if we focus our efforts on understanding what the phenotypes of these models are telling us and take full advantage of the power of genetics to lead us down unexpected, and even unintuitive, paths.

Acknowledgments

I wish to thank the members of my laboratory for their time, effort, and commitment to these projects. I would also like to thank Leo Pallanck for his generous support in establishing my own group and for continued stimulating discussions. The work in my lab is supported by grants from the Wellcome Trust (WT081987 and the Neurodegeneration Disease Initiative WT089698), Parkinson's UK (G-0713), and the European Commission FP7 (241791). The MRC Centre for Developmental and Biomedical Genetics is supported by Grant G070091.

References

Adams, M. D., Celniker, S. E., Holt, R. A., Evans, C. A., Gocayne, J. D., Amanatides, P. G., Scherer, S. E., Li, P. W., Hoskins, R. A., Galle, R. F., *et al.* (2000). The genome sequence of *Drosophila melanogaster*. *Science* **287**, 2185–2195.
Andrews, H. K., Zhang, Y. Q., Trotta, N., and Broadie, K. (2002). Drosophila sec10 is required for hormone secretion but not general exocytosis or neurotransmission. *Traffic* **3**, 906–921.

Ashburner, M., and Novitski, E. (1976). The Genetics and Biology of Drosophila. London, Academic Press.

Auluck, P. K., and Bonini, N. M. (2002). Pharmacological prevention of Parkinson disease in Drosophila. *Nat. Med.* **8**, 1185–1186.

Auluck, P. K., Chan, H. Y., Trojanowski, J. Q., Lee, V. M., and Bonini, N. M. (2002). Chaperone suppression of alpha-synuclein toxicity in a Drosophila model for Parkinson's disease. *Science* **295**, 865–868.

Auluck, P. K., Meulener, M. C., and Bonini, N. M. (2005). Mechanisms of suppression of {alpha}-synuclein neurotoxicity by geldanamycin in Drosophila. *J. Biol. Chem.* **280**, 2873–2878.

Bayersdorfer, F., Voigt, A., Schneuwly, S., and Botella, J. A. (2010). Dopamine-dependent neuro-degeneration in Drosophila models of familial and sporadic Parkinson's disease. *Neurobiol. Dis.* **40** (1), 113–119.

Bellen, H. J., Levis, R. W., Liao, G., He, Y., Carlson, J. W., Tsang, G., Evans-Holm, M., Hiesinger, P. R., Schulze, K. L., Rubin, G. M., *et al.* (2004). The BDGP gene disruption project: Single transposon insertions associated with 40% of Drosophila genes. *Genetics* **167**, 761–781.

Bender, A., Krishnan, K. J., Morris, C. M., Taylor, G. A., Reeve, A. K., Perry, R. H., Jaros, E., Hersheson, J. S., Betts, J., Klopstock, T., *et al.* (2006). High levels of mitochondrial DNA deletions in substantia nigra neurons in aging and Parkinson disease. *Nat. Genet.* **38**, 515–517.

Bennett, M. C. (2005). The role of alpha-synuclein in neurodegenerative diseases. *Pharmacol. Ther.* **105**, 311–331.

Bernal, A., and Kimbrell, D. A. (2000). Drosophila Thor participates in host immune defense and connects a translational regulator with innate immunity. *Proc. Natl. Acad. Sci. USA* **97**, 6019–6024.

Bier, E. (2005). Drosophila, the golden bug, emerges as a tool for human genetics. *Nat. Rev.* **6**, 9–23.

Birman, S., Morgan, B., Anzivino, M., and Hirsh, J. (1994). A novel and major isoform of tyrosine hydroxylase in Drosophila is generated by alternative RNA processing. *J. Biol. Chem.* **269**, 26559–26567.

Blander, G., and Guarente, L. (2004). The Sir2 family of protein deacetylases. *Annu. Rev. Biochem.* **73**, 417–435.

Bogaerts, V., Nuytemans, K., Reumers, J., Pals, P., Engelborghs, S., Pickut, B., Corsmit, E., Peeters, K., Schymkowitz, J., De Deyn, P. P., *et al.* (2008). Genetic variability in the mitochondrial serine protease HTRA2 contributes to risk for Parkinson disease. *Hum. Mutat.* **29**, 832–840.

Bonifati, V., Rizzu, P., van Baren, M. J., Schaap, O., Breedveld, G. J., Krieger, E., Dekker, M. C., Squitieri, F., Ibanez, P., Joosse, M., *et al.* (2003). Mutations in the DJ-1 gene associated with autosomal recessive early-onset parkinsonism. *Science* **299**, 256–259.

Botella, J. A., Bayersdorfer, F., and Schneuwly, S. (2008). Superoxide dismutase overexpression protects dopaminergic neurons in a Drosophila model of Parkinson's disease. *Neurobiol. Dis.* **30**, 65–73.

Boutros, M., Kiger, A. A., Armknecht, S., Kerr, K., Hild, M., Koch, B., Haas, S. A., Paro, R., and Perrimon, N. (2004). Genome-wide RNAi analysis of growth and viability in Drosophila cells. *Science* **303**, 832–835.

Bradham, C. A., and McClay, D. R. (2006). p38 MAPK is essential for secondary axis specification and patterning in sea urchin embryos. *Development* **133**, 21–32.

Brand, A. H., and Perrimon, N. (1993). Targeted gene expression as a means of altering cell fates and generating dominant phenotypes. *Development* **118**, 401–415.

Budnik, V., and White, K. (1988). Catecholamine-containing neurons in *Drosophila melanogaster*: Distribution and development. *J. Comp. Neurol.* **268**, 400–413.

Cha, G. H., Kim, S., Park, J., Lee, E., Kim, M., Lee, S. B., Kim, J. M., Chung, J., and Cho, K. S. (2005). Parkin negatively regulates JNK pathway in the dopaminergic neurons of Drosophila. *Proc. Natl. Acad. Sci. USA* **102**, 10345–10350.

Challa, M., Malladi, S., Pellock, B. J., Dresnek, D., Varadarajan, S., Yin, Y. W., White, K., and Bratton, S. B. (2007). Drosophila Omi, a mitochondrial-localized IAP antagonist and proapoptotic serine protease. *EMBO J.* **26,** 3144–3156.

Chaudhuri, A., Bowling, K., Funderburk, C., Lawal, H., Inamdar, A., Wang, Z., and O'Donnell, J. M. (2007). Interaction of genetic and environmental factors in a Drosophila parkinsonism model. *J. Neurosci.* **27,** 2457–2467.

Chen, H., and Chan, D. C. (2009). Mitochondrial dynamics—Fusion, fission, movement, and mitophagy—In neurodegenerative diseases. *Hum. Mol. Genet.* **18,** R169–R176.

Chen, L., and Feany, M. B. (2005). Alpha-synuclein phosphorylation controls neurotoxicity and inclusion formation in a Drosophila model of Parkinson disease. *Nat. Neurosci.* **8,** 657–663.

Chen, H., McCaffery, J. M., and Chan, D. C. (2007). Mitochondrial fusion protects against neurodegeneration in the cerebellum. *Cell* **130,** 548–562.

Chen, L., Periquet, M., Wang, X., Negro, A., McLean, P. J., Hyman, B. T., and Feany, M. B. (2009). Tyrosine and serine phosphorylation of alpha-synuclein have opposing effects on neurotoxicity and soluble oligomer formation. *J. Clin. Invest.* **119,** 3257–3265.

Clark, I. E., Dodson, M. W., Jiang, C., Cao, J. H., Huh, J. R., Seol, J. H., Yoo, S. J., Hay, B. A., and Guo, M. (2006). Drosophila pink1 is required for mitochondrial function and interacts genetically with parkin. *Nature* **441,** 1162–1166.

Conway, K. A., Harper, J. D., and Lansbury, P. T. (1998). Accelerated in vitro fibril formation by a mutant alpha-synuclein linked to early-onset Parkinson disease. *Nat. Med.* **4,** 1318–1320.

Dagda, R. K., Cherra, S. J., 3rd, Kulich, S. M., Tandon, A., Park, D., and Chu, C. T. (2009). Loss of PINK1 function promotes mitophagy through effects on oxidative stress and mitochondrial fission. *J. Biol. Chem.* **284,** 13843–13855.

Deng, H., Dodson, M. W., Huang, H., and Guo, M. (2008). The Parkinson's disease genes pink1 and parkin promote mitochondrial fission and/or inhibit fusion in Drosophila. *Proc. Natl. Acad. Sci. USA* **105,** 14503–14508.

Dietzl, G., Chen, D., Schnorrer, F., Su, K. C., Barinova, Y., Fellner, M., Gasser, B., Kinsey, K., Oppel, S., Scheiblauer, S., et al. (2007). A genome-wide transgenic RNAi library for conditional gene inactivation in Drosophila. *Nature* **448,** 151–156.

Drobysheva, D., Ameel, K., Welch, B., Ellison, E., Chaichana, K., Hoang, B., Sharma, S., Neckameyer, W., Srinakevitch, I., Murphy, K. J., et al. (2008). An optimized method for histological detection of dopaminergic neurons in Drosophila melanogaster. *J. Histochem. Cytochem.* **56,** 1049–1063.

Du, G., Liu, X., Chen, X., Song, M., Yan, Y., Jiao, R., and Wang, C. C. (2010). Drosophila histone deacetylase 6 protects dopaminergic neurons against {alpha}-synuclein toxicity by promoting inclusion formation. *Mol. Biol. Cell* **21,** 2128–2137.

Eriksen, J. L., Przedborski, S., and Petrucelli, L. (2005). Gene dosage and pathogenesis of Parkinson's disease. *Trends Mol. Med.* **11,** 91–96.

Exner, N., Treske, B., Paquet, D., Holmstrom, K., Schiesling, C., Gispert, S., Carballo-Carbajal, I., Berg, D., Hoepken, H. H., Gasser, T., et al. (2007). Loss-of-function of human PINK1 results in mitochondrial pathology and can be rescued by parkin. *J. Neurosci.* **27,** 12413–12418.

Faust, K., Gehrke, S., Yang, Y., Yang, L., Beal, M. F., and Lu, B. (2009). Neuroprotective effects of compounds with antioxidant and anti-inflammatory properties in a Drosophila model of Parkinson's disease. *BMC Neurosci.* **10,** 109.

Feany, M. B., and Bender, W. W. (2000). A Drosophila model of Parkinson's disease. *Nature* **404,** 394–398.

Fernandez-Funez, P., Nino-Rosales, M. L., de Gouyon, B., She, W. C., Luchak, J. M., Martinez, P., Turiegano, E., Benito, J., Capovilla, M., Skinner, P. J., et al. (2000). Identification of genes that modify ataxin-1-induced neurodegeneration. *Nature* **408,** 101–106.

Flinn, L., Mortiboys, H., Volkmann, K., Koster, R. W., Ingham, P. W., and Bandmann, O. (2009). Complex I deficiency and dopaminergic neuronal cell loss in parkin-deficient zebrafish (*Danio rerio*). *Brain* **132**, 1613–1623.

Forno, L. S. (1996). Neuropathology of Parkinson's disease. *J. Neuropathol. Exp. Neurol.* **55**, 259–272.

Freeman, M. (2008). Rhomboid proteases and their biological functions. *Annu. Rev. Genet.* **42**, 191–210.

Freeman, M. (2009). Rhomboids: 7 years of a new protease family. *Semin. Cell Dev. Biol.* **20**, 231–239.

Frezza, C., Cipolat, S., Martins de Brito, O., Micaroni, M., Beznoussenko, G. V., Rudka, T., Bartoli, D., Polishuck, R. S., Danial, N. N., De Strooper, B., *et al.* (2006). OPA1 controls apoptotic cristae remodeling independently from mitochondrial fusion. *Cell* **126**, 177–189.

Friggi-Grelin, F., Coulom, H., Meller, M., Gomez, D., Hirsh, J., and Birman, S. (2003). Targeted gene expression in Drosophila dopaminergic cells using regulatory sequences from tyrosine hydroxylase. *J. Neurobiol.* **54**, 618–627.

Fujiwara, H., Hasegawa, M., Dohmae, N., Kawashima, A., Masliah, E., Goldberg, M. S., Shen, J., Takio, K., and Iwatsubo, T. (2002). alpha-Synuclein is phosphorylated in synucleinopathy lesions. *Nat. Cell Biol.* **4**, 160–164.

Gandhi, S., and Wood, N. W. (2010). Genome-wide association studies: The key to unlocking neurodegeneration? *Nat. Neurosci.* **13**, 789–794.

Gandhi, S., Wood-Kaczmar, A., Yao, Z., Plun-Favreau, H., Deas, E., Klupsch, K., Downward, J., Latchman, D. S., Tabrizi, S. J., Wood, N. W., *et al.* (2009). PINK1-associated Parkinson's disease is caused by neuronal vulnerability to calcium-induced cell death. *Mol. Cell* **33**, 627–638.

Gasser, T. (2009). Molecular pathogenesis of Parkinson disease: Insights from genetic studies. *Expert Rev. Mol. Med.* **11**, e22.

Gegg, M. E., Cooper, J. M., Schapira, A. H., and Taanman, J. W. (2009). Silencing of PINK1 expression affects mitochondrial DNA and oxidative phosphorylation in dopaminergic cells. *PLoS ONE* **4**, e4756.

Gehrke, S., Imai, Y., Sokol, N., and Lu, B. (2010). Pathogenic LRRK2 negatively regulates micro-RNA-mediated translational repression. *Nature* **466**, 637–641.

Geisler, S., Holmstrom, K. M., Skujat, D., Fiesel, F. C., Rothfuss, O. C., Kahle, P. J., and Springer, W. (2010). PINK1/Parkin-mediated mitophagy is dependent on VDAC1 and p62/SQSTM1. *Nat. Cell Biol.* **12**, 119–131.

Gong, W. J., and Golic, K. G. (2003). Ends-out, or replacement, gene targeting in Drosophila. *Proc. Natl. Acad. Sci. USA* **100**, 2556–2561.

Greene, J. C., Whitworth, A. J., Kuo, I., Andrews, L. A., Feany, M. B., and Pallanck, L. J. (2003). Mitochondrial pathology and apoptotic muscle degeneration in Drosophila parkin mutants. *Proc. Natl. Acad. Sci. USA* **100**, 4078–4083.

Greene, J. C., Whitworth, A. J., Andrews, L. A., Parker, T. J., and Pallanck, L. J. (2005). Genetic and genomic studies of Drosophila parkin mutants implicate oxidative stress and innate immune responses in pathogenesis. *Hum. Mol. Genet.* **14**, 799–811.

Groth, A. C., Fish, M., Nusse, R., and Calos, M. P. (2004). Construction of transgenic Drosophila by using the site-specific integrase from phage phiC31. *Genetics* **166**, 1775–1782.

Gupta, A., Dawson, V. L., and Dawson, T. M. (2008). What causes cell death in Parkinson's disease? *Ann. Neurol.* **64**(Suppl. 2), S3–S15.

Hao, L. Y., Giasson, B. I., and Bonini, N. M. (2010). DJ-1 is critical for mitochondrial function and rescues PINK1 loss of function. *Proc. Natl. Acad. Sci. USA* **107**, 9747–9752.

Haque, M. E., Thomas, K. J., D'Souza, C., Callaghan, S., Kitada, T., Slack, R. S., Fraser, P., Cookson, M. R., Tandon, A., and Park, D. S. (2008). Cytoplasmic Pink1 activity protects neurons from dopaminergic neurotoxin MPTP. *Proc. Natl. Acad. Sci. USA* **105**, 1716–1721.

Hattori, N., Wanga, M., Taka, H., Fujimura, T., Yoritaka, A., Kubo, S., and Mochizuki, H. (2009). Toxic effects of dopamine metabolism in Parkinson's disease. *Parkinsonism Relat. Disord.* **15** (Suppl. 1), S35–S38.

Hayes, J. D., Flanagan, J. U., and Jowsey, I. R. (2005). Glutathione transferases. *Annu. Rev. Pharmacol. Toxicol.* **45**, 51–88.

Herlan, M., Vogel, F., Bornhovd, C., Neupert, W., and Reichert, A. S. (2003). Processing of Mgm1 by the rhomboid-type protease Pcp1 is required for maintenance of mitochondrial morphology and of mitochondrial DNA. *J. Biol. Chem.* **278**, 27781–27788.

Hiesinger, P. R., and Bellen, H. J. (2004). Flying in the face of total disruption. *Nat. Genet.* **36**, 211–212.

Igaki, T., Suzuki, Y., Tokushige, N., Aonuma, H., Takahashi, R., and Miura, M. (2007). Evolution of mitochondrial cell death pathway: Proapoptotic role of HtrA2/Omi in Drosophila. *Biochem. Biophys. Res. Commun.* **356**, 993–997.

Imai, Y., Gehrke, S., Wang, H. Q., Takahashi, R., Hasegawa, K., Oota, E., and Lu, B. (2008). Phosphorylation of 4E-BP by LRRK2 affects the maintenance of dopaminergic neurons in Drosophila. *EMBO J.* **27**, 2432–2443.

Jackson, S. M., Whitworth, A. J., Greene, J. C., Libby, R. T., Baccam, S. L., Pallanck, L. J., and La Spada, A. R. (2005). A SCA7 CAG/CTG repeat expansion is stable in *Drosophila melanogaster* despite modulation of genomic context and gene dosage. *Gene* **347**, 35–41.

Jimenez-Del-Rio, M., Guzman-Martinez, C., and Velez-Pardo, C. (2010). The effects of polyphenols on survival and locomotor activity in *Drosophila melanogaster* exposed to iron and paraquat. *Neurochem. Res.* **35**, 227–238.

Kalidas, S., and Smith, D. P. (2002). Novel genomic cDNA hybrids produce effective RNA interference in adult Drosophila. *Neuron* **33**, 177–184.

Kapahi, P., Zid, B. M., Harper, T., Koslover, D., Sapin, V., and Benzer, S. (2004). Regulation of lifespan in Drosophila by modulation of genes in the TOR signaling pathway. *Curr. Biol.* **14**, 885–890.

Karpinar, D. P., Balija, M. B., Kugler, S., Opazo, F., Rezaei-Ghaleh, N., Wender, N., Kim, H. Y., Taschenberger, G., Falkenburger, B. H., Heise, H., *et al.* (2009). Pre-fibrillar alpha-synuclein variants with impaired beta-structure increase neurotoxicity in Parkinson's disease models. *EMBO J.* **28**, 3256–3268.

Kawajiri, S., Saiki, S., Sato, S., Sato, F., Hatano, T., Eguchi, H., and Hattori, N. (2010). PINK1 is recruited to mitochondria with parkin and associates with LC3 in mitophagy. *FEBS Lett.* **584**, 1073–1079.

Khan, F. S., Fujioka, M., Datta, P., Fernandes-Alnemri, T., Jaynes, J. B., and Alnemri, E. S. (2008). The interaction of DIAP1 with dOmi/HtrA2 regulates cell death in Drosophila. *Cell Death Differ.* **15**(6), 1073–1083.

Kim, P. K., Hailey, D. W., Mullen, R. T., and Lippincott-Schwartz, J. (2008a). Ubiquitin signals autophagic degradation of cytosolic proteins and peroxisomes. *Proc. Natl. Acad. Sci. USA* **105**, 20567–20574.

Kim, Y., Park, J., Kim, S., Song, S., Kwon, S. K., Lee, S. H., Kitada, T., Kim, J. M., and Chung, J. (2008b). PINK1 controls mitochondrial localization of Parkin through direct phosphorylation. *Biochem. Biophys. Res. Commun.* **377**, 975–980.

Kirkin, V., McEwan, D. G., Novak, I., and Dikic, I. (2009). A role for ubiquitin in selective autophagy. *Mol. Cell* **34**, 259–269.

Kramer, J. M., and Staveley, B. E. (2003). GAL4 causes developmental defects and apoptosis when expressed in the developing eye of *Drosophila melanogaster*. *Genet. Mol. Res.* **2**, 43–47.

Kraytsberg, Y., Kudryavtseva, E., McKee, A. C., Geula, C., Kowall, N. W., and Khrapko, K. (2006). Mitochondrial DNA deletions are abundant and cause functional impairment in aged human substantia nigra neurons. *Nat. Genet.* **38**, 518–520.

Kumar, A., Greggio, E., Beilina, A., Kaganovich, A., Chan, D., Taymans, J. M., Wolozin, B., and Cookson, M. R. (2010). The Parkinson's disease associated LRRK2 exhibits weaker in vitro phosphorylation of 4E-BP compared to autophosphorylation. *PLoS ONE* **5,** e8730.

Lavara-Culebras, E., and Paricio, N. (2007). Drosophila DJ-1 mutants are sensitive to oxidative stress and show reduced lifespan and motor deficits. *Gene* **400,** 158–165.

Lavara-Culebras, E., Munoz-Soriano, V., Gomez-Pastor, R., Matallana, E., and Paricio, N. (2010). Effects of pharmacological agents on the lifespan phenotype of Drosophila DJ-1beta mutants. *Gene* **462**(1–2), 26–33.

Lawal, H. O., Chang, H. Y., Terrell, A. N., Brooks, E. S., Pulido, D., Simon, A. F., and Krantz, D. E. (2010). The Drosophila vesicular monoamine transporter reduces pesticide-induced loss of dopaminergic neurons. *Neurobiol. Dis.* **40**(1), 102–112.

Lee, Y. S., and Carthew, R. W. (2003). Making a better RNAi vector for Drosophila: Use of intron spacers. *Methods* **30,** 322–329.

Lee, S. B., Kim, W., Lee, S., and Chung, J. (2007). Loss of LRRK2/PARK8 induces degeneration of dopaminergic neurons in Drosophila. *Biochem. Biophys. Res. Commun.* **358,** 534–539.

Lee, F. K., Wong, A. K., Lee, Y. W., Wan, O. W., Chan, H. Y., and Chung, K. K. (2009). The role of ubiquitin linkages on alpha-synuclein induced-toxicity in a Drosophila model of Parkinson's disease. *J. Neurochem.* **110,** 208–219.

Lessing, D., and Bonini, N. M. (2009). Maintaining the brain: Insight into human neurodegeneration from *Drosophila melanogaster* mutants. *Nat. Rev.* **10,** 359–370.

Lima, S. Q., and Miesenbock, G. (2005). Remote control of behavior through genetically targeted photostimulation of neurons. *Cell* **121,** 141–152.

Liu, Z., Wang, X., Yu, Y., Li, X., Wang, T., Jiang, H., Ren, Q., Jiao, Y., Sawa, A., Moran, T., *et al.* (2008). A Drosophila model for LRRK2-linked parkinsonism. *Proc. Natl. Acad. Sci. USA* **105,** 2693–2698.

Lu, B., and Vogel, H. (2009). Drosophila models of neurodegenerative diseases. *Annu. Rev. Pathol.* **4,** 315–342.

Lundell, M. J., and Hirsh, J. (1994). Temporal and spatial development of serotonin and dopamine neurons in the Drosophila CNS. *Dev. Biol.* **165,** 385–396.

Marin, I. (2006). The Parkinson disease gene LRRK2: Evolutionary and structural insights. *Mol. Biol. Evol.* **23,** 2423–2433.

Marin, I. (2008). Ancient origin of the Parkinson disease gene LRRK2. *J. Mol. Evol.* **67,** 41–50.

Matsuda, N., Sato, S., Shiba, K., Okatsu, K., Saisho, K., Gautier, C. A., Sou, Y. S., Saiki, S., Kawajiri, S., Sato, F., *et al.* (2010). PINK1 stabilized by mitochondrial depolarization recruits Parkin to damaged mitochondria and activates latent Parkin for mitophagy. *J. Cell Biol.* **189,** 211–221.

McQuibban, G. A., Saurya, S., and Freeman, M. (2003). Mitochondrial membrane remodelling regulated by a conserved rhomboid protease. *Nature* **423,** 537–541.

McQuibban, G. A., Lee, J. R., Zheng, L., Juusola, M., and Freeman, M. (2006). Normal mitochondrial dynamics requires rhomboid-7 and affects Drosophila lifespan and neuronal function. *Curr. Biol.* **16,** 982–989.

Menzies, F. M., Yenisetti, S. C., and Min, K. T. (2005). Roles of Drosophila DJ-1 in survival of dopaminergic neurons and oxidative stress. *Curr. Biol.* **15,** 1578–1582.

Meulener, M., Whitworth, A. J., Armstrong-Gold, C. E., Rizzu, P., Heutink, P., Wes, P. D., Pallanck, L. J., and Bonini, N. M. (2005). Drosophila DJ-1 mutants are selectively sensitive to environmental toxins associated with Parkinson's disease. *Curr. Biol.* **15,** 1572–1577.

Meulener, M. C., Xu, K., Thompson, L., Ischiropoulos, H., and Bonini, N. M. (2006). Mutational analysis of DJ-1 in Drosophila implicates functional inactivation by oxidative damage and aging. *Proc. Natl. Acad. Sci. USA* **103,** 12517–12522.

Morais, V., Verstreken, P., Roethig, A., Smet, J., Snellinx, A., Vanbrabant, M., Haddad, D., Frezza, C., Mandemakers, W., Vogt-Weisenhorn, D., *et al.* (2009). Parkinson's disease mutations in PINK1 results in decreased Complex I activity and deficient synaptic function. *EMBO Mol. Med.* **1,** 99–111.

Mortiboys, H., Thomas, K. J., Koopman, W. J., Klaffke, S., Abou-Sleiman, P., Olpin, S., Wood, N. W., Willems, P. H., Smeitink, J. A., Cookson, M. R., *et al.* (2008). Mitochondrial function and morphology are impaired in parkin-mutant fibroblasts. *Ann. Neurol.* **64,** 555–565.

Muftuoglu, M., Elibol, B., Dalmizrak, O., Ercan, A., Kulaksiz, G., Ogus, H., Dalkara, T., and Ozer, N. (2004). Mitochondrial complex I and IV activities in leukocytes from patients with parkin mutations. *Mov. Disord.* **19,** 544–548.

Murphy, C. T., McCarroll, S. A., Bargmann, C. I., Fraser, A., Kamath, R. S., Ahringer, J., Li, H., and Kenyon, C. (2003). Genes that act downstream of DAF-16 to influence the lifespan of *Caenorhabditis elegans*. *Nature* **424,** 277–283.

Narendra, D., Tanaka, A., Suen, D. F., and Youle, R. J. (2008). Parkin is recruited selectively to impaired mitochondria and promotes their autophagy. *J. Cell Biol.* **183,** 795–803.

Narendra, D. P., Jin, S. M., Tanaka, A., Suen, D.-F., Gautier, C. A., Shen, J., Cookson, M. R., and Youle, R. J. (2010). PINK1 is selectively stabilized on impaired mitochondria to activate parkin. *PLoS Biol.* **8,** e1000298.

Nassel, D. R., and Elekes, K. (1992). Aminergic neurons in the brain of blowflies and Drosophila: Dopamine- and tyrosine hydroxylase-immunoreactive neurons and their relationship with putative histaminergic neurons. *Cell Tissue Res.* **267,** 147–167.

Neckameyer, W. S. (1996). Multiple roles for dopamine in Drosophila development. *Dev. Biol.* **176,** 209–219.

Neckameyer, W. S. (1998a). Dopamine and mushroom bodies in Drosophila: Experience-dependent and -independent aspects of sexual behavior. *Learn. Mem.* **5,** 157–165.

Neckameyer, W. S. (1998b). Dopamine modulates female sexual receptivity in *Drosophila melanogaster*. *J. Neurogenet.* **12,** 101–114.

Neckameyer, W. S., and White, K. (1993). Drosophila tyrosine hydroxylase is encoded by the pale locus. *J. Neurogenet.* **8,** 189–199.

Ng, C. H., Mok, S. Z., Koh, C., Ouyang, X., Fivaz, M. L., Tan, E. K., Dawson, V. L., Dawson, T. M., Yu, F., and Lim, K. L. (2009). Parkin protects against LRRK2 G2019S mutant-induced dopaminergic neurodegeneration in Drosophila. *J. Neurosci.* **29,** 11257–11262.

Nguyen, T., Sherratt, P. J., and Pickett, C. B. (2003). Regulatory mechanisms controlling gene expression mediated by the antioxidant response element. *Annu. Rev. Pharmacol. Toxicol.* **43,** 233–260.

Okochi, M., Walter, J., Koyama, A., Nakajo, S., Baba, M., Iwatsubo, T., Meijer, L., Kahle, P. J., and Haass, C. (2000). Constitutive phosphorylation of the Parkinson's disease associated alpha-synuclein. *J. Biol. Chem.* **275,** 390–397.

Outeiro, T. F., Kontopoulos, E., Altmann, S. M., Kufareva, I., Strathearn, K. E., Amore, A. M., Volk, C. B., Maxwell, M. M., Rochet, J. C., McLean, P. J., *et al.* (2007). Sirtuin 2 inhibitors rescue alpha-synuclein-mediated toxicity in models of Parkinson's disease. *Science* **317,** 516–519.

Palacino, J. J., Sagi, D., Goldberg, M. S., Krauss, S., Motz, C., Wacker, M., Klose, J., and Shen, J. (2004). Mitochondrial dysfunction and oxidative damage in parkin-deficient mice. *J. Biol. Chem.* **279,** 18614–18622.

Park, J., Kim, S. Y., Cha, G. H., Lee, S. B., Kim, S., and Chung, J. (2005). Drosophila DJ-1 mutants show oxidative stress-sensitive locomotive dysfunction. *Gene* **361,** 133–139.

Park, J., Lee, S. B., Lee, S., Kim, Y., Song, S., Kim, S., Bae, E., Kim, J., Shong, M., Kim, J. M., *et al.* (2006). Mitochondrial dysfunction in Drosophila PINK1 mutants is complemented by parkin. *Nature* **441,** 1157–1161.

Park, S. S., Schulz, E. M., and Lee, D. (2007). Disruption of dopamine homeostasis underlies selective neurodegeneration mediated by alpha-synuclein. *Eur. J. Neurosci.* **26,** 3104–3112.

Park, J., Lee, G., and Chung, J. (2009). The PINK1-Parkin pathway is involved in the regulation of mitochondrial remodeling process. *Biochem. Biophys. Res. Commun.* **378,** 518–523.

Paterna, J. C., Leng, A., Weber, E., Feldon, J., and Bueler, H. (2007). DJ-1 and Parkin modulate dopamine-dependent behavior and inhibit MPTP-induced nigral dopamine neuron loss in mice. *Mol. Ther.* **15,** 698–704.

Pendleton, R. G., Rasheed, A., Sardina, T., Tully, T., and Hillman, R. (2002). Effects of tyrosine hydroxylase mutants on locomotor activity in Drosophila: A study in functional genomics. *Behav. Genet.* **32,** 89–94.

Pesah, Y., Pham, T., Burgess, H., Middlebrooks, B., Verstreken, P., Zhou, Y., Harding, M., Bellen, H., and Mardon, G. (2004). Drosophila parkin mutants have decreased mass and cell size and increased sensitivity to oxygen radical stress. *Development* **131,** 2183–2194.

Pesah, Y., Burgess, H., Middlebrooks, B., Ronningen, K., Prosser, J., Tirunagaru, V., Zysk, J., and Mardon, G. (2005). Whole-mount analysis reveals normal numbers of dopaminergic neurons following misexpression of alpha-synuclein in Drosophila. *Genesis* **41,** 154–159.

Plun-Favreau, H., Klupsch, K., Moisoi, N., Gandhi, S., Kjaer, S., Frith, D., Harvey, K., Deas, E., Harvey, R. J., McDonald, N., *et al.* (2007). The mitochondrial protease HtrA2 is regulated by Parkinson's disease-associated kinase PINK1. *Nat. Cell Biol.* **9,** 1243–1252.

Poewe, W. (2009). Treatments for Parkinson disease—Past achievements and current clinical needs. *Neurology* **72,** S65–S73.

Poole, A. C., Thomas, R. E., Andrews, L. A., McBride, H. M., Whitworth, A. J., and Pallanck, L. J. (2008). The PINK1/Parkin pathway regulates mitochondrial morphology. *Proc. Natl. Acad. Sci. USA* **105,** 1638–1643.

Poole, A. C., Thomas, R. E., Yu, S., Vincow, E. S., and Pallanck, L. (2010). The mitochondrial fusion-promoting factor mitofusin is a substrate of the PINK1/parkin pathway. *PLoS ONE* **5,** e10054.

Quik, M., O'Leary, K., and Tanner, C. M. (2008). Nicotine and Parkinson's disease: Implications for therapy. *Mov. Disord.* **23,** 1641–1652.

Riparbelli, M. G., and Callaini, G. (2007). The Drosophila parkin homologue is required for normal mitochondrial dynamics during spermiogenesis. *Dev. Biol.* **303,** 108–120.

Roman, G., Endo, K., Zong, L., and Davis, R. L. (2001). P[Switch], a system for spatial and temporal control of gene expression in *Drosophila melanogaster*. *Proc. Natl. Acad. Sci. USA* **98,** 12602–12607.

Rong, Y. S., and Golic, K. G. (2000). Gene targeting by homologous recombination in Drosophila. *Science* **288,** 2013–2018.

Rorth, P. (1996). A modular misexpression screen in Drosophila detecting tissue-specific phenotypes. *Proc. Natl. Acad. Sci. USA* **93,** 12418–12422.

Rosen, K. M., Veereshwarayya, V., Moussa, C. E., Fu, Q., Goldberg, M. S., Schlossmacher, M. G., Shen, J., and Querfurth, H. W. (2006). Parkin protects against mitochondrial toxins and beta-amyloid accumulation in skeletal muscle cells. *J. Biol. Chem.* **281,** 12809–12816.

Ross, O. A., Soto, A. I., Vilarino-Guell, C., Heckman, M. G., Diehl, N. N., Hulihan, M. M., Aasly, J. O., Sando, S., Gibson, J. M., Lynch, T., *et al.* (2008). Genetic variation of Omi/HtrA2 and Parkinson's disease. *Parkinsonism Relat. Disord.* **14,** 539–543.

Rubin, G. M., Yandell, M. D., Wortman, J. R., Gabor Miklos, G. L., Nelson, C. R., Hariharan, I. K., Fortini, M. E., Li, P. W., Apweiler, R., Fleischmann, W., *et al.* (2000). Comparative genomics of the eukaryotes. *Science* **287,** 2204–2215.

Ryder, E., and Russell, S. (2003). Transposable elements as tools for genomics and genetics in Drosophila. *Brief Funct. Genomic. Proteomic.* **2,** 57–71.

Saini, N., and Schaffner, W. (2010). Zinc supplement greatly improves the condition of parkin mutant Drosophila. *Biol. Chem.* **391,** 513–518.

Saini, N., Oelhafen, S., Hua, H., Georgiev, O., Schaffner, W., and Bueler, H. (2010). Extended lifespan of Drosophila parkin mutants through sequestration of redox-active metals enhancement of anti-oxidative pathways. *Neurobiol. Dis.* **40**(1), 82–92.

Samaranch, L., Lorenzo-Betancor, O., Arbelo, J. M., Ferrer, I., Lorenzo, E., Irigoyen, J., Pastor, M. A., Marrero, C., Isla, C., Herrera-Henriquez, J., *et al.* (2010). PINK1-linked parkinsonism is associated with Lewy body pathology. *Brain* **133,** 1128–1142.

Sang, T. K., Chang, H. Y., Lawless, G. M., Ratnaparkhi, A., Mee, L., Ackerson, L. C., Maidment, N. T., Krantz, D. E., and Jackson, G. R. (2007). A Drosophila model of mutant human parkin-induced toxicity demonstrates selective loss of dopaminergic neurons and dependence on cellular dopamine. *J. Neurosci.* **27,** 981–992.

Sarov, M., and Stewart, A. F. (2005). The best control for the specificity of RNAi. *Trends Biotechnol.* **23,** 446–448.

Sayre, L. M., Perry, G., and Smith, M. A. (2008). Oxidative stress and neurotoxicity. *Chem. Res. Toxicol.* **21,** 172–188.

Scherzer, C. R., Jensen, R. V., Gullans, S. R., and Feany, M. B. (2003). Gene expression changes presage neurodegeneration in a Drosophila model of Parkinson's disease. *Hum. Mol. Genet.* **12,** 2457–2466.

Schwarzschild, M. A., Agnati, L., Fuxe, K., Chen, J. F., and Morelli, M. (2006). Targeting adenosine A2A receptors in Parkinson's disease. *Trends Neurosci.* **29,** 647–654.

Shulman, J. M., and Feany, M. B. (2003). Genetic modifiers of tauopathy in Drosophila. *Genetics* **165,** 1233–1242.

Silvestri, L., Caputo, V., Bellacchio, E., Atorino, L., Dallapiccola, B., Valente, E. M., and Casari, G. (2005). Mitochondrial import and enzymatic activity of PINK1 mutants associated to recessive parkinsonism. *Hum. Mol. Genet.* **14,** 3477–3492.

Simon-Sanchez, J., and Singleton, A. B. (2008). Sequencing analysis of OMI/HTRA2 shows previously reported pathogenic mutations in neurologically normal controls. *Hum. Mol. Genet.* **17,** 1988–1993.

Spillantini, M. G., Schmidt, M. L., Lee, V. M., Trojanowski, J. Q., Jakes, R., and Goedert, M. (1997). Alpha-synuclein in Lewy bodies. *Nature* **388,** 839–840.

St Johnston, D. (2002). The art and design of genetic screens: *Drosophila melanogaster. Nat. Rev.* **3,** 176–188.

Steffan, J. S., Bodai, L., Pallos, J., Poelman, M., McCampbell, A., Apostol, B. L., Kazantsev, A., Schmidt, E., Zhu, Y. Z., Greenwald, M., *et al.* (2001). Histone deacetylase inhibitors arrest polyglutamine-dependent neurodegeneration in Drosophila. *Nature* **413,** 739–743.

Steinbrink, S., and Boutros, M. (2008). RNAi screening in cultured Drosophila cells. *Methods Mol. Biol.* **420,** 139–153.

Strauss, K. M., Martins, L. M., Plun-Favreau, H., Marx, F. P., Kautzmann, S., Berg, D., Gasser, T., Wszolek, Z., Muller, T., Bornemann, A., *et al.* (2005). Loss of function mutations in the gene encoding Omi/HtrA2 in Parkinson's disease. *Hum. Mol. Genet.* **14,** 2099–2111.

Sutovsky, P., Moreno, R. D., Ramalho-Santos, J., Dominko, T., Simerly, C., and Schatten, G. (1999). Ubiquitin tag for sperm mitochondria. *Nature* **402,** 371–372.

Tain, L. S., Chowdhury, R. B., Tao, R. N., Plun-Favreau, H., Moisoi, N., Martins, L. M., Downward, J., Whitworth, A. J., and Tapon, N. (2009a). Drosophila HtrA2 is dispensable for apoptosis but acts downstream of PINK1 independently from Parkin. *Cell Death Differ.* **16**(8), 1118–1125.

Tain, L. S., Mortiboys, H., Tao, R. N., Ziviani, E., Bandmann, O., and Whitworth, A. J. (2009b). Rapamycin activation of 4E-BP prevents parkinsonian dopaminergic neuron loss. *Nat. Neurosci.* **12,** 1129–1135.

Takahashi, M., Kanuka, H., Fujiwara, H., Koyama, A., Hasegawa, M., Miura, M., and Iwatsubo, T. (2003). Phosphorylation of alpha-synuclein characteristic of synucleinopathy lesions is recapitulated in alpha-synuclein transgenic Drosophila. Neurosci. Lett. 336, 155–158.

Teleman, A. A., Chen, Y. W., and Cohen, S. M. (2005). 4E-BP functions as a metabolic brake used under stress conditions but not during normal growth. Genes Dev. 19, 1844–1848.

Tempel, B. L., Livingstone, M. S., and Quinn, W. G. (1984). Mutations in the dopa decarboxylase gene affect learning in Drosophila. Proc. Natl. Acad. Sci. USA 81, 3577–3581.

Tettweiler, G., Miron, M., Jenkins, M., Sonenberg, N., and Lasko, P. F. (2005). Starvation and oxidative stress resistance in Drosophila are mediated through the eIF4E-binding protein, d4E-BP. Genes Dev. 19, 1840–1843.

Tofaris, G. K., and Spillantini, M. G. (2007). Physiological and pathological properties of alpha-synuclein. Cell. Mol. Life Sci. 64, 2194–2201.

Trinh, K., Moore, K., Wes, P. D., Muchowski, P. J., Dey, J., Andrews, L., and Pallanck, L. J. (2008). Induction of the phase II detoxification pathway suppresses neuron loss in Drosophila models of Parkinson's disease. J. Neurosci. 28, 465–472.

Trinh, K., Andrews, L., Krause, J., Hanak, T., Lee, D., Gelb, M., and Pallanck, L. (2010). Decaffeinated coffee and nicotine-free tobacco provide neuroprotection in Drosophila models of Parkinson's disease through an NRF2-dependent mechanism. J. Neurosci. 30, 5525–5532.

Twig, G., Elorza, A., Molina, A. J., Mohamed, H., Wikstrom, J. D., Walzer, G., Stiles, L., Haigh, S. E., Katz, S., Las, G., et al. (2008a). Fission and selective fusion govern mitochondrial segregation and elimination by autophagy. EMBO J. 27, 433–446.

Twig, G., Hyde, B., and Shirihai, O. S. (2008b). Mitochondrial fusion, fission and autophagy as a quality control axis: The bioenergetic view. Biochim. Biophys. Acta 1777, 1092–1097.

Valente, E. M., Abou-Sleiman, P. M., Caputo, V., Muqit, M. M., Harvey, K., Gispert, S., Ali, Z., Del Turco, D., Bentivoglio, A. R., Healy, D. G., et al. (2004). Hereditary early-onset Parkinson's disease caused by mutations in PINK1. Science 304, 1158–1160.

Vaux, D. L., and Silke, J. (2003). HtrA2/Omi, a sheep in wolf's clothing. Cell 115, 251–253.

Ved, R., Saha, S., Westlund, B., Perier, C., Burnam, L., Sluder, A., Hoener, M., Rodrigues, C. M., Alfonso, A., Steer, C., et al. (2005). Similar patterns of mitochondrial vulnerability and rescue induced by genetic modification of alpha-synuclein, parkin, and DJ-1 in Caenorhabditis elegans. J. Biol. Chem. 280, 42655–42668.

Venderova, K., Kabbach, G., Abdel-Messih, E., Zhang, Y., Parks, R. J., Imai, Y., Gehrke, S., Ngsee, J., Lavoie, M. J., Slack, R. S., et al. (2009). Leucine-Rich Repeat Kinase 2 interacts with Parkin, DJ-1 and PINK-1 in a Drosophila melanogaster model of Parkinson's disease. Hum. Mol. Genet. 18, 4390–4404.

Venken, K. J., and Bellen, H. J. (2005). Emerging technologies for gene manipulation in Drosophila melanogaster. Nat. Rev. 6, 167–178.

Verstreken, P., Ly, C. V., Venken, K. J., Koh, T. W., Zhou, Y., and Bellen, H. J. (2005). Synaptic mitochondria are critical for mobilization of reserve pool vesicles at Drosophila neuromuscular junctions. Neuron 47, 365–378.

Vives-Bauza, C., Zhou, C., Huang, Y., Cui, M., de Vries, R. L., Kim, J., May, J., Tocilescu, M. A., Liu, W., Ko, H. S., et al. (2010). PINK1-dependent recruitment of Parkin to mitochondria in mitophagy. Proc. Natl. Acad. Sci. USA 107, 378–383.

Wang, D., Qian, L., Xiong, H., Liu, J., Neckameyer, W. S., Oldham, S., Xia, K., Wang, J., Bodmer, R., and Zhang, Z. (2006). Antioxidants protect PINK1-dependent dopaminergic neurons in Drosophila. Proc. Natl. Acad. Sci. USA 103, 13520–13525.

Wang, D., Tang, B., Zhao, G., Pan, Q., Xia, K., Bodmer, R., and Zhang, Z. (2008). Dispensable role of Drosophila ortholog of LRRK2 kinase activity in survival of dopaminergic neurons. Mol. Neurodegener. 3, 3.

Warrick, J. M., Paulson, H. L., Gray-Board, G. L., Bui, Q. T., Fischbeck, K. H., Pittman, R. N., and Bonini, N. M. (1998). Expanded polyglutamine protein forms nuclear inclusions and causes neural degeneration in Drosophila. *Cell* **93**, 939–949.

Wassef, R., Haenold, R., Hansel, A., Brot, N., Heinemann, S. H., and Hoshi, T. (2007). Methionine sulfoxide reductase A and a dietary supplement S-methyl-L-cysteine prevent Parkinson's-like symptoms. *J. Neurosci.* **27**, 12808–12816.

Weihofen, A., Ostaszewski, B., Minami, Y., and Selkoe, D. J. (2008). Pink1 Parkinson mutations, the Cdc37/Hsp90 chaperones and Parkin all influence the maturation or subcellular distribution of Pink1. *Hum. Mol. Genet.* **17**, 602–616.

Whitworth, A. J., and Pallanck, L. J. (2009). The PINK1/Parkin pathway: A mitochondrial quality control system? *J. Bioenerg. Biomembr.* **41**, 499–503.

Whitworth, A. J., Theodore, D. A., Greene, J. C., Benes, H., Wes, P. D., and Pallanck, L. J. (2005). Increased glutathione S-transferase activity rescues dopaminergic neuron loss in a Drosophila model of Parkinson's disease. *Proc. Natl. Acad. Sci. USA* **102**, 8024–8029.

Whitworth, A. J., Lee, J., Ho, V.-W., Chaudhury, R., Flick, R., and McQuibban, G. (2008). Rhomboid-7 and HtrA2/Omi act in a common pathway with the Parkinson's disease factors Pink1 and Parkin. *Dis. Model. Mech.* **1**, 168–174.

Willingham, S., Outeiro, T. F., DeVit, M. J., Lindquist, S. L., and Muchowski, P. J. (2003). Yeast genes that enhance the toxicity of a mutant huntingtin fragment or alpha-synuclein. *Science* **302**, 1769–1772.

Yang, Y., Nishimura, I., Imai, Y., Takahashi, R., and Lu, B. (2003). Parkin suppresses dopaminergic neuron-selective neurotoxicity induced by Pael-R in Drosophila. *Neuron* **37**, 911–924.

Yang, Y., Gehrke, S., Haque, M. E., Imai, Y., Kosek, J., Yang, L., Beal, M. F., Nishimura, I., Wakamatsu, K., Ito, S., *et al.* (2005). Inactivation of Drosophila DJ-1 leads to impairments of oxidative stress response and phosphatidylinositol 3-kinase/Akt signaling. *Proc. Natl. Acad. Sci. USA* **102**(38), 13670–13675.

Yang, Y., Gehrke, S., Imai, Y., Huang, Z., Ouyang, Y., Wang, J. W., Yang, L., Beal, M. F., Vogel, H., and Lu, B. (2006). Mitochondrial pathology and muscle and dopaminergic neuron degeneration caused by inactivation of Drosophila Pink1 is rescued by Parkin. *Proc. Natl. Acad. Sci. USA* **103**, 10793–10798.

Yang, Y., Ouyang, Y., Yang, L., Beal, M. F., McQuibban, A., Vogel, H., and Lu, B. (2008). Pink1 regulates mitochondrial dynamics through interaction with the fission/fusion machinery. *Proc. Natl. Acad. Sci. USA* **105**, 7070–7075.

Yun, J., Cao, J. H., Dodson, M. W., Clark, I. E., Kapahi, P., Chowdhury, R. B., and Guo, M. (2008). Loss-of-function analysis suggests that Omi/HtrA2 is not an essential component of the PINK1/ PARKIN pathway in vivo. *J. Neurosci.* **28**, 14500–14510.

Zadikoff, C., Rogaeva, E., Djarmati, A., Sato, C., Salehi-Rad, S., St. George-Hyslop, P., Klein, C., and Lang, A. E. (2006). Homozygous and heterozygous PINK1 mutations: Considerations for diagnosis and care of Parkinson's disease patients. *Mov. Disord.* **21**, 875–879.

Zhou, C., Huang, Y., Shao, Y., May, J., Prou, D., Perier, C., Dauer, W., Schon, E. A., and Przedborski, S. (2008). The kinase domain of mitochondrial PINK1 faces the cytoplasm. *Proc. Natl. Acad. Sci. USA* **105**, 12022–12027.

Zid, B. M., Rogers, A. N., Katewa, S. D., Vargas, M. A., Kolipinski, M. C., Lu, T. A., Benzer, S., and Kapahi, P. (2009). 4E-BP extends lifespan upon dietary restriction by enhancing mitochondrial activity in Drosophila. *Cell* **139**, 149–160.

Ziviani, E., and Whitworth, A. J. (2010). How could Parkin-mediated ubiquitination of mitofusin promote mitophagy? *Autophagy* **6**, 660–662.

Ziviani, E., Tao, R. N., and Whitworth, A. J. (2010). Drosophila parkin requires PINK1 for mitochondrial translocation and ubiquitinates mitofusin. *Proc. Natl. Acad. Sci. USA* **107**, 5018–5023.

2

Cell Cycle Regulated Gene Expression in Yeasts

Christopher J. McInerny
Division of Molecular and Cellular Biology, Faculty of Biomedical and Life Sciences, Davidson Building, University of Glasgow, Glasgow, United Kingdom

I. Introduction
   A. Cell cycle controls
   B. Gene expression
II. Cell Cycle Gene Expression in Yeasts
   A. Budding yeast
   B. Fission yeast
III. Regulatory Networks
IV. Conservation of Controls
V. Importance of Cell Cycle Regulated Gene Expression
VI. Unimportance of Cell Cycle Regulated Gene Expression
VII. Related Areas of Interest and Topics for Further Study
   A. Histone transcription in budding yeast and G2–M transcription in fission yeast
   B. Integration into general cell cycle controls
   C. Evolution of cell cycle control mechanisms
   D. Meiosis
   E. Mediator complex
   References

Advances in Genetics, Vol. 73
0065-2660/11 $35.00
DOI: 10.1016/B978-0-12-380860-8.00002-1

ABSTRACT

The regulation of gene expression through the mitotic cell cycle, so that genes are transcribed at particular cell cycle times, is widespread among eukaryotes. In some cases, it appears to be important for control mechanisms, as deregulated expression results in uncontrolled cell divisions, which can cause cell death, disease, and malignancy. In this review, I describe the current understanding of such regulated gene expression in two established simple eukaryotic model organisms, the budding yeast *Saccharomyces cerevisiae* and the fission yeast *Schizosaccharomyces pombe*. In these two yeasts, the global pattern of cell cycle gene expression has been well described, and most of the transcription factors that control the various waves of gene expression, and how they are in turn themselves regulated, have been characterized. As related mechanisms occur in all other eukaryotes, including humans, yeasts offer an excellent paradigm to understand this important molecular process. © 2011, Elsevier Inc.

I. INTRODUCTION

A. Cell cycle controls

How a cell duplicates and divides is a major area of interest in biology, not only because the process is so incredibly accurate and complex, and so of inherent fascination in its own right, but also because defects in the process are the basis for many human diseases.

Traditionally, the eukaryotic mitotic cell cycle is divided into four separate, consecutive, and distinct stages: S phase (where DNA replication occurs) and M phase (where the chromosomes separate), with these two separated by gap phases called G1 and G2. Research in many organisms, from simple yeasts to much more complex eukaryotes, including humans, has revealed that the cell division cycle is controlled in many ways and at many levels. As a generalization, this complexity of control seems to be to ensure that cell division occurs in a highly reproducible and accurate way, with multiple levels of controls introducing "double check" and "fail-safe" processes. The various types of control mechanisms include changes in protein activity through posttranslational modification (such as phosphorylation), changes in protein stability (in some cases through specific, targeted degradation), and changes in protein distribution. It is important to emphasize that although the regulation of gene expression through transcription offers yet another important layer of control, it is just part of an overall complex array of multiple control mechanisms.

The study of cell cycle regulated gene expression in yeasts has been ongoing since the early 1980s, when it was discovered that the transcript levels of the budding yeast histone genes fluctuated dramatically through the cell cycle (Hereford *et al.*, 1981, 1982). Since then many cell cycle regulated genes in both budding and fission yeasts have been identified by traditional methods, and in many cases the transcriptional control mechanisms characterized (Bähler, 2005a; Breeden, 1996; Futcher, 2000, 2002; McInerny, 2004; Tyers, 2004). Furthermore, the advent of global analyses of transcript levels by microarrays and the binding of cell cycle transcription factors throughout the genome by ChIP-on-chip technologies has consolidated the field (Banerjee and Zhang, 2003; Cho *et al.*, 1998; Iyer *et al.*, 2001; Kato *et al.*, 2004; Oliva *et al.*, 2005; Peng *et al.*, 2005; Rustici *et al.*, 2004; Spellman *et al.*, 1998). Such studies have also permitted complex hierarchies and networks to be proposed, suggesting how cell cycle regulated transcription fits into general cycle controls, and how each wave of gene expression may control, and be controlled by, those before and after (de Lichtenberg *et al.*, 2005; Horak *et al.*, 2002; Jensen *et al.*, 2006; Lee *et al.*, 2002; Simon *et al.*, 2001) (Fig. 2.3).

This accumulated knowledge means that the field is now at a mature stage, with a reasonably comprehensive picture of the process in both model yeast species (Bähler, 2005a; Breeden, 2003; McInerny, 2004; Toone *et al.*, 1997; Tyers, 2004; Wittenberg and Reed, 2005). These paradigms are of wide interest, as related mechanisms operate in all eukaryotes, including humans, where deregulated cell cycle gene expression is a signature of many diseases.

In this review, I first describe in detail the various waves of gene expression in the two yeast species, budding and fission yeasts, and the different transcriptional control mechanisms that regulate them. Particular emphasis will be made on those genes and mechanisms that are conserved between the two species, as in many cases these have orthologs in higher eukaryotes. Having described the basic architecture of cell cycle regulated gene expression, the mechanisms by which these various waves are integrated to each other and to more general cell cycle controls will be described.

B. Gene expression

In the context of cell cycle gene expression, we will be focussing on genes whose transcription is controlled by RNA polymerase II. These include most genes whose transcription produces messenger RNAs (mRNAs), which when translated produce polypeptides or proteins with functions within the cell. The molecular mechanisms by which these genes are controlled during transcription have been well studied (Fig. 2.1). Upstream of the start methionine of the open reading frames of genes, lie DNA regions described as promoters. Within these promoter regions are found short, conserved DNA sequence motifs, to which

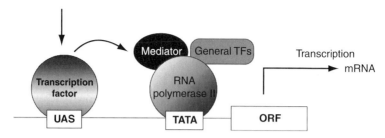

Figure 2.1. Molecular architecture of regulated gene expression controlled by RNA polymerase II. The open reading frame (ORF) of a gene is transcribed into mRNA by a combination of RNA polymerase II, general transcription factors (TFs), a Mediator complex, and specific transcription factors that bind to particular DNA sequences present in the gene's promoter region. RNA polymerase with the general TFs and the Mediator complex can bind to TATA box motifs to stimulate unregulated, constitutive expression. But for specific, regulated gene expression additional transcription factors bind to the promoter. These contact upstream activating sequences (UAS) in the promoter and upon stimulation activate RNA polymerase II through the Mediator complex. Thus, a combination of specific transcription factors and RNA polymerase together controls regulated gene expression. (See Color Insert.)

various transcription factors bind that control the genes' expression. The first of these motifs is the TATA box, to which RNA polymerase II binds, the enzyme that directly converts the gene to mRNA during transcription. For some genes, binding of RNA polymerase II along with general transcription factors and the Mediator complex to the promoter is sufficient to allow continuous, basal transcription to occur. But for controlled, regulated gene expression, a more complex promoter structure is observed. Instead, further DNA motifs are present, usually upstream of the TATA box, to which bind additional transcription factors which specifically regulate the activity of RNA polymerase II through the Mediator complex, to ensure that gene expression only occurs at the required time, or in the correct cell type. These extra motifs are known collectively in yeasts as upstream activating sequences (UASs), and enhancers in humans. Many different UASs have been identified, and in each case the UAS is bound by a particular transcription factor or complex, which modulates the activity of RNA polymerase II and consequent gene expression.

As will be seen in this review, in the context of cell cycle regulated gene expression in yeasts, a large number of UASs have been described, and the specific cell cycle transcription factor complexes that bind to them. It is the combination of these two that results in gene expression in these organisms at certain cell cycle times. Therefore, a primary aim of this review is to describe the various UASs and transcription factor complexes that control regulated, cell cycle-specific gene expression in yeasts.

II. CELL CYCLE GENE EXPRESSION IN YEASTS

A. Budding yeast

In budding yeast, traditional and genome-wide experiments have identified a relatively large number of genes whose transcript levels fluctuate significantly through the cell cycle. The two original transcriptome experiments identified up to 800 genes, with approximately 300 strongly periodically expressed in both studies (Cho et al., 1998; Spellman et al., 1998), though subsequent high-resolution mapping has identified more, with up to 1100 genes (Pramila et al., 2006; Rowicka et al., 2007). These analyses described periodic genes that fall into groups expressed at four cell cycle stages: those at the start of S phase, others at the G1–S transition, more during S phase, and those at the start and the end of M phase, the G2–M and M–G1 transitions (Table 2.1; Fig. 2.2).

1. G1–S waves

One of the first mechanisms of cell cycle gene expression in yeasts to be identified whose transcription mechanisms are well characterized was a group of two waves of genes whose peak transcript levels occur at the so-called G1–S *transition*, the start of S phase, in budding yeast (Breeden, 1996). This gene expression occurs at a time in the cell cycle where cells decide whether to commit to cell division or to exit from the cell cycle, which is called Start in yeasts, or the restriction point in mammalian cells.

In budding yeast, the two waves of gene expression are controlled by two different, but overlapping, transcription factor complexes, named MCB-binding factor (MBF; also sometimes called DSC1), and SCB-binding factor (SBF). MBF is composed of at least two proteins, Swi6p and Mbp1p, whereas SBF contains Swi6p and another protein, Swi4p (Andrews and Herskowitz, 1989a,b; Dirick et al., 1992; Koch et al., 1993; Lowndes et al., 1991, 1992a; Primig et al., 1992). In each case, the transcription factors bind to different short DNA sequence motifs, sometimes repeated, present in the promoters of the genes that they control, which are the UASs. One of the motifs is called an MCB, for *Mlu* I cell cycle box as, coincidentally, it is identical to the *Mlu* I restriction enzyme recognition site (McIntosh et al., 1991); and the other motif is called an SCB, for Swi4/Swi6-dependent cell cycle box. It is believed that Mbp1p and Swi4p have DNA-binding properties, whereas Swi6p, in both cases, is thought instead to have a regulatory role. All three proteins share sequence similarities with each other, including ankyrin motifs thought to be required for protein–protein interactions, suggesting that they evolved from a common ancestor (Bork, 1993).

Broadly speaking, SBF and MBF control the expression of groups of genes with different functions at G1–S. SBF-controlled genes function mostly in cell wall biosynthesis and cell budding, whereas MBF targets genes

Table 2.1. Cell Cycle Control Mechanisms in Budding and Fission Yeasts

Cell cycle phase	UAS name	Budding yeast			Fission yeast		
		UAS DNA motif	Transcription factor complex	Components	UAS DNA motif	Transcription factor complex	Components
G1–S	SCB	CNCGAAA	SBF	Swi4p, Swi6p	Not present	Not present	Not present
G1–S	MCB	ACGCGT	MBF (DSC1)	Mbp1p, Swi6p	ACGCGT	MBF (DSC1)	Cdc10p, Res1p, Res2p, Rep1p, Rep2p
S	Not named (histones)	GCGAAAANNGAAC	Not characterized	Not characterized	AACCCT	Asm2p	Asm2p
S	HCM	TAAACAA	Hcm1p	Hcm1p	Not characterized	Not characterized	Not characterized
G2–M	SFF/FKH	TTNCCNAANNNGG +GTAAACAA	SFF	Mcm1p, Fkh1p, Fkh2p, Nddl1p	Not present	Not present	Not present
G2–M	New 3	Not present	Not present	Not present	ACCNCGCT	Not characterized	Not characterized
M–G1	PCB	Not present	Not present	Not present	GNAACN +GTAAACAA	PBF	Mbx1p, Fkh2p, Sep1p
M–G1	ECB	TTNCCNAANNNGG +NAATTA	EBF	Mcm1p, Yox1p, Yhp1p	Not characterized	Not characterized	Not characterized
M–G1	SWI5/ACE2	ACCAGCN	Swi5p-Ace2p	Swi5p-Ace2p	CCAGCC	Ace2p	Ace2p

Abbreviated names for transcription factor complexes, UAS (upstream activating sequence) DNA motifs, and constituent proteins. For details of each, see text.

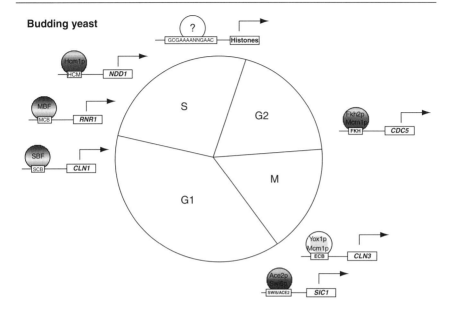

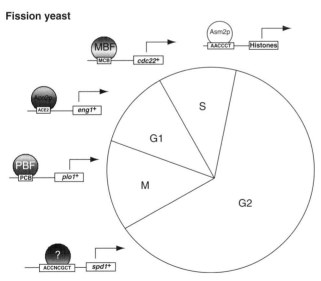

Figure 2.2. Cell cycle regulated gene expression in budding and fission yeasts. For each yeast species, a single gene is shown to represent the group of genes that are transcribed at particular cell cycle times. Where known the name or DNA sequence of the upstream activating sequence and the transcription factor are indicated. Related transcription factors in the two yeasts are shown by the same color. For details of each, see text. (See Color Insert.)

predominately involved in DNA replication. However, it should be emphasized that there is significant functional overlap between these two transcription factor complexes, and they do not account for all G1–S-specific transcription in this organism (Bean et al., 2005; Horak et al., 2002; Partridge et al., 1997).

Much research has been done to understand how the activity of the MBF and SBF transcription factor complexes is regulated, and this has revealed complex, multiple levels of control. Primary control occurs through the Cdc28p/Cln3p cyclin-dependent kinase (CDK) complex (Stuart and Wittenberg, 1995; Tyers et al., 1993): both MBF and SBF bind to their respective promoters in early G1, but activation of Cln3p–Cdc28p in later G1 results in the binding of the RNA polymerase II machinery (Cosma et al., 1999, 2001; Dirick et al., 1995; Wijnen and Futcher, 1999; Wijnen et al., 2002). The link between Cln3p and MBF/SBF is provided by the Whi5p protein, which acts as an inhibitor of G1–S gene expression through binding to SBF; this is reversed through its phosphorylation by Cln3p–Cdc28p which causes it to vacate the nucleus so relieving its inhibition (Costanzo et al., 2004; de Bruin et al., 2004). Other proteins, with apparent activating roles for G1–S transcription, have been identified, such as Bck2p and Stb1p (Costanzo et al., 2003; Ho et al., 1999; Wijnen and Futcher, 1999). Another protein, Nrm1p, has been shown to have a repressing role. The NRM1 gene is under MCB control, and encodes a component of MBF; as its absence results in overexpression of MBF genes, it acts as a repressor in a negative feedback loop inhibiting gene expression at G1–S (de Bruin and Wittenberg, 2009; de Bruin et al., 2006).

As well as these posttranslational control mechanisms, Swi4p expression is controlled at the level of transcription, being regulated through both ECB promoter sequence motifs in its promoter resulting in gene expression at M–G1 (Section II.A.4), and through SBF- and MBF-mediated transcription at G1–S, as part of a positive feedback loop (Foster et al., 1993; McInerny et al., 1997). MBF- and SBF-mediated transcription is turned off in G2 through another CDK complex containing Cdc28p–Clb1p/Clb2p, which is itself in turn activated by transcriptional activation of CLN1 and CLN2 by MBF and SBF (Amon et al., 1993; Breeden, 2000; Koch et al., 1996). Yet another CDK complex, Cdc28p–Clb6p, which is also activated transcriptionally by MBF and SBF, controls the distribution of Swi6p in cells, causing it by direct phosphorylation to exit the nucleus, so repressing gene expression; this event is antagonized by the phosphatase Cdc14p which, through dephosphorylating Swi6p, causes its nuclear entry during G1 (Geymonat et al., 2004; Sidorova et al., 1995).

2. S-phase waves

The first wave of gene expression to be identified in either yeast species was the histone gene wave in budding yeast (Hereford et al., 1981, 1982). Though potential transcription factor sites have been identified that activate this cell cycle expression during S phase, frustratingly, the transcription factors have eluded characterization (Freeman *et al.*, 1992). However, global transcription factor binding experiments have indicated that SBF binds to histone promoters (Iyer *et al.*, 2001), so it may have a role in the regulation of their expression (Simon *et al.*, 2001). Furthermore, two potential transcriptional repressors have been identified, and named Hir1p and Hir2p (Sherwood *et al.*, 1993; Spector *et al.*, 1997).

Another wave of gene expression during S phase is controlled by the Hcm1p forkhead-like transcription factor (Pramila *et al.*, 2006), which belongs to a class of conserved transcription factors found throughout the animal kingdom, all containing a so-called *forkhead DNA-binding motif* (Carlsson and Mahlapuu, 2002; Durocher and Jackson, 2002; Durocher *et al.*, 1999; Kaufmann and Knöchel, 1996). The *HCM1* gene is transcribed during G1–S under both MBF and SBF controls (Rowicka *et al.*, 2007). In addition to activating S-phase expression of genes involved in budding and chromosome segregation, Hcm1p controls expression of Whi5p, repressing G1–S transcription in the following cycle. It also activates *YHP1*, which represses ECB elements later in the cycle (Section II.A.4). In addition, Hcm1p also directly controls the transcription of *FKH1*, *FKH2*, and *NDD1* that encode transcription factors which control G2–M transcription. Thus, this transcription factor has a central role in cell cycle controls, regulating gene expression both directly and indirectly throughout the cell cycle.

3. G2–M waves

The G2–M phase in budding yeast is a major cell cycle transition in this organism, and correspondingly a large group of cell cycle regulated genes, sometimes known as the CLB2 cluster, is expressed exclusively at this time. The pioneer gene in this class was *SWI5*, with Mcm1p identified as the transcription factor that controls its expression and the rest of the group (Althoefer *et al.*, 1995; Lydall *et al.*, 1991; Maher *et al.*, 1995). Mcm1p is a member of a class of so-called *MADS box transcription factors* that are found in all eukaryotic organisms, from yeasts to humans, where it is known as serum response factor (Shore and Sharrocks, 1995; Treisman, 1994; Wynne and Treisman, 1992). Mcm1p had been much studied in budding yeast and shown to have many roles in controlling gene expression of a range of genes, including those for mating-type switching (Gelli, 2002). The ability of Mcm1p to control various groups of gene expression occurs through it having differing binding partners that confer specificity, and is the modus operandi of MADS box proteins in all organisms (Acton *et al.*, 1997; Messenguy and Dubois, 2003; Tuch *et al.*, 2008a).

The associated binding partner to Mcm1p that controls G2–M transcription was first characterized biochemically, and named Swi five factor (SFF), with two forkhead-like transcription factors, Fkh1p and Fkh2p, and another protein Ndd1p, subsequently identified as important components (Jorgensen and Tyers, 2000, 2004; Knapp et al., 1996; Koranda et al., 2000; Pic et al., 2000; Zhu et al., 2000). The two forkhead transcription factors are required to ensure correct gene expression at the G2–M interval (Boros et al., 2003; Darieva et al., 2003; Hollenhorst et al., 2000, 2001; Kumar et al., 2000; Lim et al., 2003; Reynolds et al., 2003). Both bind to forkhead DNA motifs present in the promoters of genes in the CLB2 cluster. The two proteins are partially redundant, though they appear to have separate and distinct roles. Fkh2p binds to Mcm1p on gene promoters, whereas Fkh1p binds in the absence of Mcm1p, with changes in DNA conformation thought to be important for the recruitment of SFF to promoter DNA.

Transcriptional activation is not imparted through changes in binding of either Mcm1p or Fkh2p to promoter DNA, as both remain in contact throughout the cell cycle. Instead, another transcription activator, Ndd1p, has an important role in this process (Koranda et al., 2000; Loy et al., 1999). Ndd1p is required for G2–M transcription, with it binding to CLB2 cluster gene promoters dependent on Fkh2p and Mcm1p. Ndd1p protein levels peak in S and G2 phases, through both regulation of its transcription and protein stability. Importantly, binding of Ndd1p to promoters is controlled by the regulated phosphorylation of Fkh2p by the Clb5p–Cdc28p CDK complex (Pic-Taylor et al., 2004). Once Ndd1p is bound to Fkh2p, this interaction is stabilized by another CDK complex, Clb2p–Cdc28p, phosphorylating Ndd1p on threonine 319 and resulting in it binding specifically to the FHA domain of Fkh2p. The complex of Ndd1p–Fkh2p–Mbx1p results in the transcription of CLB1 and CLB2, thus creating a positive feedback loop. Ndd1p is also directly phosphorylated on serine 85 by another cell cycle protein kinase, the polo-like kinase encoded by the CDC5 gene, which is required for the normal temporal expression of the CLB2 cluster (Darieva et al., 2006). Part of this mechanism occurs through Cdc5p binding specifically to promoter DNA at the time when it activates Ndd1p. As CDC5 is part of the CLB2 cluster, this process offers yet another positive feedback mechanism to control its expression.

A separate layer of control involving protein stability appears to also operate, at least for one gene in the CLB2 cluster, CDC20. This occurs through Cks1p recruiting the proteasome to its promoter, which results in gene expression at G2–M (Morris et al., 2003; Tempé et al., 2007; Yu et al., 2005).

4. M–G1 waves

Two waves of gene expression occur late in the budding yeast cell cycle and are controlled by two different transcription systems. One of these involves Mcm1p along with two homeodomain proteins, Yox1p and Yhp1p, and the other the Swi5p/Ace2p transcription factor pair.

Mcm1p is fascinating and unique in cell cycle regulated transcription in that it is involved in at least two waves of gene expression at different times of the cell cycle (and indeed other important transcription control mechanisms). As already described, these differences are conferred by ancillary transcription factors that bind alongside Mcm1p on promoters with, in the case of M–G1 gene expression, these being two homeodomain proteins, Yox1p and Yhp1p. Here, Mcm1p binds to a promoter element named an early cell cycle box (ECB) in the promoters of important cell cycle genes such as *SWI4* and *CLN3*, themselves rate limiting for cell cycle progression, and essential for the control of subsequent waves of gene expression at G1–S (MacKay *et al.*, 2001; McInerny *et al.*, 1997). Regulated expression is not conferred through changes in Mcm1p binding to DNA, as it is in contact with promoters throughout the cell cycle (Mai *et al.*, 2002), instead through changes in binding of Yox1p and Yhp1p, with their removal from promoters allowing gene expression at M–G1 (Pramila *et al.*, 2002). As *YOX1* is itself under transcriptional regulation by SBF, and Mcm1p controls the expression of *SWI4* encoding a component of SBF, this sets up a negative feedback loop.

A separate wave of gene expression at M–G1 is controlled by the paralogous pair of transcription factors, Swi5p and Ace2p, with these two having overlapping, but distinct, roles in these processes, reminiscent of SBF/MBF at G1–S (Dohrmann *et al.*, 1992, 1996; Doolin *et al.*, 2001; Laabs *et al.*, 2003; McBride *et al.*, 1999). Both transcription factors bind to the same promoter sequence, which has been called a Swi5-binding site and is present in a number of gene promoters for genes that encode products required for cytokinesis and cell separation.

In a crucial linking mechanism between two waves of gene expression, the *ACE2* and *SWI5* genes are transcriptionally regulated by the Mcm1p–Fkh2p system at G2–M. Swi5p protein is also negatively regulated by Clb2p–Cdc28p, as the specific degradation of Clb2p during anaphase results in the dephosphory-lation of Swi5p, so causing its retention in the nucleus and allowing it to activate target genes (Moll *et al.*, 1991; Nasmyth *et al.*, 1990). In contrast, phosphorylation of Ace2p by the RAM network causes its activation to stimulate gene expression (Nelson *et al.*, 2003).

Swi5p and Ace2p also have other important roles in cell cycle gene expression, being part of the process which controls mother–daughter cell identity. This occurs through the accumulation of the product of one of their target genes, *ASH1*: Ash1p, along with Ace2p, accumulates only in daughter cells, thus ensuring the specific control of gene expression in this cell type (Bertrand *et al.*, 1998; Bobola *et al.*, 1996; Cosma, 2004; McBride *et al.*, 1999; Paquin and Chartrand, 2008; Shen *et al.*, 2009; Sil and Herskowitz, 1996). Another essential function for Swi5p is the control of mating-type switching through the control of the *HO* endonuclease, which is also mediated by SBF (Andrews and Herskowitz, 1989a; Breeden and Nasmyth, 1987).

B. Fission yeast

In fission yeast, four main waves of gene expression have been described, with the genes identified both by traditional methods and through transcriptome experiments. These fall across various stages of the cell cycle, at G1–S, S phase, G2–M, and M–G1 (Table 2.1; Fig. 2.2).

1. G1–S waves

The G1–S wave was the first to be identified in this organism and has been the most well characterized. It contains up to 20 genes whose encoded products have roles in DNA synthesis and cell cycle controls. Similar to that in budding yeast, the genes all contain MCB sequence motifs in their promoters, which are the UASs that control their G1–S-specific transcription (Lowndes et al., 1992b; Maqbool et al., 2003). Indeed, the MCB DNA sequence motifs in budding and fission yeasts are interchangeable, showing that this promoter motif is evolutionarily conserved.

The transcription factor complex that binds to MCB sequences in fission yeast is also called MBF (originally DSC1), with a number of components identified. First among these was Cdc10p, which appears to have a central role in controlling G1–S gene expression with, like Swi6p, both positive and negative functions (Lowndes et al., 1992b; McInerny et al., 1995). Cdc10p is similar to Swi4p and Swi6p in peptide sequence, predicted structure, and also contains ankyrin motifs, suggesting a common ancestry for this class of transcription factor (Aves et al., 1985). Other components of MBF include Res1p (also named Pct1p), Res2p, Rep1p, and Rep2p, each with apparently differing functions, but altogether contributing to ensure the correct expression at the beginning of S phase, with some having mitotic and others meiotic specific roles (Ayté et al., 1995, 1997; Baum et al., 1997, 1998; Caligiuri and Beach, 1993; Miyamoto et al., 1994; Tahara et al., 1998; Tanaka et al., 1992; White et al., 2001; Whitehall et al., 1999; Zhu et al., 1994). For example, Rep2p has an important regulatory role in controlling MBF activity in mitosis (Nakashima et al., 1995); and Res2p and Rep1p have significant roles in controlling meiotic gene expression (Cunliffe et al., 2004; Ding and Smith, 1998). Other proteins to which Cdc10p bind have been identified, and include Pol5p, which controls rRNA production (Nadeem et al., 2006), and the MBF complex also has a role in DNA damage response (Chu et al., 2007).

How MBF is itself controlled is not fully understood, although this is not through cell cycle-specific binding to promoter DNA, at least of Cdc10p, as it contacts MCB motifs throughout the cell cycle (Wuarin et al., 2002). As mentioned, Cdc10p has both repressing and activating roles (McInerny et al., 1995), with further repressive functions conferred to MBF by SpNrm1p (de Bruin and Wittenberg, 2009; de Bruin *et al.*, 2006). Though it has been suggested that the Cdc2p CDK activates MBF (Connolly *et al.*, 1997; Reymond *et al.*, 1993), other experiments indicate that it has no role (Baum *et al.*, 1997) with, instead, another cyclin, Pas1p, when associated with a kinase Pef1p, activating MBF (Tanaka and Okayama, 2000). Other layers of controls also operate with, for example, the cyclin Cig2p, which itself contains MCB sequences in its promoter and whose transcription is at least in part controlled by MBF, inhibiting MBF activity through binding to Res2p and phosphorylating Res1p when associated with the CDK Cdc2p (Ayté *et al.*, 2001). Further controls may occur through specific proteolytic degradation by the SCF ubiquitin ligase system during S, G2, and M phases, when MBF is inactive (Yamano *et al.*, 2000).

2. S-phase wave

A number of genes encoding histones, like the situation in budding yeast, are periodically expressed in fission yeast during DNA synthesis (Matsumoto and Yanagida, 1985; Matsumoto *et al.*, 1987), with the promoter motif and transcription factor, Ams2p, that positively regulate their expression identified (Takayama and Takahashi, 2007). Furthermore, a repressor named Hip1p has been characterised, that is similar to the Hir1p/Hir2p repressors in budding yeast (Blackwell *et al.*, 2004). Not all histone genes are expressed at the same cell cycle time, with the centromere histone variant $cnp1^+$ (CENP-A) expressed slightly earlier in the cell cycle that the other histones, which is likely to be important for function (Takahashi *et al.*, 2000). Presumably, $cnp1^+$ is under different transcriptional control mechanisms, though these remain unknown.

3. G2–M waves

Despite G2 contributing to a significant proportion of the mitotic cell cycle in fission yeast, and the G2–M transition being the major control point, only a smaller group of weakly regulated genes has been identified that are expressed at this cell cycle stage. These include $spd1^+$, $psu1^+$, and $rds1^+$, with a putative common promoter UAS proposed, named "New 3" (Rustici *et al.*, 2004). This UAS has neither been confirmed experimentally nor has the transcription factor that binds to it been identified.

4. M–G1 waves

An important wave of transcription occurs in fission yeast at the M–G1 interval. This group of genes numbers at least 20 genes, with the first identified being $cdc15^+$ (Anderson *et al.*, 2002; Fankhauser *et al.*, 1995; Rustici *et al.*, 2004).

Most encode products required for processes at the end of the cell cycle, such as chromosome separation, cytokinesis, and septation. The promoter sequences and transcription factors required for their expression have been identified, named pombe cell cycle boxes (Pcbs) and PCB-binding factor (PBF), respectively (Anderson *et al.*, 2002). At least three components of PBF have been identified, and these include two forkhead-like transcription factors, Fkh2p and Sep1p, and a MADS box-like protein, Mbx1p (Buck *et al.*, 2004; Bulmer *et al.*, 2004; Szilagyi *et al.*, 2005; Zilahi *et al.*, 2000).

The two forkhead transcription factors have complementary and opposing roles in regulating gene expression: Fkh2p, which is only bound to PCB promoters when M–G1 genes are not being expressed, appears to have a repressive role; in contrast, Sep1p is only bound when genes are being expressed, and thus has a positive role (Papadopoulou *et al.*, 2008). These proposed functions are supported by experiments where either gene has been deleted from the chromosome in cells, with the absence of Fkh2p resulting is unnaturally high transcription of M–G1 genes and, in contrast, the absence of Sep1p resulting in low gene expression (Buck *et al.*, 2004; Bulmer *et al.*, 2004; Rustici *et al.*, 2004). Thus, the stepwise replacement of one forkhead transcription factor by another appears to be a crucial part of the control process. How this change in binding is mediated is not yet understood, although Fkh2p and Sep1p bind to each other in cells, suggesting that a transient interaction between the two may be part of the process, and Sep1p is required for Fkh2p function (Buck *et al.*, 2004; Papadopoulou *et al.*, 2008). Fkh2p also binds to its own promoter and is periodically expressed as part of the cluster at M–G1, so forming a negative feedback loop. Furthermore, both proteins are phosphorylated by as yet unknown protein kinases, although genetic experiments imply a role for Cdc2p with Sep1p (Grallert *et al.*, 1998).

Another level of control of M–G1 gene expression occurs through Mbx1p. This, too, is a cell cycle regulated phosphoprotein, but in this case the protein kinase has been identified as being Plo1p, a polo-like kinase (Papadopoulou *et al.*, 2008). Plo1p phosphorylates Mbx1p at M–G1, with this occurring in part through Plo1p only binding to PCB promoters at this cell cycle time. As the $plo1^+$ gene itself is expressed at M–G1 and has PCB sequences in its promoter, Plo1p acts in a positive feedback loop controlling its own transcription.

Similar to the situation in budding yeast, a group of genes whose function is required for cell separation is controlled by an Ace2p transcription factor (Alonso-Nunez *et al.*, 2005; Petit *et al.*, 2005). This group, which includes $eng1^+$, encoding Eng1p, which controls degradation of the cell septum, is also expressed at M–G1 under the control of an ACE2 promoter motif (Dekker *et al.*, 2006; Martín-Cuadrado *et al.*, 2003). Ace2p is itself regulated by PBF, so these two transcription waves are functionally linked (Rustici *et al.*, 2004).

This dependency between two waves of gene expression is the only example in fission yeast, but is strikingly similar to that seen in budding yeast (Bähler, 2005a). It has been suggested that it is part of the mechanism that allowed fission to evolve into a single-celled organism (Bähler, 2005b; Borup, 2006).

III. REGULATORY NETWORKS

Now that reasonably comprehensive pictures of cell cycle regulated gene expression in both budding and fission yeasts have been created, attempting to link together the various waves to each other in each organism, and to integrate these into more general cell cycle controls has been possible.

In budding yeast, a combination of traditional methods and genome-wide approaches has created an integrated "cycle within a cycle" whereby the consecutive waves of gene expression are functionally linked to one another with, in the simplest example, one wave of gene expression containing the transcription factor that controls transcription of the next wave of gene expression (Fig. 2.3; Breeden, 2003; Ihmels et al., 2002; Lee et al., 2002; Simon et al., 2001; Tyers, 2004; Wittenberg and Reed, 2005). Arbitrarily beginning the cell cycle at G1–S, SBF and MBF control expression of HCM1, with Hcm1p controlling the expression of NDD1 and with Ndd1p activating the Fkh2p–Mcm1p complex, which, in turn, stimulates transcription of SWI5 and ACE2. Swi5p and Ace2p, along with Mcm1p, stimulate the production of both Swi4p and Cln3p which, either directly or indirectly, activate SBF and MBF. Thus, the loop is completed. On top of this "loop of transcription" is another layer of posttranslational control, with certain gene products activating or repressing the activity of transcription factor complexes (de Lichtenberg et al., 2007).

In contrast, such a "loop of transcription" appears in fission yeast not to be complete (Fig. 2.3). So far, the only direct link between two consecutive cell cycle waves of transcription occurs with the Ace2p transcription factor being under transcriptional control by PBF at M–G1, with it controlling the Ace2 wave of gene expression (Bähler, 2005a; McInerny, 2004; Nachman and Regev, 2009). Consistent with the suggestion that the other waves of gene expression are unlinked, deregulation of PBF–PCB gene expression has no effect on MBF–MCB controls, and vice versa (Anderson et al., 2002). Instead, it appears that most waves are controlled separately by cell cycle molecules such as the CDK Cdc2p or the polo kinase Plo1p. Quite how the global mechanism will operate in higher eukaryotes, and whether a loop of transcription operates, remains to be seen.

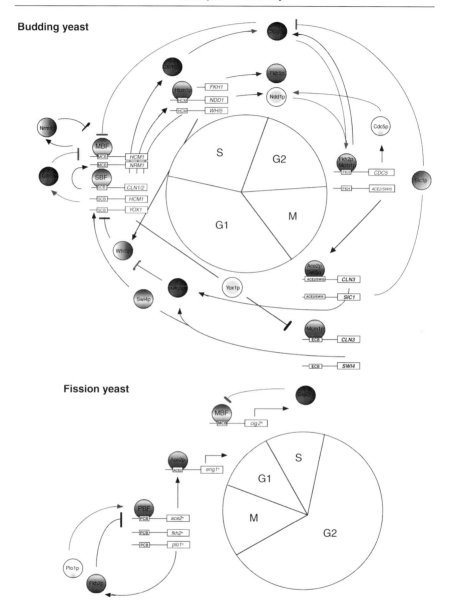

Figure 2.3. Molecular networks integrating and controlling cell cycle regulated gene expression in budding and fission yeasts. Transcriptional regulation is shown by black lines and posttranslational in red, with positive control by arrows and negative control by bars. Related transcription factors in the two yeasts are shown by the same color. For details of each, see text. (See Color Insert.)

IV. CONSERVATION OF CONTROLS

A major reason for using yeasts to study fundamental biological problems is to exploit the acquired knowledge to understand more complex organisms, especially humans. In many cases, molecular processes are so complicated in higher eukaryotes that it is impossible to unravel and understand them, without prior knowledge from more simple systems. In contrast, in yeasts, the relatively simplicity of processes permits their characterization, with this information permitting subsequent analysis and understanding in higher eukaryotes. It has also been very useful to study the same process in both budding and fission yeasts, as it has been observed that mechanisms conserved between these two distantly related organisms, are often conserved in other eukaryotes (Côte et al., 2009; Forsburg, 1999; Forsburg and Nurse, 1991; Ukil et al., 2008).

Among the various genes transcribed at particular cell cycle times, a number of genes have been shown to be periodically expressed in both budding and fission yeasts (Table 2.2). Computational analyses have identified up to 20% of cycling genes encode orthologs that are periodically expressed in both species, with 5–7% in other eukaryotic organisms (Bähler, 2005a; Lu et al., 2007).

Table 2.2. Cell Cycle Regulated Genes in Both Budding and Fission Yeasts

Cell cycle phase	Budding yeast	Fission yeast	Function	Human ortholog
G1–S	RNR1	cdc22$^+$	Ribonucleotide reductase	Yes
	CDC6	cdc18$^+$	Regulator of DNA replication	Yes
	SWE1	mik1$^+$	Protein kinase regulating mitotic entry	Yes
	CLB1-CLB6	cig2$^+$	B-type cyclins	Yes
	POL1	pol1$^+$	DNA polymerase α	Yes
	POL2	cdc20$^+$	DNA polymerase ε	Yes
S phase	Histone genes	Histone genes	Histones H2A, H2B, H3, and H4	Yes
G2–M–G1 phases	CDC5	plo1$^+$	Polo kinase	Yes
	IPL1	ark1$^+$	AuroraB kinase	Yes
	KIN3	fin1$^+$	NimA kinase	Yes
	CDC20	slp1$^+$	APC regulator	Yes
	ACE2	ace2$^+$	Cell cycle transcription factor	Yes
	BUD4	mid2$^+$	Cytokinesis	Yes
	SPO12	spo12$^+$	Cell cycle regulator	Not known

Representative genes from each cluster are shown, which have significant roles in cell cycle controls and whose orthologs are also periodically expressed in mammalian cells.

Notably, most of these genes encode products that have important and significant roles in controlling progression through the cell division cycle, and many are also periodically transcribed in humans by related transcriptional mechanisms (Cho et al., 2001; Whitfield et al., 2002; Zwicker and Muller, 1995).

Among the waves of gene expression described in the two yeast species, three show striking similarities between the two organisms. These are the SCB/MCB waves at G1–S, the SFF/PCB waves at G2–M/M–G1, and the Swi5/Ace2 wave at M–G1.

In budding yeast, gene expression at G1–S is controlled by two transcription factor complexes working in parallel, with some overlap in their composition. SBF contains Swi4p and Swi6p, whereas MBF contains Swi6p and Mbp1p. In fission yeast, MBF instead contains Cdc10p along with a number of ancillary proteins. All three transcription factor complexes are related to each other containing conserved structural motifs. Furthermore, the MCB DNA motif to which MBF binds is conserved between the two organisms. One significant difference between the organisms, however, is that in fission yeast, there is just one transcription factor complex operating at G1–S, which encompasses both the functions of SBF and MBF. Despite these differences, related transcriptional mechanisms operate in human cells with a wave of gene expression at G1–S controlled by the E2F–Dp transcription factor complex (Genovese et al., 2006; La Thangue and Taylor, 1993; Sherr and McCormick, 2002; Taylor et al., 1997).

Other waves of gene expression occurring later in the cell cycle also show similarities between the two yeasts and humans. These include those that regulate various genes required late in the cell cycle for M phase and cytokinesis. The waves of gene expression are all controlled by forkhead-like transcription factors in both yeasts and humans (Alvarez et al., 2001; Laoukili et al., 2005; Park et al., 2008). What is also striking is that these transcription factors control the expression of and are controlled by polo kinases in both yeasts and humans (Darieva et al., 2006; Fu et al., 2008; Papadopoulou et al., 2008).

V. IMPORTANCE OF CELL CYCLE REGULATED GENE EXPRESSION

A significant question to consider is the importance of cell cycle regulated gene expression. The fact that so many genes' transcription varies through the cell cycle, and that in some cases this is conserved among organisms, suggests that such regulation is required and important. Why is the expression of some genes cell cycle regulated? A number of explanations have been suggested, which fall into three broad categories: cell economy, impairment of cell cycle progression, and cell cycle controls.

When cells progress through the cell cycle, a very large number of different cellular structures and processes are made or activated, in many cases for a short period to last only for the cell cycle phase for which they are required. For example, during the S phase when DNA replication occurs, proteins required for making DNA must be present, and during M phase proteins needed for chromosome separation need to be made. Therefore, an efficient and economical way to use cellular constituents would be to produce proteins only at the cell cycle stage when they are required, using regulated gene transcription as one mechanism among others to achieve this.

In other instances, it may be deleterious for cell cycle progression for certain proteins to be present at a particular cell cycle phase when they are usually not present. Such proteins may inhibit or prevent cell cycle progression, or result in faulty processes occurring. Cell cycle-specific expression is one way of ensuring that proteins are only present at the cell cycle phase when they are meant to be present. Examples of this are the $cdc18^+$ gene in fission yeast, and the BUD8 and BUD9 gene pair in budding yeast. $cdc18^+$ is a critical controller of S phase, being a component of the origin of replication complex (ORC) that is under MCB–MBF control and normally only transcribed at G1–S (Kelly et al., 1993). Its artificial expression at other times in the cell cycle causes multiple S phases which are lethal to cells. Indeed, the restriction of the presence of Cdc18p to the start of S phase is so critical to cells that it is controlled at multiple levels, including transcription and protein stability (Jallepalli et al., 1997). The Bub8p and Bub9p proteins are related membrane proteins that control localization of cell poles, partly through their expression at different cell cycle times (Schenkman et al., 2002). This was shown by the swapping of their promoters resulting in each gene product being produced at the wrong cell cycle time and consequent mis-localization to the opposite pole.

Another interesting function for cell cycle-specific expression is when it plays a role in regulating or controlling cell cycle progression. For this small number of genes, their phase-specific expression is crucial for controlling the timing of, and the orderly, correct passage into the next cell cycle phase. A good example of this is shown by the polo kinases, Plo1p and Cdc5p, protein kinases with central roles in controlling M-phase progression and cytokinesis, conserved between the two yeast species, and also present in humans (Archambault and Glover, 2009). In all three eukaryotes, the genes are cell cycle regulated, and this regulation is very important, as its deregulation has serious effects. The most dramatic example of this is shown in fission yeast where the expression of $plo1^+$ during interphase results in premature and lethal cytokinesis (Ohkura et al., 1995). This is in part because Plo1p controls its own and other genes' transcription at the M–G1 interval by directly binding to promoters and phosphorylating at least one component of the PBF transcription factor complex that controls M–G1 gene expression in this organism (Anderson et al., 2002; Ng et al., 2006;

Papadopoulou *et al.*, 2008). Cdc5p has similar regulatory roles in budding yeast, binding to, and phosphorylating Ndd1p to stimulate its own and other genes' cell cycle expression (Darieva *et al.*, 2006). Related mechanisms also occur in humans with polo kinases phosphorylating FoxM1 forkhead-like transcription factors (Fu *et al.*, 2008) which control cyclin and polo kinase gene expression (Alvarez *et al.*, 2001; Laoukili *et al.*, 2005). Thus, polo kinases, known to have primary roles in controlling events through mitosis and cytokinesis, cause this to occur in part through regulating expression of genes required at this cell cycle time.

A different example of phase-specific expression being important for cell cycle control is shown by the fission yeast gene *mik1*$^+$. *mik1*$^+$ encodes a protein kinase that phosphorylates the fission yeast CDK, Cdc2p, to inhibit its activity and so prevent mitotic entry, thus functioning at the G2–M boundary (Lee *et al.*, 1994). It was a surprise, therefore, when it was found that *mik1*$^+$ mRNA and protein levels peak at the G1–S boundary, and that it is under MBF/MCB controls (Baber-Furnari *et al.*, 2000; Christensen *et al.*, 2000; Ng *et al.*, 2001). The explanation for this is that Mik1p protein is produced during S phase as part of the mechanism to prevent M phase from occurring before DNA synthesis is complete. Thus, the specific production of Mik1p protein during S phase by transcriptional regulation is part of the control mechanism that ensures orderly entry into the following M phase, later in the cell cycle.

VI. UNIMPORTANCE OF CELL CYCLE REGULATED GENE EXPRESSION

Despite the fact that some genes are periodically expressed in both yeast species, and in a few cases this is significant for cell cycle controls, it is important to realize that in many other cases, indeed the majority, periodic expression of genes appears to be irrelevant to cell cycle progression, at least under experimental conditions. This conclusion can be stated because of observation from two types of experimental approaches.

The first approach is to ask the simple question whether an individual gene's periodic expression is significant, by replacing its promoter with another constitutively active one, resulting in the gene being expressed throughout the cell cycle. It is surprising on how many occasions this experiment has little or no effect on cell division. Perhaps even more striking, however, is when cell cycle transcription factors are either deleted from the chromosome or modified to become constitutively activated, resulting in gross effects on the expression of their whole target group. In many cases when this is done, not only are the cells still alive but often they also complete reasonably normal cell cycles. Examples of this include from budding yeast, deleting the genes encoding Swi4p, Swi6p, Fkh1p, and Fkh2p (Pic *et al.*, 2000); and from fission yeast, Res1p, Res2p,

Fkh2p, and Sep1p (Buck *et al.*, 2004; Caligiuri and Beach, 1993; Ribár *et al.*, 1997, 1999). So, not only manipulating individual genes but also altering the expression of whole groups of genes can have little effect. In a few cases, this can be explained by genetic redundancy, with double deletions of transcription factor pairs having far more severe phenotypes. But it is a fact that the removal of apparently important cell cycle transcription factors affecting whole groups of target genes has, in some cases, very little effect.

However, at least two important points need to be considered in such discussions about the relevance of cell cycle gene expression. The first is that most, if not all, experiments studying this phenomenon in yeasts, are with cells growing under optimal conditions during exponential growth phase, and so going through multiple, consecutive cell cycles. It is likely that these are not usual growth conditions that yeast experience in the wild where, instead, they spend most time in stationary phase, with short bursts of division and growth, when nutrients sporadically appear. This means that perhaps more relevant to the biology of these single-celled organisms is their first divisions immediately after exiting from stationary phase. Under such conditions, the G1 phase is highly extended when cells reenter the cell cycle, and it is perhaps then that the *de novo* production of proteins required for cell cycle progression may be critical.

Another essential point to reemphasize is that the presence or absence of critical cell cycle proteins within cells is almost always controlled at multiple levels, transcriptionally and posttranscriptionally. Such systems introduce double-check mechanisms, whose success is exemplified by the fact that experimentally manipulating transcription of the gene can have little effect on cell cycle progression.

VII. RELATED AREAS OF INTEREST AND TOPICS FOR FURTHER STUDY

Despite much research allowing a reasonably comprehensive picture of cell cycle regulated gene expression during mitosis in both yeast species, there are still some large gaps that need to be filled, and much detail to be unraveled. Here, I highlight some of the major areas and also describe related areas of interest.

A. Histone transcription in budding yeast and G2–M transcription in fission yeast

Perhaps, the most notable gap in budding yeast is the lack of understanding of how the periodic expression of the histone genes is controlled, despite these being the first cell cycle regulated genes to be identified. Histones fall into many different subtypes, characterized by differing posttranslational modifications, with these having important roles in controlling general gene expression, and their correct relative stoichiometric levels are important for correct chromosome transmission to daughter cells (Meeks-Wagner and Hartwell, 1986). Thus, their periodic expression is important and the transcription factors that regulate this need to be identified.

In fission yeast, a group of genes was identified by transcriptome analysis that is expressed at the G2–M interval, with a putative promoter sequence, "New 3" proposed from computer analyses (Rustici *et al.*, 2004). This sequence neither has been confirmed experimentally nor has the transcription factor complex been identified, although it has been suggested that the transcript levels of some of the genes may be controlled by changes in mRNA stability (Bähler, 2005a).

B. Integration into general cell cycle controls

For both yeast species, frameworks suggesting how the various waves of genes are controlled through the cell division cycle have been created. At least for budding yeast, this framework describes a transcription cycle within a cycle, though this is not the case for fission yeast, where the cycle is incomplete. Even so, in both yeasts, the waves of transcription need to be integrated and controlled by general cell cycle control mechanisms, to ensure that expression of the various waves is coupled to cell cycle processes for which they are relevant.

In some cases, such links have been established (Fig. 2.4). Perhaps, the most important cell cycle control molecules are the CDKs, which in yeasts are represented by single proteins, Cdc28p in budding yeast and Cdc2p in fission yeast. Both CDKs are sequentially activated through the cell cycle, by binding to different cyclin molecules to stimulate and control various cell cycle stages, such as S phase and M phase. As such, the cyclin–CDK molecules have a preeminent role in controlling cell cycle events, and in some cases they appear to directly link these to transcriptional regulation, of which a few examples have been described in this review, in both yeasts, at various stages of the cell cycle. Additionally, it has been shown that CDKs have roles in more general transcriptional regulation in both budding and fission yeasts (Lee *et al.*, 2005; Morris *et al.*, 2003). However, global approaches have demonstrated that, at least in budding yeast, CDKs do not control all cyclic waves of gene expression, so other regulator molecules must be involved (Orlando *et al.*, 2008).

One such class of molecules is the polo kinases, which have a major function in controlling processes late in the cell cycle. Indeed, at least in fission yeast, it is known that once cells have passed the G2–M transition and entered mitosis, CDK activity must be inhibited to allow cytokinesis and cell division to occur (Le Goff *et al.*, 1999). In both yeasts and humans, it has been shown that polo kinases directly control cell cycle gene expression at this late cell cycle stage

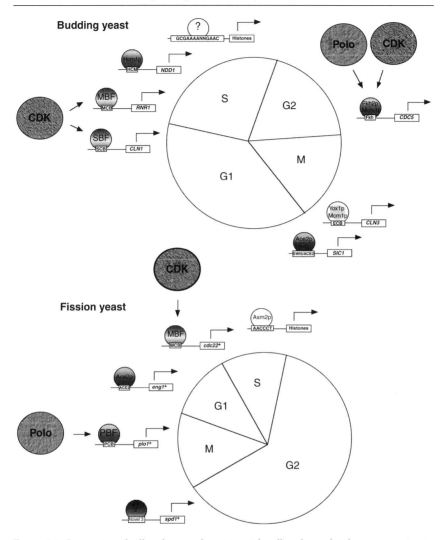

Figure 2.4. Integration of cell cycle control proteins with cell cycle regulated gene expression in budding and fission yeasts. For each yeast species, a single gene is shown to represent the group of genes that are transcribed at particular cell cycle times. Where known the name or DNA sequence of the upstream activating sequence, and the transcription factor are indicated. Related transcription factor complexes in the two yeasts are shown by the same color. For details of each, see text. (See Color Insert.)

through directly binding to and phosphorylating the transcription factors that regulate them (Anderson *et al.*, 2002; Darieva *et al.*, 2006; Fu *et al.*, 2008; Papadopoulou *et al.*, 2008).

It is likely that other protein kinases are involved in controlling cell cycle transcription factors as many are phosphoproteins, and these need to be identified. And of course, the fact that transcription factors are reversibly phosphorylated means that the antagonistic phosophatases require identification as well.

C. Evolution of cell cycle control mechanisms

An interesting question is how cell cycle transcriptional control mechanisms have evolved. Now that we have models in different organisms, comparative studies can permit this to be approached. Global analyses have already made interesting inroads into this area, suggesting how relatively simple mutations can have profound implications for expression of groups of genes, and so can account for major differences in circuitry between species (de Lichtenberg et al., 2007; Jensen et al., 2006; Tuch et al., 2008a,b).

D. Meiosis

Though this review has deliberately focused on mitotic control, it is important to state that some of the mechanisms described also operate during the alternative life cycle, meiosis. In a few cases, related or the same transcription factor complexes operate, with forkhead transcription factors and MBF being good examples (Abe and Shimoda, 2000; Cunliffe et al., 2004; Horie et al,. 1998; Raithatha and Stuart, 2005). But in many other cases, completely different transcription factors and promoter sequences specific to meiotic gene expression are present, although the same simple paradigm operates with various waves of gene expression being controlled by specific transcription factors and promoter sequences (Chu et al., 1998; Mata et al., 2002).

E. Mediator complex

There has been much interest in characterizing the Mediator complex in both budding and fission yeasts, with the complex now well described and its pattern of binding to DNA in a global context characterized (Holstege et al., 1998; Linder et al., 2008; Zhu et al., 2006). As well as its more general role in forming the link between RNA polymerase II and the specific transcription factors that control regulated gene expression, it has been found that the Mediator complex has a more specific role in controlling cell cycle gene expression in both yeasts. For example, mutations in components of the Mediator complex in fission yeast have specific effects on the transcription of genes required late in the cell cycle for cell separation under Ace2p and possibly Sep1p control (Lee et al., 2005;

Miklos *et al.*, 2008; Sharma *et al.*, 2006), with related observations in budding yeast (Porter *et al.*, 2002). Why this should be is not yet understood, but will be a fascinating area of enquiry in the future.

References

Abe, H., and Shimoda, C. (2000). Autoregulated expression of *Schizosaccharomyces pombe* meiosis specific transcription factor Mei4 and a genome-wide search for its target genes. *Genetics* **154,** 1497–1508.

Acton, T. B., Zhong, H., and Vershon, A. K. (1997). DNA-binding specificity of Mcm1: Operator mutations that alter DNA-bending and transcriptional activities by a MADS box protein. *Mol. Cell. Biol.* **17,** 1881–1889.

Alonso-Nunez, M. L., An, H., Martin-Cuadrado, A. B., Mehta, S., Petit, C., Sipiczki, M., del Rey, F., Gould, K. L., and de Aldana, C. R. (2005). Ace2p controls the expression of genes required for cell separation in *Schizosaccharomyces pombe*. *Mol. Biol. Cell* **16,** 2003–2017.

Althoefer, H., Schleiffer, A., Wassmann, K., Nordheim, A., and Ammerer, G. (1995). Mcm1 is required to coordinate G2-specific transcription in *Saccharomyces cerevisiae*. *Mol. Cell. Biol.* **15,** 5917–5928.

Alvarez, B., Martínez, A. C., Burgering, B. M., and Carrera, A. C. (2001). Forkhead transcription factors contribute to execution of the mitotic programme in mammals. *Nature* **413,** 744–747.

Amon, A., Tyers, M., Futcher, B., and Nasmyth, K. (1993). Mechanisms that help the yeast cell cycle clock tick: G2 cyclins transcriptionally activate G2 cyclins and repress G1 cyclins. *Cell* **74,** 993–1007.

Anderson, M., Ng, S. S., Marchesi, V., MacIver, F. H., Stevens, F. E., Riddell, T., Glover, D. M., Hagan, I. M., and McInerny, C. J. (2002). *plo1*⁺ regulates gene transcription at the M-G1 interval during the fission yeast mitotic cell cycle. *EMBO J.* **21,** 5745–5755.

Andrews, B. J., and Herskowitz, I. (1989a). Identification of a DNA binding factor involved in cell-cycle control of the yeast HO gene. *Cell* **57,** 21–29.

Andrews, B. J., and Herskowitz, I. (1989b). The yeast SWI4 protein contains a motif present in developmental regulators and is part of a complex involved in cell-cycle-dependent transcription. *Nature* **342,** 830–833.

Archambault, V., and Glover, D. M. (2009). Polo-like kinases: Conservation and divergence in their functions and regulation. *Nat. Rev. Mol. Cell Biol.* **10,** 265–275.

Aves, S. J., Durkacz, B. W., Carr, A., and Nurse, P. (1985). Cloning, sequencing and transcriptional control of the *Schizosaccharomyces pombe* cdc10 'start' gene. *EMBO J.* **4,** 457–463.

Ayté, J., Leis, J. F., Herrera, A., Tang, E., Yang, H., and DeCaprio, J. A. (1995). The *Schizosaccharomyces pombe* MBF complex requires heterodimerization for entry into S phase. *Mol. Cell. Biol.* **15,** 2589–2599.

Ayté, J., Leis, J. F., and DeCaprio, J. A. (1997). The fission yeast protein p73res2 is an essential component of the mitotic MBF complex and a master regulator of meiosis. *Mol. Cell. Biol.* **17,** 6246–6254.

Ayté, J., Schweitzer, C., Zarzov, P., Nurse, P., and DeCaprio, J. A. (2001). Feedback regulation of the MBF transcription factor by cyclin Cig2. *Nat. Cell Biol.* **3,** 1043–1050.

Bähler, J. (2005a). Cell cycle control of gene expression in budding and fission yeast. *Annu. Rev. Genet.* **39,** 69–94.

Bähler, J. (2005b). A transcriptional pathway for cell separation in fission yeast. *Cell Cycle* **4,** 39–41.

Banerjee, N., and Zhang, M. Q. (2003). Identifying cooperativity among transcription factors controlling the cell cycle in yeast. *Nucleic Acids Res.* **31,** 7024–7031.

Baber-Furnari, B. A., Rhind, N., Boddy, M. N., Shanahan, P., Lopez-Girona, A., and Russell, P. (2000). Regulation of mitotic inhibitor Mik1 helps to enforce the DNA damage checkpoint. Mol. Biol. Cell 11, 1–11.

Baum, B., Wuarin, J., and Nurse, P. (1997). Control of S-phase periodic transcription in the fission yeast mitotic cycle. EMBO J. 16, 4676–4688.

Baum, B., Nishitani, H., Yanow, S., and Nurse, P. (1998). Cdc18 transcription and proteolysis couple S phase to passage through mitosis. EMBO J. 17, 5689–5698.

Bean, J. M., Siggia, E. D., and Cross, F. R. (2005). High functional overlap between MluI cell-cycle box binding factor and Swi4/6 cell-cycle box binding factor in the G1/S transcriptional program in Saccharomyces cerevisiae. Genetics 171, 49–61.

Bertrand, E., Chartrand, P., Schaefer, M., Shenoy, S. M., Singer, R. H., and Long, R. M. (1998). Localization of ASH1 mRNA particles in living yeast. Mol. Cell 2, 437–445.

Blackwell, C., Martin, K. A., Greenall, A., Pidoux, A., Allshire, R. C., and Whitehall, S. K. (2004). The Schizosaccharomyces pombe HIRA-like protein Hip1 is required for the periodic expression of histone genes and contributes to the function of complex centromeres. Mol. Cell. Biol. 24, 4309–4320.

Bobola, N., Jansen, R. P., Shin, T. H., and Nasmyth, K. (1996). Asymmetric accumulation of Ash1p in postanaphase nuclei depends on a myosin and restricts yeast mating-type switching to mother cells. Cell 84, 699–709.

Bork, P. (1993). Hundreds of ankyrin-like repeats in functionally diverse proteins: Mobile modules that cross phyla horizontally? Proteins 17, 363–374.

Boros, J., Lim, F. L., Darieva, Z., Pic-Taylor, A., Harman, R., Morgan, B. A., and Sharrocks, A. D. (2003). Molecular determinants of the cell-cycle regulated Mcm1p-Fkh2p transcription factor complex. Nucleic Acids Res. 31, 2279–2288.

Borup, M. T. (2006). Stress-induced switch to pseudohyphal growth in S. pombe. Cell Cycle 5, 2138–2145.

Breeden, L. (1996). Start-specific transcription in yeast. Curr. Top. Microbiol. Immunol. 208, 95–127.

Breeden, L., and Nasmyth, K. (1987). Cell cycle control of the yeast HO gene: Cis- and trans-acting regulators. Cell 48, 389–397.

Breeden, L. L. (2000). Cyclin transcription: Timing is everything. Curr. Biol. 10, R586–R588.

Breeden, L. L. (2003). Periodic transcription: A cycle within a cycle. Curr. Biol. 13, R31–R38.

Buck, V., Ng, S. S., Ruiz-Garcia, A. B., Papadopoulou, K., Bhatti, S., Samuel, J. M., Anderson, M., Millar, J. B. A., and McInerny, C. J. (2004). Fkh2p andSep1p regulate mitotic gene transcription in fission yeast. J. Cell Sci. 117, 5623–5632.

Bulmer, R., Pic-Taylor, A., Whitehall, S. K., Martin, K. A., Millar, J. B., Quinn, J., and Morgan, B. A. (2004). The forkhead transcription factor Fkh2 regulates the cell division cycle of Schizosaccharomyces pombe. Eukaryot. Cell 3, 944–954.

Caligiuri, M., and Beach, D. (1993). Sct1 functions in partnership with Cdc10 in a transcription complex that activates cell cycle START and inhibits differentiation. Cell 72, 607–619.

Carlsson, P., and Mahlapuu, M. (2002). Forkhead transcription factors: Key players in development and metabolism. Dev. Biol. 250, 1–23.

Cho, R. J., Campbell, M. J., Winzeler, E. A., Steinmetz, L., Conway, A., Wodicka, L., Wolfsberg, T. G., Gabrielian, A. E., Landsman, D., Lochart, D. J., et al. (1998). A genome-wide transcriptional analysis of the mitotic cell cycle. Mol. Cell 2, 65–73.

Cho, R. J., Huang, M., Campbell, M. J., Dong, H., Steinmetz, L., Sapinoso, L., Hampton, G., Elledge, S. J., Davis, R. W., and Lockhart, D. J. (2001). Transcriptional regulation and function during the human cell cycle. Nat. Genet. 27, 48–54.

Chu, S., DeRisi, J., Eisen, M., Mulholland, J., Botstein, D., Brown, P. O., and Herskowitz, I. (1998). The transcriptional program of sporulation in budding yeast. Science 282, 699–705.

Chu, Z., Li, J., Eshaghi, M., Peng, X., Karuturi, R. K., and Liu, J. (2007). Modulation of cell cycle-specific gene expressions at the onset of S phase arrest contributes to the robust DNA replication checkpoint response in fission yeast. *Mol. Biol. Cell* **18,** 1756–1767.

Christensen, P. U., Bentley, N. J., Martinho, R. G., Nielsen, O., and Carr, A. M. (2000). Mik1 levels accumulate in S phase and may mediate an intrinsic link between S phase and mitosis. *Proc. Natl. Acad. Sci. USA* **97,** 2579–2584.

Connolly, T., Caligiuri, M., and Beach, D. (1997). The Cdc2 protein kinase controls Cdc10/Sct1 complex formation. *Mol. Biol. Cell* **8,** 1105–1115.

Cosma, M. P., Tanaka, T., and Nasmyth, K. (1999). Ordered recruitment of transcription and chromatin remodeling factors to a cell cycle-and developmentally regulated promoter. *Cell* **97,** 299–311.

Cosma, M. P., Panizza, S., and Nasmyth, K. (2001). Cdk1 triggers association of RNA polymerase to cell cycle promoters only after recruitment of the mediator by SBF. *Mol. Cell* **7,** 1213–1220.

Cosma, M. P. (2004). Daughter-specific repression of *Saccharomyces cerevisiae* HO: Ash1 is the commander. *EMBO Rep.* **5,** 953–957.

Costanzo, M., Nishikawa, J. L., Tang, X., Millman, J. S., Schub, O., Breitkreuz, K., Dewar, D., Rupes, I., Andrews, B., and Tyers, M. (2004). CDK activity antagonizes Whi5, an inhibitor of G1/S transcription in yeast. *Cell* **117,** 899–913.

Costanzo, M., Schub, O., and Andrews, B. (2003). G1 transcription factors are differentially regulated in *Saccharomyces cerevisiae* by the Swi6-binding protein Stb1. *Mol. Cell. Biol.* **23,** 5064–5077.

Côte, P., Hogues, H., and Whiteway, M. (2009). Transcriptional analysis of the *Candida albicans* cell cycle. *Mol. Biol. Cell* **20,** 3363–3373.

Cunliffe, L., White, S., and McInerny, C. J. (2004). DSC1-MCB regulation of meiotic transcription in *Schizosaccharomyces pombe*. *Mol. Genet. Genomics* **271,** 60–71.

Darieva, Z., Pic-Taylor, A., Boros, J., Spanos, A., Geymonat, M., Sedwick, S. G., Morgan, B. A., and Sharrocks, A. D. (2003). Cell cycle regulated transcription through the FHA domain of Fkh2p and the coactivator Ndd1p. *Curr. Biol.* **13,** 1740–1745.

Darieva, Z., Bulmer, R., Pic-Taylor, A., Doris, K. S., Geymonat, M., Sedgwick, S. G., Morgan, B. A., and Sharrocks, A. D. (2006). Polo kinase controls cell-cycle-dependent transcription by targeting a coactivator protein. *Nature* **444,** 494–498.

de Bruin, R. A., McDonald, W. H., Kalashnikova, T. I., Yates, J., and Wittenberg, C. (2004). Cln3 activates G1-specific transcription via phosphorylation of the SBF bound repressor Whi5. *Cell* **117,** 887–898.

de Bruin, R. A., Kalashnikova, T. I., Chahwan, C., McDonald, W. H., Wohlschlegel, J., Yates, J., III, Russell, P., and Wittenberg, C. (2006). Constraining G1-specific transcription to late G1 phase: The MBF-associated corepressor Nrm1 acts via negative feedback. *Mol. Cell* **23,** 483–496.

de Bruin, R. A., and Wittenberg, C. (2009). All eukaryotes: Before turning off G1-S transcription, please check your DNA. *Cell Cycle* **8,** 214–217.

de Lichtenberg, U., Jensen, L. J., Brunak, S., and Bork, P. (2005). Dynamic complex formation during the yeast cell cycle. *Science* **307,** 724–727.

de Lichtenberg, U., Jensen, T. S., Brunak, S., Bork, P., and Jensen, L. J. (2007). Evolution of cell cycle control: Same molecular machines, different regulation. *Cell Cycle* **6,** 1819–1825.

Dekker, N., de Haan, A., and Hochstenbach, F. (2006). Transcription regulation of the alpha-glucanase gene agn1 by cell separation transcription factor Ace2p in fission yeast. *FEBS Lett.* **580,** 3099–3106.

Ding, R., and Smith, G. R. (1998). Global control of meiotic recombination genes by *Schizosaccharomyces pombe* rec16 (rep1). *Mol. Gen. Genet.* **258,** 663–670.

Dirick, L., Moll, T., Auer, H., and Nasmyth, K. (1992). A central role for SWI6 in modulating cell cycle start-specific transcription in yeast. *Nature* **357,** 508–513.

Dirick, L., Bohm, T., and Nasmyth, K. (1995). Roles and regulation of Cln-Cdc28 kinases at the start of the cell cycle of *Saccharomyces cerevisiae*. *EMBO J.* **14**, 4803–4813.

Dohrmann, P. R., Butler, G., Tamai, K., Dorland, S., Greene, J. R., Thiele, D. J., and Stillman, D. J. (1992). Parallel pathways of gene regulation: Homologous regulators SWI5 and ACE2 differentially control transcription of HO and chitinase. *Genes Dev.* **6**, 93–104.

Dohrmann, P. R., Voth, W. P., and Stillman, D. J. (1996). Role of negative regulation in promoter specificity of the homologous transcriptional activators Ace2p and Swi5p. *Mol. Cell. Biol.* **16**, 1746–1758.

Doolin, M. T., Johnson, A. L., Johnston, L. H., and Butler, G. (2001). Overlapping and distinct roles of the duplicated yeast transcription factors Ace2p and Swi5p. *Mol. Microbiol.* **40**, 422–432.

Durocher, D., Henckel, J., Fersht, A. R., and Jackson, S. P. (1999). The FHA domain is a modular phosphopeptide recognition motif. *Mol. Cell* **4**, 387–394.

Durocher, D., and Jackson, S. P. (2002). The FHA domain. *FEBS Lett.* **513**, 58–66.

Fankhauser, C., Reymond, A., Cerutti, L., Utzig, S., Hofmann, K., and Simanis, V. (1995). The *S. pombe* cdc15 gene is a key element in the reorganization of F-actin at mitosis. *Cell* **82**, 435–444.

Forsburg, S. L. (1999). The best yeast? *Trends Genet.* **15**, 340–344.

Forsburg, S. L., and Nurse, P. (1991). Cell cycle regulation in the yeasts *Saccharomyces cerevisiae* and *Schizosaccharomyces pombe*. *Annu. Rev. Cell Dev. Biol.* **7**, 227–256.

Foster, R., Mikesell, G. E., and Breeden, L. (1993). Multiple SWI6-dependent cis-acting elements control SWI4 transcription through the cell cycle. *Mol. Cell. Biol.* **13**, 3792–3801.

Freeman, K. B., Karns, L. R., Lutz, K. A., and Smith, M. M. (1992). Histone H3 transcription in *Saccharomyces cerevisiae* is controlled by multiple cell cycle activation sites and a constitutive negative regulatory element. *Mol. Cell. Biol.* **12**, 5455–5463.

Fu, Z., Malureanu, L., Huang, J., Wang, W., Li, H., van Deursen, J. M., Tindal, D. J., and Chen, J. (2008). Plk1-dependent phosphorylation of FoxM1 regulates a transcriptional programme required for mitotic progression. *Nat. Cell Biol.* **10**, 1076–1082.

Futcher, B. (2000). Microarrays and cell cycle transcription in yeast. *Curr. Opin. Cell Biol.* **12**, 710–715.

Futcher, B. (2002). Transcriptional regulatory networks and the yeast cell cycle. *Curr. Opin. Cell Biol.* **14**, 676–683.

Gelli, A. (2002). Rst1 and Rst2 are required for the a/alpha diploid cell type in yeast. *Mol. Microbiol.* **46**, 845–854.

Genovese, C., Trani, D., Caputi, M., and Claudio, P. P. (2006). Cell cycle control and beyond: Emerging roles for the retinoblastoma gene family. *Oncogene* **25**, 5201–5209.

Geymonat, M., Spanos, A., Wells, G. P., Smerdon, S. J., and Sedgwick, S. G. (2004). Clb6/Cdc28 and Cdc14 regulate phosphorylation status and cellular localization of Swi6. *Mol. Cell. Biol.* **24**, 2277–2285.

Grallert, A., Grallert, B., Ribar, B., and Sipiczki, M. (1998). Coordination of initiation of nuclear division and initiation of cell division in *Schizosaccharomyces pombe*: Genetic interactions of mutations. *J. Bacteriol.* **180**, 892–900.

Hereford, L. M., Osley, M. A., Ludwig, T. R., and McLaughlin, C. S. (1981). Cell-cycle regulation of yeast histone mRNA. *Cell* **24**, 367–375.

Hereford, L., Bromley, S., and Osley, M. A. (1982). Periodic transcription of yeast histone genes. *Cell* **30**, 305–310.

Ho, Y., Costanzo, M., Moore, L., Kobayashi, R., and Andrews, B. J. (1999). Regulation of transcription at the *Saccharomyces cerevisiae* Start transition by Stb1, a Swi6-binding protein. *Mol. Cell. Biol.* **19**, 5267–5278.

Hollenhorst, P. C., Bose, M. E., Mielke, M. R., Muller, U., and Fox, C. A. (2000). Forkhead genes in transcriptional silencing, cell morphology and the cell cycle. Overlapping and distinct functions for FKH1 and FKH2 in *Saccharomyces cerevisiae*. *Genetics* **154**, 1533–1548.

Hollenhorst, P. C., Pietz, G., and Fox, C. A. (2001). Mechanisms controlling differential promoter occupancy by the yeast forkhead proteins Fkh1p and Fkh2p: Implications for regulating the cell cycle and differentiation. *Genes Dev.* **15,** 2445–2456.

Holstege, F. C., Jennings, E. G., Wyrick, J. J., Lee, T. I., Hengartner, C. J., Green, M. R., Golub, T. R., Lander, E. S., and Young, R. A. (1998). Dissecting the regulatory circuitry of a eukaryotic genome. *Cell* **95,** 717–728.

Horak, C. E., Luscombe, N. M., Qian, J., Bertone, P., Piccirrillo, S., Gerstein, M., and Snyder, M. (2002). Complex transcriptional circuitry at the G1/S transition in *Saccharomyces cerevisiae. Genes Dev.* **16,** 3017–3033.

Horie, S., Watanabe, Y., Tanaka, K., Nishiwaki, S., Fujioka, H., Abe, H., Yamamoto, M., and Shimoda, C. (1998). The *Schizosaccharomyces pombe mei4+* gene encodes a meiosis-specific transcription factor containing a forkhead DNA-binding domain. *Mol. Cell. Biol.* **18,** 2118–2129.

Ihmels, J., Friedlander, G., Bergmann, S., Sarig, O., Ziv, Y., and Barkai, N. (2002). Revealing modular organization in the yeast transcriptional network. *Nat. Genet.* **31,** 370–377.

Iyer, V. R., Horak, C. E., Scafe, C. S., Botstein, D., Snyder, M., and Brown, P. O. (2001). Genomic binding sites of the yeast cell-cycle transcription factors SBF and MBF. *Nature* **409,** 533–538.

Jallepalli, P. V., Brown, G. W., Muzi-Falconi, M., Tien, D., and Kelly, T. J. (1997). Regulation of the replication initiator protein p65^{cdc18} by CDK phosphorylation. *Genes Dev.* **11,** 2767–2779.

Jensen, L. J., Jensen, T. S., de Lichtenberg, L. U., Brunak, S., and Bork, P. (2006). Co-evolution of transcriptional and post-translational cell-cycle regulation. *Nature* **443,** 594–597.

Jorgensen, P., and Tyers, M. (2000). The fork'ed path to mitosis. *Genome Biol.* **1,** 1022.1–1022.4.

Jorgensen, P., and Tyers, M. (2004). How cells coordinate growth and division. *Curr. Biol.* **14,** R1014–R1027.

Kato, M., Hata, N., Banerjee, N., Futcher, B., and Zhang, M. Q. (2004). Identifying combinatorial regulation of transcription factors and binding motifs. *Genome Biol.* **5,** R56.

Kaufmann, E., and Knöchel, W. (1996). Five years on the wings of fork head. *Mech. Dev.* **57,** 3–20.

Kelly, T. J., Martin, G. S., Forsburg, S. L., Stephen, R. J., Russo, A., and Nurse, P. (1993). The fission yeast *cdc18+* gene product couples S-phase to START and Mitosis. *Cell* **74,** 1–20.

Knapp, D., Bhoite, L., Stillman, D. J., and Nasmyth, K. (1996). The transcription factor Swi5 regulates expression of the cyclin kinase inhibitor P40^{SIC1}. *Mol. Cell. Biol.* **16,** 5701–5707.

Koch, C., Moll, T., Neuberg, M., Ahorn, H., and Nasmyth, K. (1993). A role for the transcription factors Mbp1 and Swi4 in progression from G1 to S phase. *Science* **261,** 1551–1557.

Koch, C., Schleiffer, A., Ammerer, G., and Nasmyth, K. (1996). Switching transcription on and off during the yeast cell cycle: Cln/Cdc28 kinases activate bound transcription factor SBF (Swi4/Swi6) at start, whereas Clb/Cdc28 kinases displace it from the promoter in G2. *Genes Dev.* **10,** 129–141.

Koranda, M., Schleiffer, A., Endler, L., and Ammerer, G. (2000). Forkhead-like transcription factors recruit Ndd1 to the chromatin of G2/M-specific promoters. *Nature* **406,** 94–98.

Kumar, R., Reynolds, D. M., Shevchenko, A., Goldstone, S. D., and Dalton, S. (2000). Forkhead transcription factors, Fkh1p and Fkh2p, collaborate with Mcm1p to control transcription required for M-phase. *Curr. Biol.* **10,** 896–906.

La Thangue, N. B., and Taylor, W. R. (1993). A structural similarity between mammalian and yeast transcription factors for cell-cycle-regulated genes. *Trends Cell Biol.* **3,** 75–76.

Laabs, T. L., Markwardt, D. D., Slattery, M. G., Newcomb, L. L., Stillman, D. J., and Heideman, W. (2003). ACE2 is required for daughter cell-specific G1 delay in *Saccharomyces cerevisiae. Proc. Natl. Acad. Sci. USA* **100,** 10275–10280.

Laoukili, J., Kooistra, M. R., Bras, A., Kauw, J., Kerkhoven, R. M., Morrison, A., Clevers, H., and Medema, R. H. (2005). FoxM1 is required for execution of the mitotic programme and chromosome stability. *Nat. Cell Biol.* **7,** 126–136.

Lee, M. S., Enoch, T., and Piwnica-Worms, H. (1994). $mik1^+$ encodes a tyrosine kinase that phosphorylates $p34^{cdc2}$ on tyrosine 15. *J. Biol. Chem.* **269**, 30530–30537.

Lee, K. M., Miklos, I., Du, H., Watt, S., Szilagyi, Z., Saiz, J. E., Madabhushi, R., Penkett, C. J., Sipiczki, M., Bähler, J., et al. (2005). Impairment of the TFIIH associated CDK-activating kinase selectively affects cell. *Mol. Biol. Cell* **16**, 2734–2745.

Lee, T. I., Rinaldi, N. J., Robert, F., Odom, D. T., Bar-Joseph, Z., Gerber, G. K., Hannett, N. M., Harbison, C. T., Thompson, C. M., Simon, I., et al. (2002). Transcriptional regulatory networks in *Saccharomyces cerevisiae*. *Science* **298**, 799–804.

Le Goff, X., Utzig, S., and Simanis, V. (1999). Controlling septation in fission yeast: Finding the middle, and timing it right. *Curr. Genet.* **35**, 571–584.

Lim, F. L., Hayes, A., West, A. G., Pic-Taylor, A., Darieva, Z., Morgan, B. A., Oliver, S. G., and Sharrocks, A. D. (2003). Mcm1p-induced DNA bending regulates the formation of ternary transcription factor complexes. *Mol. Cell. Biol.* **23**, 450–461.

Linder, T., Rasmussen, N. N., Samuelsen, C. O., Chatzidaki, E., Baraznenok, V., Beve, J., Henriksen, P., Gustafsson, C. M., and Holmberg, S. (2008). Two conserved modules of *Schizosaccharomyces pombe* mediator regulate distinct cellular pathways. *Nucleic Acids Res.* **36**, 2489–24504.

Lowndes, N. F., Johnson, A. L., and Johnston, L. H. (1991). Coordination of expression of DNA synthesis genes in budding yeast by a cell-cycle regulated trans factor. *Nature* **350**, 247–250.

Lowndes, N. F., Johnson, A. L., Breeden, L., and Johnston, L. H. (1992a). SWI6 protein is required for transcription of the periodically expressed DNA synthesis genes in budding yeast. *Nature* **357**, 505–508.

Lowndes, N. F., McInerny, C. J., Johnson, A. L., Fantes, P. A., and Johnston, L. H. (1992b). Control of DNA synthesis genes in fission yeast by the cell-cycle gene $cdc10^+$. *Nature* **355**, 449–453.

Loy, C. J., Lydall, D., and Surana, U. (1999). *NDD1*, a high-dosage suppressor of *cdc28-1N*, is essential for expression of a subset of late-S-phase-specific genes in *Saccharomyces cerevisiae*. *Mol. Cell. Biol.* **19**, 3312–3327.

Lu, Y., Mahony, S., Benos, P. V., Rosenfeld, R., Simon, I., Breeden, L. L., and Bar-Joseph, Z. (2007). Combined analysis reveals a core set of cycling genes. *Genome Biol.* **8**, R146.

Lydall, D., Ammerer, G., and Nasmyth, K. (1991). A new role for MCM1 in yeast: Cell cycle regulation of *SWI5* transcription. *Genes Dev.* **5**, 2405–2419.

MacKay, V. L., Mai, B., Waters, L., and Breeden, L. L. (2001). Early cell cycle box-mediated transcription of *CLN3* and *SWI4* contributes to the proper timing of the G1-to-S transition in budding yeast. *Mol. Cell. Biol.* **21**, 4140–4148.

Maher, M., Cong, F., Kindelberger, D., Nasmyth, K., and Dalton, S. (1995). Cell cycle-regulated transcription of the *CLB2* gene is dependent on Mcm1 and a ternary complex factor. *Mol. Cell. Biol.* **15**, 3129–3137.

Mai, B., Miles, S., and Breeden, L. L. (2002). Characterization of the ECB binding complex responsible for the M/G1-specific transcription of *CLN3* and *SWI4*. *Mol. Cell. Biol.* **22**, 430–441.

Maqbool, Z., Kersey, P. J., Fantes, P. A., and McInerny, C. J. (2003). MCB-mediated regulation of cell cycle-specific $cdc22^+$ transcription in fission yeast. *Mol. Genet. Genomics* **269**, 765–775.

Martín-Cuadrado, A. B., Dueñas, E., Sipiczki, M., Vázquez de Aldana, C. R., and Del Rey, F. (2003). The endo-beta-1,3-glucanase eng1p is required for dissolution of the primary septum during cell separation in *Schizosaccharomyces pombe*. *J. Cell Sci.* **116**, 1689–1698.

Mata, J., Lyne, R., Burns, G., and Bähler, J. (2002). The transcriptional program of meiosis and sporulation in fission yeast. *Nat. Genet.* **32**, 143–147.

Matsumoto, S., and Yanagida, M. (1985). Histone gene organization of fission yeast: A common upstream sequence. *EMBO J.* **4**, 3531–3538.

Matsumoto, S., Yanagida, M., and Nurse, P. (1987). Histone transcription in cell cycle mutants of fission yeast. *EMBO J.* **6**, 1093–1097.

McBride, H. J., Yu, Y., and Stillman, D. J. (1999). Distinct regions of the Swi5 and Ace2 transcription factors are required for specific gene activation. *J. Biol. Chem.* **274,** 21029–21036.

McInerny, C. J. (2004). Cell cycle-regulated transcription in fission yeast. *Biochem. Soc. Trans.* **32,** 967–972.

McInerny, C. J., Kersey, P. J., Creanor, J., and Fantes, P. A. (1995). Positive and negative roles for cdc10 in cell cycle gene expression. *Nucleic Acids Res.* **23,** 4761–4768.

McInerny, C. J., Partridge, J. F., Mikesell, G. E., Creemer, D. P., and Breeden, L. L. (1997). A novel Mcm1-dependent element in the *SWI4, CLN3, CDC6,* and *CDC47* promoters activates M/G1-specific transcription. *Genes Dev.* **11,** 1277–1288.

McIntosh, E. M., Atkinson, T., Storms, R. K., and Smith, M. (1991). Characterization of a short, cis-acting DNA sequence which conveys cell cycle stage-dependent transcription in *Saccharomyces cerevisiae. Mol. Cell. Biol.* **11,** 329–337.

Meeks-Wagner, D., and Hartwell, L. H. (1986). Normal stoichiometry of histone dimer sets is necessary for high fidelity of mitotic chromosome transmission. *Cell* **44,** 43–52.

Messenguy, F., and Dubois, E. (2003). Role of MADS box proteins and their cofactors in combinatorial control of gene expression and cell development. *Gene* **316,** 1–21.

Miklos, I., Szilagyi, Z., Watt, S., Zilahi, E., Batta, G., Antunovics, Z., Enczi, K., Bähler, J., and Sipiczki, M. (2008). Genomic expression patterns in cell separation mutants of *Schizosaccharomyces pombe* defective in the genes *sep10⁺* and *sep15⁺* coding for the Mediator subunits Med31 and Med8. *Mol. Genet. Genomics* **279,** 225–238.

Miyamoto, M., Tanaka, K., and Okayama, H. (1994). *res2⁺,* a new member of the *cdc10⁺/SWI4* family, controls the 'start' of mitotic and meiotic cycles in fission yeast. *EMBO J.* **13,** 1873–1880.

Moll, T., Tebb, G., Surana, U., Robitsch, H., and Nasmyth, K. (1991). The role of phosphorylation and the CDC28 protein kinase in cell cycle-regulated nuclear import of the *S. cerevisiae* transcription factor SWI5. *Cell* **66,** 743–758.

Morris, M. C., Kaiser, P., Rudyak, S., Baskerville, C., Watson, M. H., and Reed, S. I. (2003). Cks1-dependent proteasome recruitment and activation of *CDC20* transcription in budding yeast. *Nature* **424,** 1009–1013.

Nachman, I., and Regev, A. (2009). BRNI: Modular analysis of transcriptional regulatory programs. *BMC Bioinform.* **10,** 155.

Nadeem, F. K., Blair, D., and McInerny, C. J. (2006). Pol5p, a novel binding partner to Cdc10p in fission yeast involved in rRNA production. *Mol. Genet. Genomics* **276,** 391–401.

Nakashima, N., Tanaka, K., Sturm, S., and Okayama, H. (1995). Fission yeast Rep2 is a putative transcriptional activator subunit for the cell cycle 'start' function of Res2-Cdc10. *EMBO J.* **14,** 4794–4802.

Nasmyth, K., Adolf, G., Lydall, D., and Seddon, A. (1990). The identification of a second cell cycle control on the *HO* promoter in yeast: Cell cycle regulation of SWI5 nuclear entry. *Cell* **62,** 631–647.

Nelson, B., Kurischko, C., Horecka, J., Mody, M., Nair, P., Pratt, L., Zougman, A., McBroom, L. D., Hughes, T. R., Boone, C., *et al.* (2003). RAM: A conserved signaling network that regulates Ace2p transcriptional activity and polarized morphogenesis. *Mol. Biol. Cell* **14,** 3782–3803.

Ng, S. S., Anderson, M., White, S., and McInerny, C. J. (2001). *mik1⁺* G1-S transcription regulates mitotic entry in fission yeast. *FEBS Lett.* **503,** 131–134.

Ng, S. S., Papadopoulou, K., and McInerny, C. J. (2006). Regulation of gene expression and cell division by Polo-like kinases. *Curr. Genet.* **50,** 73–80.

Ohkura, H., Hagan, I. M., and Glover, D. M. (1995). The conserved *Schizosaccharomyces pombe* kinase Plo1, required to form a bipolar spindle, the actin ring, and septum, can drive septum formation in G1 and G2 cells. *Genes Dev.* **9,** 1059–1073.

Oliva, A., Rosebrock, A., Ferrezuelo, F., Pyne, S., Chen, H., Skiena, S., Futcher, B., and Leatherwood, J. (2005). The cell cycle-regulated genes of *Schizosaccharomyces pombe*. *PLoS Biol.* **3**, e225.

Orlando, D. A., Lin, C. Y., Bernard, A., Wang, J. Y., Socolar, J. E., Iversen, E. S., Hartemink, A. J., and Haase, S. B. (2008). Global control of cell-cycle transcription by coupled CDK and network oscillators. *Nature* **453**, 944–947.

Papadopoulou, K., Ng, S. S., Ohkura, H., Geymonat, M., Sedgwick, S. G., and McInerny, C. J. (2008). Regulation of gene expression during M-G1-phase in fission yeast through Plo1p and forkhead transcription factors. *J. Cell Sci.* **121**, 38–47.

Paquin, N., and Chartrand, P. (2008). Local regulation of mRNA translation: New insights from the bud. *Trends Cell Biol.* **18**, 105–111.

Partridge, J. F., Mikesell, G. E., and Breeden, L. L. (1997). Cell cycle-dependent transcription of CLN1 involves Swi4 binding to MCB-like elements. *J. Biol. Chem.* **272**, 9071–9077.

Park, H. J., Costa, R. H., Lau, L. F., Tyner, A. L., and Raychaudhuri, P. (2008). Anaphase-promoting complex/cyclosome-CDH1-mediated proteolysis of the forkhead box M1 transcription factor is critical for regulated entry into S phase. *Mol. Cell. Biol.* **28**, 5162–5171.

Peng, X., Karuturi, R. K., Miller, L. D., Lin, K., Jia, Y., Kondu, P., Wang, L., Wong, L. S., Liu, E. T., Balasubramanian, M. K., *et al.* (2005). Identification of cell cycle-regulated genes in fission yeast. *Mol. Biol. Cell* **16**, 1026–1042.

Petit, C. S., Mehta, S., Roberts, R. H., and Gould, K. L. (2005). Ace2p contributes to fission yeast septin ring assembly by regulating *mid2+* expression. *J. Cell Sci.* **118**, 5731–5742.

Pic-Taylor, A., Darieva, Z., Morgan, B. A., and Sharrocks, A. D. (2004). Regulation of cell cyclespecific gene expression through cyclin-dependent kinase-mediated phosphorylation of the forkhead transcription factor Fkh2p. *Mol. Cell. Biol.* **24**, 10036–10046.

Pic, A., Lim, F. L., Ross, S. J., Veal, E. A., Johnson, A. L., Sultan, A. R., West, A. G., Johnston, L. H., Sharrocks, A. D., and Morgan, B. A. (2000). The forkhead protein Fkh2 is a component of the yeast cell cycle transcription factor SFF. *EMBO J.* **19**, 3750–3761.

Porter, S. E., Washburn, T. M., Chang, M., and Jaehning, J. A. (2002). The yeast pafl-RNA polymerase II complex is required for full expression of a subset of cell cycle-regulated genes. *Eukaryot. Cell* **1**, 830–842.

Pramila, T., Miles, S., GuhaThakurta, D., Jemiolo, D., and Breeden, L. L. (2002). Conserved homeodomain proteins interact with MADS box protein Mcm1 to restrict ECB-dependent transcription to the M/G1 phase of the cell cycle. *Genes Dev.* **16**, 3034–3045.

Pramila, T., Wu, W., Miles, S., Noble, W. S., and Breeden, L. L. (2006). The forkhead transcription factor Hcm1 regulates chromosome segregation genes and fills the S-phase gap in the transcriptional circuitry of the cell cycle. *Genes Dev.* **20**, 2266–2278.

Primig, M., Sockanathan, S., Auer, H., and Nasmyth, K. (1992). Anatomy of a transcription factor important for the start of the cell cycle in *Saccharomyces cerevisiae*. *Nature* **358**, 593–597.

Raithatha, S. A., and Stuart, D. T. (2005). Meiosis-specific regulation of the *S. cerevisiae* S-phase cyclin CLB5 is dependent upon MCB elements in its promoter but is independent of MBF activity. *Genetics* **169**, 1329–1342.

Reymond, A., Marks, J., and Simanis, V. (1993). The activity of *S. pombe* DSC-1-like factor is cell cycle regulated and dependent on the activity of p34cdc2. *EMBO J.* **12**, 4325–4334.

Reynolds, D., Shi, B. J., McLean, C., Katsis, F., Kemp, B., and Dalton, S. (2003). Recruitment of Thr319-phosphorylated Ndd1p to the FHA domain of Fkh2p requires Clb kinase activity: A mechanism for CLB cluster gene activation. *Genes Dev.* **17**, 1789–1802.

Ribár, B., Banrevi, A., and Sipiczki, M. (1997). *sep1+* encodes a transcription-factor homologue of the HNF-3/forkhead DNA-binding-domain family in *Schizosaccharomyces pombe*. *Gene* **202**, 1–5.

Ribár, B., Grallert, A., Oláh, E., and Szállási, Z. (1999). Deletion of the *sep1⁺* forkhead transcription factor homologue is not lethal but causes hyphal growth in *Schizosaccharomyces pombe*. *Biochem. Biophys. Res. Commun.* **263**, 465–474.

Rowicka, M., Kudlicki, A., Tu, B. P., and Otwinowski, Z. (2007). High-resolution timing of cell cycle-regulated gene expression. *Proc. Natl. Acad. Sci. USA* **104**, 16892–16897.

Rustici, G., Mata, J., Kivinen, K., Lió, P., Penkett, C. J., Burns, G., Hayles, J., Brazma, A., Nurse, P., and Bähler, J. (2004). Periodic gene expression program of the fission yeast cell cycle. *Nat. Genet.* **36**, 809–817.

Sharma, N., Marguerat, S., Mehta, S., Watt, S., and Bähler, J. (2006). The fission yeast Rpb4 subunit of RNA polymerase II plays a specialized role in cell separation. *Mol. Genet. Genomics* **276**, 545–554.

Schenkman, L. R., Caruso, C., Pagé, N., and Pringle, J. R. (2002). The role of cell cycle-regulated expression in the localization of spatial landmark proteins in yeast. *J. Cell Biol.* **156**, 829–841.

Shen, Z., Paquin, N., Forget, A., and Chartrand, P. (2009). Nuclear shuttling of She2p couples *ASH1* mRNA localization to its translational repression by recruiting Loc1p and Puf6p. *Mol. Biol. Cell* **20**, 2265–2275.

Sherr, C. J., and McCormick, F. (2002). The RB and p53 pathways in cancer. *Cancer Cell* **2**, 103–112.

Sherwood, P. W., Tsang, S. V., and Osley, M. A. (1993). Characterization of *HIR1* and *HIR2*, two genes required for regulation of histone gene transcription in *Saccharomyces cerevisiae*. *Mol. Cell. Biol.* **13**, 28–38.

Shore, P., and Sharrocks, A. D. (1995). The MADS-box family of transcription factors. *Eur. J. Biochem.* **229**, 1–13.

Sidorova, J. M., Mikesell, G. E., and Breeden, L. L. (1995). Cell cycle-regulated phosphorylation of Swi6 controls its nuclear localization. *Mol. Biol. Cell* **6**, 1641–1658.

Sil, A., and Herskowitz, I. (1996). Identification of asymmetrically localized determinant, Ash1p, required for lineage-specific transcription of the yeast *HO* gene. *Cell* **84**, 711–722.

Simon, I., Barnett, J., Hannett, N., Harbison, C. T., Rinaldi, N. J., Volkert, T. L., Wyrick, J. J., Zeitlinger, J., Gifford, D. K., Jaakkola, T. S., *et al.* (2001). Serial regulation of transcriptional regulators in the yeast cell cycle. *Cell* **106**, 697–708.

Spector, M. S., Raff, A., DeSilva, H., Lee, K., and Osley, M. A. (1997). Hir1p and Hir2p function as transcriptional corepressors to regulate histone gene transcription in the *Saccharomyces cerevisiae* cell cycle. *Mol. Cell. Biol.* **17**, 545–552.

Spellman, P. T., Sherlock, G., Zhang, M. Q., Iyer, V. R., Anders, K., Eisen, M. B., Brown, P. O., Botstein, D., and Futcher, B. (1998). Comprehensive identification of cell cycle-regulated genes of the yeast *Saccharomyces cerevisiae* by microarray hybridization. *Mol. Biol. Cell* **9**, 3273–3297.

Stuart, D., and Wittenberg, C. (1995). *CLN3*, not positive feedback, determines the timing of *CLN2* transcription in cycling cells. *Genes Dev.* **9**, 2780–2794.

Szilagyi, Z., Batta, G., Enczi, K., and Sipiczki, M. (2005). Characterisation of two novel fork-head gene homologues of *Schizosaccharomyces pombe*: Their involvement in cell cycle and sexual differentiation. *Gene* **348**, 101–109.

Tahara, S., Tanaka, K., Yuasa, Y., and Okayama, H. (1998). Functional domains of rep2, a transcriptional activator subunit for Res2-Cdc10, controlling the cell cycle "start". *Mol. Biol. Cell* **9**, 1577–1588.

Takahashi, K., Chen, E. S., and Yanagida, M. (2000). Requirement of Mis6 centromere connector for localizing a CENP-A-like protein in fission yeast. *Science* **288**, 2215–2219.

Takayama, Y., and Takahashi, K. (2007). Differential regulation of repeated histone genes during the fission yeast cell cycle. *Nucleic Acids Res.* **35**, 3223–3237.

Tanaka, K., and Okayama, H. (2000). A pcl-like cyclin activates the Res2p-Cdc10p cell cycle "start" transcriptional factor complex in fission yeast. *Mol. Biol. Cell* **11**, 2845–2862.

Tanaka, K., Okazaki, K., Okazaki, N., Ueda, T., Sugiyama, A., Nojima, H., and Okayama, H. (1992). A new cdc gene required for S phase entry of *Schizosaccharomyces pombe* encodes a protein similar to the $cdc10^+$ and *SWI4* gene products. *EMBO J.* **11,** 4923–4932.

Taylor, I. A., Treiber, M. K., Olivi, L., and Smerdon, S. J. (1997). The X-ray structure of the DNA binding domain from the *Saccharomyces cerevisiae* cell-cycle transcription factor Mbp1 at 2.1 Ao resolution. *J. Mol. Biol.* **272,** 1–8.

Tempé, D., Brengues, M., Mayonove, P., Bensaad, H., Lacrouts, C., and Morris, M. C. (2007). The alpha helix of ubiquitin interacts with yeast cyclin-dependent kinase subunit CKS1. *Biochemistry* **46,** 45–54.

Toone, W. M., Aerne, B. L., Morgan, B. A., and Johnston, L. H. (1997). Getting started: Regulating the initiation of DNA replication in yeast. *Annu. Rev. Microbiol.* **51,** 125–149.

Treisman, R. (1994). Ternary complex factors: Growth factor regulated transcriptional activators. *Curr. Opin. Genet. Dev.* **4,** 96–101.

Tuch, B. B., Galgoczy, D. J., Hernday, A. D., Li, H., and Johnson, A. D. (2008a). The evolution of combinatorial gene regulation in fungi. *PLoS Biol.* **6,** e38.

Tuch, B. B., Li, H., and Johnson, A. D. (2008b). Evolution of eukaryotic transcription circuits. *Science* **319,** 1797–1799.

Tyers, M. (2004). Cell cycle goes global. *Curr. Opin. Cell Biol.* **16,** 602–613.

Tyers, M., Tokiwa, G., and Futcher, B. (1993). Comparison of the *Saccharomyces cerevisiae* G1 cyclins: Cln3 may be an upstream activator of Cln1, Cln2 and other cyclins. *EMBO J.* **12,** 1955–1968.

Ukil, L., Varadaraj, A., Govindaraghavan, M., Liu, H. L., and Osmani, S. A. (2008). Copy number suppressors of the *Aspergillus nidulans* nimA1 mitotic kinase display distinctive and highly dynamic cell cycle-regulated locations. *Eukaryot. Cell* **7,** 2087–2099.

White, S., Khaliq, F., Sotiriou, S., and McInerny, C. J. (2001). The role of DSC1 components $cdc10^+$, $rep1^+$ and $rep2^+$ in MCB gene transcription at the mitotic G1-S boundary in fission yeast. *Curr. Genet.* **40,** 251–259.

Whitehall, S., Stacey, P., Dawson, K., and Jones, N. (1999). Cell cycle-regulated transcription in fission yeast: Cdc10-Res protein interactions during the cell cycle and domains required for regulated transcription. *Mol. Biol. Cell* **10,** 3705–3715.

Whitfield, M. L., Sherlock, G., Saldanha, A. J., Murray, J. I., Ball, C. A., Alexander, K. E., Matese, J. C., Perou, C. M., Hurt, M. M., Brown, P. O., et al. (2002). Identification of genes periodically expressed in the human cell cycle and their expression in tumors. *Mol. Biol. Cell* **13,** 1977–2000.

Wijnen, H., and Futcher, B. (1999). Genetic analysis of the shared role of CLN3 and BCK2 at the G1-S transition in *Saccharomyces cerevisiae*. *Genetics* **153,** 1131–1143.

Wijnen, H., Landman, A., and Futcher, B. (2002). The G1 cyclin Cln3 promotes cell cycle entry via the transcription factor Swi6. *Mol. Cell. Biol.* **22,** 4402–4418.

Wittenberg, C., and Reed, S. I. (2005). Cell cycle-dependent transcription in yeast: Promoters, transcription factors, and transcriptomes. *Oncogene* **24,** 2746–2755.

Wuarin, J., Buck, V., Nurse, P., and Millar, J. B. (2002). Stable association of mitotic cyclin B/Cdc2 to replication origins prevents endoreduplication. *Cell* **111,** 419–431.

Wynne, J., and Treisman, R. (1992). SRF and MCM1 have related but distinct DNA binding specificities. *Nucleic Acids Res.* **20,** 3297–3303.

Yamano, H., Kitamura, K., Kominami, K., Lehmann, A., Katayama, S., Hunt, T., and Toda, T. (2000). The spike of S phase cyclin Cig2 expression at the G1-S border in fission yeast requires both APC and SCF ubiquitin ligases. *Mol. Cell* **6,** 1377–1387.

Yu, V. P., Baskerville, C., Grünenfelder, B., and Reed, S. I. (2005). A kinase-independent function of Cks1 and Cdk1 in regulation of transcription. *Mol. Cell* **17,** 145–151.

Zhu, G., Spellman, P. T., Volpe, T., Brown, P. O., Botstein, D., Davis, T. N., and Futcher, B. (2000). Two yeast forkhead genes regulate the cell cycle and pseudohyphal growth. *Nature* **406**, 90–94.

Zhu, X., Wirén, M., Sinha, I., Rasmussen, N. N., Linder, T., Holmberg, S., Ekwall, K., and Gustafsson, C. M. (2006). Genome-wide occupancy profile of mediator and the Srb8-11 module reveals interactions with coding regions. *Mol. Cell* **22**, 169–178.

Zhu, Y., Takeda, T., Nasmyth, K., and Jones, N. (1994). *pct1*⁺, which encodes a new DNA-binding partner of p85cdc10, is required for meiosis in the fission yeast *Schizosaccharomyces pombe*. *Genes Dev.* **8**, 885–898.

Zilahi, E., Salimova, E., Simanis, V., and Sipiczki, M. (2000). The *S. pombe* sep1 gene encodes a nuclear protein that is required for periodic expression of the *cdc15* gene. *FEBS Lett.* **481**, 105–108.

Zwicker, J., and Muller, R. (1995). Cell cycle-regulated transcription in mammalian cells. *Prog. Cell Cycle Res.* **1**, 91–99.

3

RNA Editing by Mammalian ADARs

Marion Hogg, Simona Paro, Liam P. Keegan, and Mary A. O'Connell
MRC Human Genetics Unit, Institute of Genetics and Molecular Medicine, Western General Hospital, Edinburgh, United Kingdom

Advances in Genetics, Vol. 73   0065-2660/11 $35.00
Copyright 2011, Elsevier Inc. All rights reserved.   DOI: 10.1016/B978-0-12-380860-8.00003-3

ABSTRACT

The main type of RNA editing in mammals is the conversion of adenosine to inosine which is translated as if it were guanosine. The enzymes that catalyze this reaction are ADARs (adenosine deaminases that act on RNA), of which there are four in mammals, two of which are catalytically inactive. ADARs edit transcripts that encode proteins expressed mainly in the CNS and editing is crucial to maintain a correctly functioning nervous system. However, the majority of editing has been found in transcripts encoding *Alu* repeat elements and the biological role of this editing remains a mystery. This chapter describes in detail the different ADAR enzymes and the phenotype of animals that are deficient in their activity. Besides being enzymes, ADARs are also double-stranded RNA-binding proteins, so by binding alone they can interfere with other processes such as RNA interference. Lack of editing by ADARs has been implicated in disorders such as forebrain ischemia and Amyotrophic Lateral Sclerosis (ALS) and this will also be discussed.   © 2011, Elsevier Inc.

I. INTRODUCTION

The term "RNA editing" was originally coined in 1986 to describe the insertion of uridine residues into trypanosome mitochondrial mRNA encoding cytochrome oxidase subunit II (Benne *et al.*, 1986). Since then, the term has been used to describe insertion or deletion of bases, as well as conversion of one base to another. RNA editing by insertion or deletion of bases has not been found in higher eukaryotes, whereas base conversion is very widespread.

Uridine to cytidine (U-to-C) conversion by amination is very rare and there has been only example found, in the transcript encoding the Wilms' tumor susceptibility gene in mammals (Sharma *et al.*, 1994). Deamination or removal of an amine group to release ammonia is the most common type and can occur on several bases within DNA or RNA: adenosine to inosine (A-to-I) in double-stranded (ds) RNA or cytosine to uracil (C-to-U) in DNA or RNA (Fig. 3.1). These deamination events are mediated by cytidine deaminase (CDA) superfamily of enzymes, which include the adenosine deaminases that act on RNA (ADARs) family and the apolipoprotein B mRNA-editing enzyme catalytic polypeptide (APOBEC) family.

The consequences of base conversion can be dramatic. When occurring in DNA, base conversion can induce chromosomal rearrangements as occurs in class switch recombination of immunoglobins (for review see Honjo *et al.*, 2002). In messenger RNA (mRNA), nucleotide conversion can change the coding capacity of the transcript, alter splice donor, acceptor or regulatory sequences, or create alternative start or stop codons. These editing enzymes are regulators of protein diversity and function, in particular, within the immune system and the

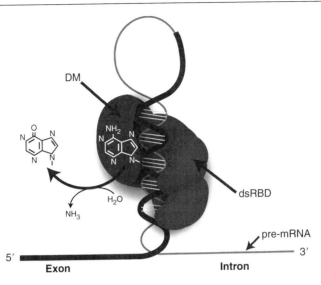

Figure 3.1. The dsRNA-binding domains of ADARs bind to a duplex that is formed between an exon and a downstream intron. ADAR's catalytic deaminase domain (DM) catalyses the conversion of a specific adenosine to inosine by hydrolytic deamination.

central nervous system (CNS) and it is possible that these regions of the body require or tolerate a greater diversity of proteins. This chapter describes RNA editing catalyzed by mammalian ADARs.

II. STRUCTURE AND EVOLUTION OF THE CDA DOMAIN

The crystal structure of the deaminase domain from two key enzymes was solved: the murine adenosine deaminase (ADA) (Wilson *et al.*, 1991) and *Escherichia coli* CDA (Betts *et al.*, 1994). Both structures reveal a zinc atom within the active site and four conserved residues that coordinate it and are essential for catalytic activity. Both ADA and CDA enzymes deaminate free nucleosides through hydrophilic attack, the difference is that ADA targets carbon 6 of a purine ring, whereas CDA targets carbon 4 of a pyrimidine ring.

Sequence comparisons revealed conserved motifs involved in zinc binding that show both the ADAR and APOBEC proteins, regardless of substrate, are members of the CDA family. The conserved residues within the CDA domain are: a cysteine (C) or a histidine (H) residue followed by a glutamic acid (E) residue which acts as a proton donor during the nucleophilic deamination reaction; a proline (P) followed by two conserved cysteine residues which coordinate the zinc atom in the active site; and a conserved secondary structure of alternating α-helices and β-sheets. These conserved residues are organized into three motifs: motif I (H/CxE), motif II (PC), and motif III (C), where x indicates any amino acid. The distance between the two C motifs (motifs II and III) varies; within the APOBEC and ADAT family it is two amino acids, whereas within the ADAR and ADAT1 family it is more variable.

III. THE ADAS THAT ACT ON RNA FAMILY

ADARs were originally identified as an activity in *Xenopus laevis* embryo extracts that appeared to unwind dsRNA so that it migrated differently when electrophoresised on a native polyacrylamide gel (Bass and Weintraub, 1987). Originally, it was thought to be a helicase activity but then it was found to modify dsRNA so that up to 50% of the adenosine residues in the dsRNA were deaminated to inosines (Bass, 1988, p. 3).

The ADAR family evolved from tRNA ADA (Tad1/ADAT1) by the acquisition of several dsRNA-binding domains (dsRBDs) at the amino terminus (Gerber and Keller, 2001; Fig. 3.2). There are four members of the ADAR family in mammals; however, alternatively, splicing generates different isoforms with either altered activity or cellular localization. Expression of ADAR1 and ADAR2 is highest in the CNS, where most of the specifically edited transcripts have been identified. ADAR3, which is catalytically inactive, is only expressed in the brain (Melcher *et al.*, 1996). Another member of the family known as testis-expressed nuclear RNA-binding protein (TENR) is only expressed in the testis (Schumacher *et al.*, 1995). It is likely to be catalytically inactive as it lacks crucial conserved residues involved in catalysis (Connolly *et al.*, 2005).

A. ADAR1

ADAR1 was the original protein to be purified to homogeneity from different species (Hough and Bass, 1994; Kim *et al.*, 1994; O'Connell and Keller, 1994). The activity assay employed by different groups was the conversion of A-to-I in long dsRNA so using this assay ADAR2 should also have been isolated. However, as this did not occur, it suggests that ADAR2 protein is either not as abundant as ADAR1 or is less stable. ADAR1 protein is ubiquitously expressed and is highly conserved in many organisms but is not found in *Drosophila*; however, the possibility exists that it has been lost from the species (Keegan *et al.*, 2004).

ADAR1 is alternatively spliced at the amino terminus to generate two different isoforms that differ in their translation start sites and subsequently in their subcellular localization (Kawakubo and Samuel, 2000; Patterson and Samuel, 1995). The longer ADAR1 isoform (p150) is interferon-inducible,

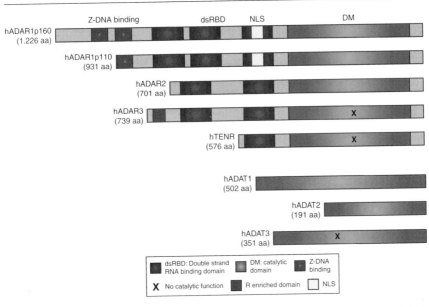

Figure 3.2. Proteins domains of ADARs consist of dsRNA (black) which bind dsRNA and the catalytic deaminase domain (DM; gray box). ADAT1, ADAT2, and ADAT3 have a DM domain which they also use to bind tRNA. ADAR3, TENR, and ADAT3 are inactive and that is represented by an (X) in the DM domain. ADAR1 has Z-DNA binding at the N-terminus domains and ADAR3 contains an R-enriched region at the N-terminus.

whereas the shorter isoform p110 is constitutively expressed (Fig. 3.2). Expression of ADAR1p150 is regulated by an interferon-inducible promoter which generates a transcript that utilizes the first methionine, whereas utilization of an internal start site at methionine 296 generates the p110 isoform (Kawakubo and Samuel, 2000; Patterson and Samuel, 1995). Both proteins have three dsRBDs and an unusual NLS that overlaps with the third dsRBD (Strehblow *et al.*, 2002). The p150 isoform is a nucleocytoplasmic shuttling protein; however, its localization is mainly cytoplasmic due to a nuclear export signal (NES) signal within the first Z-DNA-binding domain at the amino terminus (Poulsen *et al.*, 2001). The Z-DNA-binding domains are at the amino terminus so ADAR1p150 has two, however as translation of ADARp110 starts at methionine 296 it only has one (Fig. 3.2). These Z-DNA-binding domains allow interaction with non-B-form RNA (Herbert, 2001, p. 599).

ADAR1 p110 is also a nucleocytoplasmic shuttling protein despite lacking an NES. Instead, it interacts with the export factor exportin-5 (Exp-5) and export of ADAR1 is enhanced by binding to dsRNA (Fritz *et al.*, 2009). Transportin-1 (TRN 1) is the nuclear import factor that interacts with the NSL present in the third dsRBD and binding to dsRNA abolishes import of ADAR1.

Both TRN 1 and Exp-5 bind to an overlapping region of ADAR1 so binding is mutually exclusive. As export is enhanced by binding to dsRNA and import is abolished, this gives directionality to export of dsRNA and prevents dsRNA from being reimported. ADAR1 has been shown to localize to the nucleolus in some mammalian cell lines (Desterro *et al.*, 2003; Sansam *et al.*, 2003). Photobleaching experiments indicate this localization is dynamic, and the protein leaves the nucleolus to interact with substrates; therefore, its localization to the nucleolus may be for storage purposes (Desterro *et al.*, 2003).

1. ADAR1-null mice

Unlike ADAR2, there is no known essential editing event that depends exclusively on ADAR1. ADAR1 does edit specific sites such as three of the five editing sites in exon 5 of the serotonin 5-hydroxytryptamine subtype 2C (5-HT_{2C}) receptor (Burns *et al.*, 1997). It was therefore surprising when *Adar1*$^{-/-}$ mice displayed a severe embryonic phenotype of massive liver disintegration and died at embryonic day (E) 12.5 (Hartner *et al.*, 2004; Wang *et al.*, 2004) as no ascribed role for ADARs during development and no edited transcripts has been found in the liver. Characterization of the *Adar1*$^{-/-}$ liver phenotype revealed a defect in proliferation of hematopoietic cells, with those present showing fragmented DNA indicative of apoptotic cell death. The heterozygous mice were indistinguishable from wild type.

Mouse embryonic fibroblast cells derived from *Adar1*$^{-/-}$ mice displayed elevated apoptosis in response to serum starvation indicative of a widespread predisposition to apoptotic cell death (Jiang *et al.*, 2004). The onset of the *Adar1*$^{-/-}$ apoptotic phenotype correlated with an induction of the interferon-inducible ADAR1 p150 isoform following serum starvation implying a role for ADAR1 in mediating the induction of apoptosis in response to stress. Interestingly, analysis of the transcript encoding the HT_{2C} isolated from E12 *Adar1*$^{-/-}$ embryos showed that exon 5, which contains all five editing sites, was aberrantly spliced out. This implies that splicing and editing at these sites in the HT_{2C} transcript are coupled (Hartner *et al.*, 2004).

Further characterization of the liver phenotype in mice with induced disruption of *Adar1* in the hematopoietic system revealed that ADAR1 is crucial for the maintenance of hematopoietic stem cells (Hartner *et al.*, 2009). Another group found that ADAR1 was instead required for the survival of hematopoietic progenitor cells (XuFeng *et al.*, 2009). During liver development, ADAR1 somehow acts to suppress interferon signaling and in the absence of ADAR1 elevated apoptosis occurs with aberrant activation of interferon-inducible genes. Therefore, the hypothesis has been proposed that some unknown substrate of ADAR1 is critical for correct interferon response during embryonic liver development (Hartner *et al.*, 2009). The *Adar1*$^{-/-}$ phenotype could be rescued by the expression of ADAR1 and rescue was largely dependent on RNA-editing activity (XuFeng *et al.*, 2009). Recent work (Vitali and Scadden, 2010) has shown that I–U-containing dsRNA inhibits the induction of an interferon response by binding to the cytosolic dsRNA receptors RIG1 and MDA5, suggesting that ADAR1 may act to restrain or to terminate interferon responses.

2. Human mutations in the *ADAR1* gene

Several human mutations distributed across the entire ADAR1 gene cause the relatively benign dyschromatosis symmetrica hereditaria (DSH) (Liu *et al.*, 2006; Miyamura *et al.*, 2003; Zhang *et al.*, 2004, 2008) which is an autosomal dominant skin pigmentation disorder that predominantly affects the hands and feet. The discovery of 30 different point mutations in the ADAR1 gene associated with this disorder indicates that ADAR1 has an as yet undefined role in melanocyte development or pigmentation. The disease arises from haploinsuffiency of the remaining *ADAR1* gene copy. Some mutations are truncations within the N-terminal region unique to ADAR1 p150, suggesting that loss of this isoform alone is sufficient to cause disease. One individual with the disorder had a mutation, G1007R, which is in a highly conserved region that lies within the deaminase domain (Tojo *et al.*, 2006). This individual also presented with chronic dystonic posture and calcification in the brain. Functional assays demonstrated that the G1007R mutant protein is not capable of RNA editing although it retains RNA-binding activity, indicating it may behave as a dominant negative mutant with the strongest *ADAR1* loss of function phenotype yet seen in humans (Heale *et al.*, 2009a,b). This mutation has subsequently been found in another individual with DSH, dystonia, and calcinosis of the brain; however, the mother who carried the mutation did not have neurological symptoms which suggests that this mutation may have reduced penetrance (Kondo *et al.*, 2008).

B. ADAR2

ADAR2 has a very similar domain structure to ADAR1 with two dsRBDs at the amino terminus and the catalytic deaminase domain at the C-terminus (Fig. 3.2). This protein is more conserved than ADAR1 and its ortholog is found in *Drosophila* (Palladino *et al.*, 2000). It is a nuclear protein, and like ADAR1 p110, it has been shown to localize to the nucleolus in a dynamic manner (Dawson *et al.*, 2004; Desterro *et al.*, 2003). Further investigation revealed that a region encompassing the first dsRBD is required for nucleolar localization, and the ADAR2 protein relocalizes to the nucleoplasm upon expression of a substrate. However, it is unclear whether the region harbors a nucleolar localization signal or whether RNA-binding is required for the nucleolar localization. However, an ADAR2 mutant containing point mutations within both dsRBDs displayed exclusion from the nucleolus indicating RNA-binding was required for nucleolar localization (Dawson *et al.*, 2004). These results suggest that ADAR2 activity is regulated through subcellular localization.

The crystal structure of the deaminase domain of ADAR2 revealed the presence of a zinc ion in the catalytic center as had been proposed but in addition, they found an inositol hexakisphosphate molecule buried in the core of the enzyme (Macbeth *et al.*, 2005). The amino acid residues that coordinate the inositol hexakisphosphate are conserved throughout the ADAR and ADAT1 families, but absent from the ADAT2/3 family. The inositol hexakisphosphate molecule stabilizes the ADAR2 protein structure and is essential for catalytic activity.

Similar to ADAR1, there are different isoforms of ADAR2. Human ADAR2 cloned from a brain cDNA library encoded two major isoforms which differed by the inclusion or exclusion of an *Alu* cassette within the deaminase domain (Gerber *et al.*, 1997). Characterization of the two ADAR2 isoforms revealed that both had the same substrate specificity, but inclusion of the *Alu*-containing exon led to a twofold reduction in ADAR2 activity on a *GluR-B R/G* site substrate *in vitro*. Another major transcript was isolated that lacked the last two amino acids and surprisingly this isoform was catalytically inactive (Lai *et al.*, 1997).

Rat ADAR2 is alternatively spliced at the amino terminus and the inclusion of one exon of 47 bp alters the predicted reading frame of the protein (Rueter *et al.*, 1999). Use of the canonical translation start site in the presence of the 47 bp exon would generate a truncated protein lacking the dsRBDs and the catalytic domain. The inclusion of this exon is dependent upon editing of an adenosine residue at the -1 position of the $3'$ splice site, which mimics the canonical AG acceptor site. Editing at this site occurs up to 30% in ADAR2 pre-mRNA isolated from rat brain. Therefore, rat ADAR2 has evolved a complex autoregulatory feedback, such that alternative splicing of its own transcript is dependent on the presence of a functional ADAR2 protein (Rueter *et al.*, 1999). ADAR2 autoediting also occurs in human cell lines, and the level of autoediting and inclusion of the alternatively spliced exon have been shown to correlate, such that higher expression of ADAR2 results in increased inclusion of the 47 bp exon (Maas *et al.*, 2001). This editing is independent of ADAR1 expression. The stem–loop structure surrounding the edited splice site is created by interactions between intron 4 and the downstream exon. Despite it containing an intervening loop of 1354 nucleotides, the paired regions are highly conserved.

1. Editing of glutamate receptor transcripts by ADAR2

Although the biological role of ADAR1 is unclear, ADAR2 is a genuine RNA-editing enzyme with a clearly defined key target. ADAR2 is expressed in many tissues yet the transcripts it specifically edits are expressed within the CNS. The critical site that it edits is the Q/R site in the transcript encoding subunit B of the glutamate-gated ion channel receptor (GluR-B also referred to as GluR2). Ionotropic glutamate receptors are subdivided into three groups according to the synthetic agonists that they bind to: α-amino-3-hydroxy-5-methylisoxasole-4-propionate (AMPA), N-methyl-D-aspartic acid (NMDA), and kainite receptors. Pre-mRNAs encoding both the AMPA and kainite receptor subunits are edited leading to variation at key residues within the receptor subunits (Table 3.1). AMPA receptors mediate fast excitatory synaptic transmission and characteristically have low calcium permeability. The glutamate receptor is a tetramer composed of GluR subunits which are GluR-A, GluR-B, GluR-C, or GluR-D. If the GluR-B subunit is present, then the receptor is impermeable to calcium.

The position that regulates the calcium permeability is the Q/R site and this is edited by ADAR2. Editing changes the genomically encoded glutamine (Q) at position 607 to an arginine (R) residue (Higuchi et al., 1993; Sommer et al., 1991). Editing of the Q/R site occurs to >99% in the mouse and human CNS. Thus, one site-specific RNA-editing event has great functional importance for neuronal calcium homeostasis. Despite sharing considerable sequence and structural homology, the other three GluR subunit transcripts A, C, and D (also referred to as GluR1, 3, and 4) do not undergo editing at the Q/R site and retain a glutamine at the corresponding amino acid and are calcium permeable. Editing at the Q/R site also results in retention of the GluR-B subunit with the endoplasmic reticulum ER, while the unedited GluR-B (Q) subunit is efficiently assembled into receptors and transported to the cell surface (Greger et al., 2002, 2003). ER retention is thought to promote inclusion of the GluR-B (R) subunit into heterotetramers, whereas GluR-B (Q) subunits more readily form homotetramers (Greger et al., 2003). Therefore, RNA processing events determine the kinetics and assembly of AMPA receptors at the synapse and if editing is prevented at this position then this leads to increased intracellular levels of calcium and neuronal cell death.

There is editing at another position in the *GluR-B* transcript which is adjacent to the alternatively spliced FLIP/FLOP exons. Editing results in a change from the genomically encoded arginine to a glycine residue (R/G site). The resulting glycine substitution increases the rate of recovery of the channel from desensitization (Lomeli et al., 1994). Editing at the R/G site is performed by both ADAR1 and ADAR2, and occurs at low levels during development increasing to approximately 75% in adult mouse brain. The same position (R/G) is also edited in *GluR-C* and *GluR-D* transcripts; however, it is absent from the *GluR-A*

Table 3.1. Mammalian Transcripts That Are Edited by ADARs

Gene	Amino acid change and editing site	Tissue	Reference
5-Hydroxytryptamine receptor 2C (5-HT$_{2C}$)	I > M	CNS	Burns et al. (1997)
5-Hydroxytryptamine receptor 2C (5-HT$_{2C}$)	I > V	CNS	Burns et al. (1997)
5-Hydroxytryptamine receptor 2C (5-HT$_{2C}$)	N > S	CNS	Burns et al. (1997)
GRIK1 or glutamate receptor, ionotropic kainate 1 (GLUR5)	Q > R	CNS	Nutt and Kamboj (1994)
GRIK2 or glutamate receptor, ionotropic kainate 2 (GLUR6)	Q > R	CNS	Köhler et al. (1993)
GRIK2 or glutamate receptor, ionotropic kainate 2 (GLUR6)	I > V	CNS	Köhler et al. (1993)
GRIK2 or glutamate receptor, ionotropic kainate 2 (GLUR6)	Y > C	CNS	Köhler et al. (1993)
GRIA2 or glutamate receptor ionotropic, AMPA B (GLURB)	Q > R	CNS	Paschen and Djuricic (1994) and Sommer et al. (1991)
GRIA2 or glutamate receptor ionotropic, AMPA B (GLURB)	R > G	CNS	Lomeli et al. (1994) and Sun et al. (1994)
GRIA3 or glutamate receptor ionotropic, AMPA C (GLURC)	R > G	CNS	Lomeli et al. (1994)
GRIA4 or glutamate receptor ionotropic, AMPA D (GLURD)	R > G	CNS	Lomeli et al. (1994)
KCNA1 or potassium voltage-gated channel subfamily A member 1	I > V	CNS	Bhalla et al. (2004)
GABRA3 or γ-aminobutyric acid receptor subunit alpha-3	I > M	CNS	Ohlson et al. (2007)
CYFIP2 or cytoplasmic FMR1-interacting protein 2	K > E	CNS	Levanon et al. (2005)
ARL6IP4 or SRp25	K > R	Expressed only in G1/ S phase	Gommans et al. (2008)
BLCAP (BC10) or bladder cancer-associated protein	Y > C	Brain and B cells	Clutterbuck et al. (2005)
BLCAP (BC10) or bladder cancer-associated protein	Q > R	Brain and B cells	Clutterbuck et al. (2005) and Levanon et al. (2005)
BLCAP (BC10) or bladder cancer-associated protein	K > R	Brain and B cells	Clutterbuck et al. (2005) and Levanon et al. (2005)
FLNA or filamin A	Q > R		Levanon et al. (2005)
IGFBP7 or insulin-like growth factor binding protein 7	R > G		Levanon et al. (2005)
IGFBP7 or insulin-like growth factor binding protein 7	K > R		Levanon et al. (2005)

The gene name, the edited position, and where the transcripts are expressed are summarized in this table.

transcript due to loss of a functional ECS region (Lomeli *et al.*, 1994). Therefore, the range of receptor subunits available is considerably increased through alternative splicing and editing.

The calcium permeability of kainate receptors is also regulated by editing at the position which corresponds to the Q/R site in the *GluR-B* transcript. Editing of the transcripts encoding the kainate receptor subunits GluR-5 and GluR-6 is both spatially and developmentally regulated, and editing of each subunit transcript is independent of the other (Bernard and Khrestchatisky, 1994; Köhler *et al.*, 1993).

2. ADAR2 knockout and transgenic mice

In contrast to the embryonic lethal phenotype of the $Adar1^{-/-}$ gene, $Adar2^{-/-}$ mice survive to birth but die by postnatal day 20 with an increasing predisposition to epileptic seizures (Higuchi *et al.*, 2000). The ADAR2-null phenotype could be completely rescued by a pre-edited version of the endogenous GluR-B gene in which the genome-encoded glutamine is mutated to arginine. This indicates that the primary function of ADAR2 is to edit the *GluR-B* Q/R site (Higuchi *et al.*, 2000). The rescue also confirmed that there is no obvious role for the unedited *GluR-B* (Q) allele as the rescued mice were phenotypically wild type. The $Adar2^{-/-}$ mice have reduced editing at the *GluR-B* Q/R site: 10% of pre-mRNA transcripts were edited but 40% of mature messages were edited in these mice due to facilitated splicing of the edited *GluR-B* transcript. ADAR1 is probably responsible for the residual *GluR-B* Q/R site editing as expression of ADAR1 is unaffected by alterations in ADAR2. Inefficient editing led to accumulation of unspliced pre-mRNA in the nucleus that retained intron 11 where the ECS is located (Higuchi *et al.*, 2000), and consequentially there was a fivefold decrease in the amount of mature *GluR-B* mRNA.

Transgenic mice were generated with an asparagines (N) residue at the Q/R site to allow mice to live longer and develop more extensive neurodegeneration (Kuner *et al.*, 2005). An asparagine residue was chosen as it conferred a twofold increase in calcium permeability but did not significantly alter the AMPA channel conductance. Selective death of motor neurons was observed that increased with age. The motor neuron phenotype was similar to that observed in a mouse model of familial amyotrophic lateral sclerosis (fALS) in which a mutation in the superoxide dismutase 1 (*SOD1*) gene results in selective motor neuron pathology (Gurney *et al.*, 1994). Indeed, crossing the *GluR-B* (*N*) mice with *SOD1* mutants enhanced the motor neuron degeneration, implicating a role of GluR-B mediated calcium homeostasis in human motoneuron diseases such as ALS (Kuner *et al.*, 2005).

Transgenic mice overexpressing *ADAR2* under the control of a CMV promoter developed adult-onset obesity (Singh *et al.*, 2007). There was no alteration in editing levels in transcripts edited normally by ADAR2. The transgenic mice also expressed the endogenous *ADAR2* and the level of transgenic ADAR2 protein in the brain was only 1.2-fold higher than the endogenous level, leading to the hypothesis that the observed phenotype was due to mis-expression of the transgene in tissues that do not normally express *ADAR2*. Interestingly, the same adult-onset obesity phenotype was observed with transgenic mice overexpressing a catalytic site mutant *ADAR2 E396A*, indicating the phenotype was not due to aberrant editing. These data suggest that the catalytically inactive ADAR2 has another function, probably mediated through RNA-binding.

C. ADAR3

The third ADAR identified ADAR3 is only expressed in specific regions of the brain including the amygdala and thalamus (Melcher *et al.*, 1996). Similar to other ADAR proteins, it is capable of binding to dsRNA *in vitro* but is unique in that it also binds to ssRNA through an N-terminal arginine and lysine-rich region (R domain) (Fig. 3.2). The protein is catalytically inactive; however, the conservation of the deaminase domain among vertebrates as shown by amino acid alignment (Keegan *et al.*, 2004) indicates it is under positive selective pressure. *In vitro* ADAR3 has been shown to act as an effective competitor by binding to the same transcripts as ADAR1 and ADAR2 and preventing deamination (Chen *et al.*, 2000). To date, no known function has been described for ADAR3; however, as it is conserved, it may act as a competitor of ADAR1 and ADAR2 in the brain and thereby regulate their editing of particular transcripts.

IV. TESTIS-EXPRESSED NUCLEAR RNA-BINDING PROTEIN

TENR is expressed in the round spermatids in the testis. TENR was identified in a screen to find RNA-binding proteins that interact with the 3′ untranslated region (UTR) of the *protamine 1* (*Prm1*) RNA (Schumacher *et al.*, 1995). *Tenr*$^{-/-}$ male mice are sterile (Connolly *et al.*, 2005). This is due to low sperm counts and a high incidence of sperm morphological defects indicating that TENR plays a role in sperm morphogenesis. TENR has dsRBDs but lacks the conserved residues within the deaminase domain that are essential for catalytic activity indicating it is not a functional deaminase (Connolly *et al.*, 2005; Fig. 3.2). Besides *Prm1*, no other transcripts have been identified that it binds to and the biological function of TENR remains unclear.

V. RNA BINDING BY ADARS

ADAR belongs to a diverse group of proteins that have one or more copies of a dsRBD of approximately 70 amino acids (reviewed in Fierro-Monti and Mathews, 2000). Despite only a few residues being conserved, all dsRBDs analyzed fold into the same secondary structure of α-helices and β-sheets, organized as $\alpha1$-$\beta1$-$\beta2$-$\beta3$-$\alpha2$, where the α-helices make contact with the same face of the RNA. Other proteins with dsRBDs members include *Staufen* a protein involved in mRNA transport; PKR, an interferon-inducible, RNA-dependent protein kinase; and many proteins involved in the RNA interference pathway which have domain arrangements resembling ribonuclease III, an RNA nuclease.

The dsRBD-containing proteins do not exhibit sequence specificity, it is the secondary structure of the A-form RNA that they recognize. A-form nucleic acid differs from B-form in that the major groove is narrow and the minor groove is wide and shallow. Highly sequence-specific interaction between proteins and DNA occurs on the major grove; however, on A-form nucleic acid this is not possible so the protein–RNA interactions occur with the 2-hydroxyl group of the ribose sugar in the minor and are therefore not sequence specific.

However, the dsRBDs of ADAR2 do display a binding selectivity to the transcripts they edit. Experiments on a short RNA encompassing the *GluR-B Q/R* site demonstrated that ADAR2 dsRBDs exhibit selective RNA binding to the *GluR-B Q/R* site and this was distinct from the binding site of a dsRBD from PKR (Stephens *et al.*, 2004). The dsRBD occupies approximately 16 bp of dsRNA but many can accommodate or prefer binding to substrates that contain bulges or loops within the dsRNA. The individual dsRBDs of ADAR2 exhibit different binding specificities when analyzed separately (Poulsen *et al.*, 2006; Stefl *et al.*, 2006). dsRBD1 preferentially binds to perfect duplex dsRNA located in the stem–loop region, whereas dsRBD2 shows preference for dsRNA containing an A–C mismatch which is usually located near to the editing site. Both dsRBDs are required for efficient editing by ADAR2 (Stefl *et al.*, 2006).

In mammals, ADAR1 contains three dsRBDs, whereas ADAR2 and ADAR3 both contain two dsRBDs. Recently, an ADAR was cloned from squid that is an ortholog of ADAR2 (Palavicini *et al.*, 2009). It encodes three dsRBD; however, the first dsRBD is in an alternatively spliced exon at the amino terminus of the protein. When both isoforms were expressed and assayed *in vitro* the isoform with three dsRBDs had increased enzymatic activity.

To investigate what was responsible for the specificity of ADAR1 and ADAR2, chimeric proteins were generated with the dsRBDs from one ADAR and the deaminase domain from another (Wong *et al.*, 2001). This revealed that the editing specificity in the chimeric proteins was provided by the deaminase domain. However, when the dsRBDs of ADAR1 and PKR were exchanged, this resulted in a reduction or a complete loss of enzymatic activity on edited transcript; however, the chimeric protein was still active on dsRNA. Therefore, specificity is also provided by the dsRBDs (Liu *et al.*, 2000). ADAR3 is the only member of the ADAR family known to bind to ssRNA, and experiments have shown this interaction is mediated through the arginine- and lysine-rich R domain in the amino terminus of the protein which is not found in other ADAR proteins (Chen *et al.*, 2000; Fig. 3.1). ADAR3 is also capable of binding dsRNA through its canonical dsRBDs but it is unable to edit either known substrates or dsRNA.

The dsRBDs have additional roles other than binding dsRNA such as nucleocytoplasmic shuttling of ADAR1 as previously described. The deletion of dsRBD1 of ADAR2 reduced editing efficiency, whereas loss of dsRBD2 abolished editing altogether (Poulsen *et al.*, 2006). The N-terminal region of ADAR2 had an autoinhibitory effect on catalytic activity as a truncated ADAR2 protein containing a deaminase domain and dsRB2 was capable of editing a 15-bp substrate, whereas the full-length protein did not. However, the addition of the N-terminal region of ADAR2 to the truncated protein in *trans* reduced editing efficiency on shorter substrates, suggesting that the RNA substrate has to be long enough for binding of both dsRBDs to relieve the autoinhibition for efficient editing (Macbeth *et al.*, 2004).

The dsRBDs have also been implicated in dimer formation as binding to RNA has been shown to be a prerequisite for dimer formation. However, there have been conflicting reports and the issue of whether dimer formation requires RNA binding for its formation has not been resolved. Studies have shown that the formation of homodimers is required for ADAR activity (Cho *et al.*, 2003; Gallo *et al.*, 2003; Jaikaran *et al.*, 2002; Poulsen *et al.*, 2006). A ternary complex can be observed when increasing amounts of ADAR are added to substrate RNA, such that one monomer binds and then another, indicating dimerization is RNA-dependent (Jaikaran *et al.*, 2002). Analysis of RNA editing *in vitro* using one wild-type monomer and one catalytically inactive monomer showed that both monomers contribute to hyperediting of dsRNA and site-specific editing of substrates (Cho *et al.*, 2003). However, fluorescence energy resonance transfer (FRET) experiments indicate that ADAR1 and ADAR2 form homodimers in an RNA-independent manner, and are capable of forming heterodimers *in vivo* (Chilibeck *et al.*, 2006).

To assess the role of RNA binding in dimerization (Valente and Nishikura, 2007), mutations were made in three conserved lysine residues (KKxxK → EAxxA) within each of the dsRBDs of ADAR1 and ADAR2. The mutations introduced were based on data from alanine-scanning mutagenesis of the *Drosophila* RNA-binding protein *Staufen*, which demonstrated that mutation of exposed lysine residues within the conserved dsRBD eliminated RNA binding without disrupting structure (Ramos *et al.*, 2000). Sequential purification of protein complexes containing both wild-type and mutant ADAR proteins demonstrated that homodimerization of ADAR proteins occurs independently of RNA binding (Valente and Nishikura, 2007). However, the dimeric ADAR proteins containing one mutant and one wild-type monomer behaved in a dominant negative manner in both an RNA-binding assay and an *in vitro* editing assay indicating that RNA binding of both monomers is required for deamination to occur.

The conflicting reports likely arise from the different experimental models, whether the experiments were performed *in vitro* or *in vivo* in cell culture. Also, posttranslational modification could be involved as some proteins were expressed in yeast and others in Sf9 insect cells. Furthermore, proteins from different species such as *Drosophila* and human ADARs were used.

In yeast, ADAT2 and ADAT3 have been shown to function as a heterodimer where ADAT2 provides the catalytic function and ADAT3 provides substrate specificity (Gerber and Keller, 1999). This raised the possibility that ADARs may function as heterodimers. Heterodimerization between ADAR proteins may reveal a role for ADAR3, where ADAR1 or ADAR2 provides the catalytic activity and ADAR3 affects the substrate specificity. FRET analysis indicates that heterodimers between ADAR1 and ADAR2 monomers form *in vivo* (Chilibeck *et al.*, 2006) and heterodimers between ADAR1 and ADAR2 were co-immunoprecipitated from astrocytoma cell lines, where it was demonstrated that increased levels of ADAR1 can inhibit editing of substrates by ADAR2 (Cenci *et al.*, 2008). It remains to be seen whether overexpression of ADAR3 can have the same effect. Dimers formed between the two isoforms of ADAR1 (p110 and p150) are readily detectable indicating that the Z-DNA binding domains absent in the p110 isoform are not required for dimerization (Cho *et al.*, 2003).

VI. SECONDARY STRUCTURES OF RNA-EDITING SUBSTRATES

Initial investigations into *GluR-B Q/R* site editing revealed that a complementary editing site complementary sequence (ECS) located in the downstream intron was required for editing to occur (Higuchi *et al.*, 1993; Fig. 3.1). These sequences form an imperfect stem–loop structure which is bound by the dsRBDs of ADAR2. Point mutations which destabilized the stem–loop decreased the editing frequency, but this could be restored by making the complementary mutation in the opposite base, indicating it is the structure not the sequence that is important.

Mammalian ADAR1 and ADAR2 edit some of the same sites yet other sites are clearly specific for one or other protein, raising questions about substrate recognition. While there is no consensus sequence surrounding edited adenosine residues, ADAR enzymes exhibit local sequence preferences. For example, the $5'$ nearest neighbor preference for ADAR1 is $U = A > C > G$, similarly for ADAR2 it is $U = A > C = G$ (Lehmann and Bass, 2000).

It is thought that the deamination reaction occurs outside the dsRNA, with the target adenosine residue rotated or flipped $180°$ allowing deamination of the base and minimizing structural disruption (reviewed in Roberts and Cheng, 1998). It was observed that ADAR2 induced changes in 2-aminopurine fluorescence of a modified substrate which is consistent with a base-flipping model (Yi-Brunozzi *et al.*, 2001).

VII. TRANSCRIPTS EDITED BY ADARS

Most of the transcripts that are specifically edited by ADARs are expressed in the CNS (Table 3.1). Pre-mRNA encoding subunits of glutamate receptors both of the AMPA and kainite classes are edited; however, there is no editing of pre-mRNA encoding subunits of the NMDA class of receptor. We have previously discussed editing of *GluR-B* Q/R and R/G sites, also *GluR-C* and *GluR-D* transcripts are both edited at the R/G site (Lomeli *et al.*, 1994). In transcripts encoding the kainate subunit GluR6, there is editing at two positions that convert amino acids in transmembrane segment 1 and transmembrane segment 2 (Köhler *et al.*, 1993). RNA editing at these positions affects calcium permeability of the receptor. Editing of transcripts encoding the kainate subunit GluR5 also affects calcium permeability (Nutt and Kamboj, 1994). Both edited channels are strongly inhibited by low concentrations of fatty acids, whereas channels that contain a mixture of edited and unedited isoforms resist inhibition (Wilding *et al.*, 2005). Transcripts encoding the serotonin receptor are also edited and up to 24 different isoforms can be generated by editing (Burns *et al.*, 1997). The fully edited isoform displays blunted G-protein coupling and reduced constitutive activity. The potassium channel transcript K(V)1.1 is also edited, and this editing is also found in the *Drosophila* ortholog of the gene and is an example of convergent evolution (Bhalla *et al.*, 2004). The consequence of editing on this protein is channel inactivation. The transcript encoding the inhibitory GABA-type A receptor (GABA$_A$) is also edited at one position in transmembrane segment 3 of α_3 where an isoleucine is converted to methionine (Ohlson *et al.*, 2007). Whole cell recordings in HEK 293 cells reveal that receptors containing the edited isoform have smaller amplitudes, slower activation, and faster deactivation than those that contain the unedited isoform. It is likely that editing also affects subunit assembly. The consequence of editing on the functioning of the other proteins (Clutterbuck *et al.*, 2005; Levanon *et al.*, 2005) listed in Table I is unknown.

VIII. SEARCHES FOR MAMMALIAN TRANSCRIPTS EDITED BY ADARS

Inosine occurs at a frequency of approximately one base in every 17,000 nucleotides in mouse brain poly(A)$^+$ RNA (Paul and Bass, 1998). As this frequency of inosine nucleotides could not be accounted for with known site-specific editing events, this prompted the search for new ADAR substrates. Before the era of deep-sequencing, many attempts were made to conclusively describe the list of ADAR editing events in different model organisms through the detection of A–G transition changes when comparing cDNA and genomic DNA sequences. However, this approach proved problematic due to the high frequency of single nucleotide polymorphisms (SNPs) and sequencing errors.

Levanon et al. used a bioinformatic approach to search for expressed sequence tags (ESTs) containing A–G mismatches (Levanon et al., 2004). As the substrate for ADAR is dsRNA, they limited their search to predicted double-stranded regions in ESTs. This approach identified 12,723 putative editing sites in 1637 genes, 26 of which were validated by sequencing genomic DNA, and cDNA from the same individual. Interestingly, 92% of editing sites identified using this approach were located within Alu elements and 1.3% were in LINE elements.

Using a similar approach, Athanasiadis et al. (2004) searched databases for clusters of multiple A–G changes within short sequences, reasoning that these were unlikely to be SNPs or sequencing artifacts. They also found A–G changes clustered at the site of Alu repeat element insertions. This approach identified 1445 mRNAs that were edited at 14,500 sites, which would be sufficient to account for the amount of inosine observed in brain mRNA. Alu insertion favors actively transcribed regions, often occurring in UTR regions and within intronic regions. The edited Alu repeat elements were found to occur in tandem in inverted repeat orientation such that when transcribed a dsRNA stem–loop structure would form between the two inverted repeats and be edited by ADAR (Fig. 3.3). Due to the extensive similarity between Alu elements, the stem–loop structures formed are inherently stable, although bulges and mismatches were edited more frequently than near-perfect duplexes reflecting the in vivo substrate preference of ADARs. Some editing events were found to alter splice sites resulting in exonization of a partial or whole Alu element, although the novel alternative splice sites were usually used at a low frequency and rarely constitutively (Athanasiadis et al., 2004). Surprisingly, editing of transcripts containing other known repeat elements was considerably lower than that of Alu elements, which is likely to be due to the high level of conservation between Alu element subgroups and therefore the ease of creating the required secondary structure.

The prevalence of editing of transcripts containing Alu elements was further confirmed with similar large-scale bioinformatic approaches (Blow et al., 2004; Kim et al., 2004). It has been postulated that editing of repetitive elements

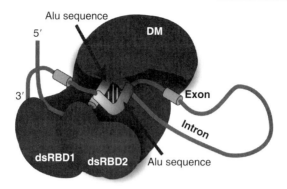

Figure 3.3. Transcripts encoding Alu repeat elements are often present in introns of transcripts. If two are present in opposite orientation less than 2 kb apart they can base pair with the other and form a duplex structure that is edited by ADARs.

may act to inhibit retrotransposition; however, there is no real evidence to either support or contradict this hypothesis and editing of *Alu*-containing transcripts may have a novel biological role.

Recently, massively parallel amplification and deep-sequencing were performed to find potential human editing sites in nonrepetitive sequences at computationally predicted editing sites. Amplification of these small regions using unique linked primers that form a "padlock" across potential editing sites and high-throughput sequencing (Li *et al.*, 2009) identified 10 known edited transcripts, along with 207 new transcripts that were edited in two tissues greater than 5%. The number of new edited transcripts increased dramatically when the stringent selection criteria were relaxed as many transcripts showed low levels of editing (<5%) and were only edited in one tissue. Of the sites identified, 55 occurred in coding regions and 38 led to amino acid changes. Editing of transcripts also varied between tissues and within brain regions. Next generation sequencing technology has made the identification and verification of novel editing sites easier; however, the functional consequences of these newly identified editing events have yet to be elucidated.

IX. CONSEQUENCES OF INOSINE IN TRANSCRIPTS

Deamination of adenosine bases in RNA results in transcripts that contain one or many inosine bases, which can have different consequences for the cell. If editing occurred within coding sequences, the inosine base is read as guanosine by the translation machinery which can lead to a recoding event so that a different amino acid is inserted that can change the properties of proteins as is the case with the GluR-B subunit, as described earlier. Deamination of adenosine residues that occur close to splicing junctions can affect the inclusion/exclusion of either introns or exons. For example, the ADAR2 autoediting site changes a splice site and consequently generates a nonfunctional protein (Rueter et al., 1999). In the transcript encoding *PTPN6*, editing within an intron was shown to alter the conserved adenosine residue at the branch point resulting in retention of the intron (Beghini et al., 2000). The retained intron contains an in-frame stop codon which is predicted to generate a truncated protein lacking catalytic activity.

Editing can also remove a stop codon altering the protein produced as occurs with the hepatitis delta virus (HDV) viral proteins p24 and p27 and this has a crucial role in the life cycle of the HDV (Polson et al., 1996). The viral genome contains an amber stop codon (UAG) which is altered by A–I editing to produce a UIG codon which is translated as tryptophan (UGG) by the translation machinery. The two proteins produced, short (p24) and long (p27), differ by 19 amino acids and have specific roles in the viral life cycle. p24 is required for replication of the viral RNA genome, whereas p27 represses replication and is required for packaging of the virus (Polson et al., 1996). Editing can also result in the exonization of *Alu* elements via generation of splice site consensus sequences (Athanasiadis et al., 2004).

Editing of noncoding regions is widespread (Li et al., 2009); however, little is known about the effects of these editing events. Editing within 3'UTR regions could potentially alter polyadenylation signal sequences; however, this has yet to be reported. Editing can also alter miRNA binding sites thereby altering their target specificity (Borchert et al., 2009). Over 3000 of the 12,723 editing events that were analyzed in this study formed 7-mer seed matches to a subset of human miRNAs. They also found that in 200 of the ESTs that editing within a specific 13 nucleotide motif created seed matches to three otherwise unrelated miRNAs.

The formation of one or many I–U base pairs in dsRNA can have structural implications as I–U base pairing is less stable than A–U pairing, and has a lower melting temperature (Serra et al., 2004). I–U base pairs can alter the stacking of the dsRNA helix and in hyperedited transcripts these changes are likely to have a more substantial effect.

Hyperediting of dsRNA transcripts, where up to 50% of the adenosine residues in a single transcript are deaminated to inosines, has been observed *in vitro* for perfect RNA duplexes. However, hyperedited transcripts can also occur *in vivo*, an example being the voltage-dependent potassium channel (sqKv2) RNA from squid, where up to 17 adenosines are modified in a 360-base region (Patton et al., 1997). Hyperedited RNAs are cleaved by Tudor staphylococcal nuclease (Tudor-SN) protein, a component of the RNA-induced silencing complex (RISC). Tudor-SN specifically binds to inosine-containing RNA and promotes cleavage at the cleavage site (5'-IIUI-3') (Scadden, 2005).

This cleavage site is a highly preferred editing site and was shown to be generated in dsRNA *in vitro* by editing with ADAR1, ADAR2, or *Drosophila* ADAR (Scadden and O'Connell, 2005).

X. INOSINE-CONTAINING RNA REGULATES GENE EXPRESSION IN *TRANS*

An affinity column of immobilized inosine-containing dsRNA was used to purify proteins that specifically interact with inosine-containing dsRNA (Scadden, 2005). This led to the isolation of factors which are present in stress granules (SGs), including poly(A)-binding protein (PABP), and components of the eukaryotic initiation complex (eIF-4A, 4G, and 4E) (Scadden, 2007). Cytoplasmic SGs are formed when cells undergo environmental stress and proteins required for survival must be rapidly synthesized, while synthesis of other nonessential proteins is halted (reviewed in Anderson and Kedersha, 2009). Therefore, SGs are comprised of a host of RNA processing proteins and cellular mRNAs that are maintained in a translationally silent state until the stress is relieved. The observation that SG-associated proteins also interact with inosine-containing dsRNA led to the hypothesis that inosine-containing dsRNA could affect gene mRNA expression levels through the induction of SGs. This hypothesis proved to be true for several different dsRNAs containing I–U base pairs and to a lesser extent dsRNAs with G–U base pairs, including an endogenously edited miRNA-142.

Inosine-containing duplexes transfected into cells resulted in a general decrease in mRNA levels of both endogenous mRNAs and reporter mRNAs. This was accompanied by a decrease in translation which was due to inhibition of translation initiation. A model was proposed whereby editing of dsRNAs by ADAR produces inosine-containing dsRNAs which are sequestered into SGs with cellular mRNAs and translation initiation factors leading to a decrease in translation within the cell (Scadden, 2007). The decrease in translation factors results in increased traffic of cellular mRNAs to SGs that are subsequently either degraded in processing (P) bodies or held in a translationally silent state. The translation block is relieved when inosine-containing dsRNAs are degraded by Tudor-SN (Scadden, 2005), a component of the RISC complex. However, further work is required to identify the endogenous inosine RNAs that are responsible for this.

XI. EDITING AND RNA INTERFERENCE

RNA interference is the process whereby double-stranded short interfering RNA (siRNA) or microRNA (miRNA) of approximately 21–23 nucleotides (nt) can target the degradation or prevent the translation of an mRNA which contains complementarity to it (for review see Carthew and Sontheimer, 2009). In summary, siRNAs are cleaved from longer dsRNAs by Dicer in the cytoplasm and have complete complementarity to their target mRNA. However, miRNAs are processed from primary miRNAs (pri-miRNAs) that are often encoded within introns of transcripts. They form stem–loop dsRNA structures that are cleaved from surrounding the sequence by the microprocessor complex in the nucleus to generate a pre-miRNA. The pre-miRNA of approximately 75 nt is then exported from the nucleus and processed further by Dicer to remove the loop region and produce a mature miRNA (Lee *et al.*, 2002). In contrast to siRNAs, miRNAs can contain mismatches although a critical "seed" region of 4–7 nt must be complementary to their target mRNA. One strand of the mature miRNA or siRNA associates with the RISC complex (Gregory *et al.*, 2005) to direct silencing of target mRNAs, which can be mediated through cleavage and degradation of the transcript or translational inhibition. siRNA and miRNA activities depend on the formation of dsRNA during their biogenesis and for functional RNA silencing, and given that ADARs bind dsRNA it has long been hypothesized that they could modulate the RNAi pathway.

An initial survey to analyze the extent of miRNA editing revealed that 13% of human pri-miRNAs were edited and that levels of editing varied across the tissues analyzed (Blow *et al.*, 2006). This corresponded to 6% of mature miRNAs being edited and target site prediction software indicated that editing of mature miRNA could change their target specificity by redirecting them to another transcript. A more comprehensive sequence analysis of pri-miRNAs revealed numerous editing sites and led to the prediction that up to 16% pri-miRNAs could be edited in human brain (Kawahara *et al.*, 2008), which if correct would imply that editing of miRNA occurs more frequently than initially thought. However, only a subset of pri-miRNAs has been found to be edited indicating ADARs specifically target certain pri-miRNAs and they recognize more than just dsRNA structure.

The reported consequences of miRNA editing are varied; however, only one example of redirection of target specificity has been reported. In the case of the human, miRNA-376 cluster editing occurred to nearly 100% at several sites (Kawahara *et al.*, 2007b). One editing site was located within the seed region of the mature miRNA and editing at this site altered the target specificity of the miRNAs so that it targeted the transcript encoding phosphoribosyl pyrophosphate synthetase 1, an enzyme involved in the uric acid synthesis pathway. The functional consequence of this redirection of the edited miRNA was confirmed as altered levels of phosphoribosyl pyrophosphate synthetase 1 protein expression was detected in $ADAR2^{-/-}$.

However, in the cases reported so far editing of miRNA inhibits biogenesis of the miRNA by preventing cleavage by either Drosha or Dicer (Heale *et al.*, 2009a; Kawahara *et al.*, 2007b; Yang *et al.*, 2006). This inhibition of processing can be mediated through binding of ADARs to pre-mirNAs thereby preventing cleavage by Drosha and does not require catalytic activity of ADAR (Heale *et al.*, 2009a). RNA editing of pri-miR-151 resulted in inhibition both of its cleavage by Dicer and accumulation of edited pre-miR-151 (Kawahara *et al.*, 2007a). Yang and colleagues found that processing of pri-miR-142 by Drosha was inhibited by editing of adenosine residues located close to the Drosha cleavage site (Yang *et al.*, 2006). Consequently, downstream Dicer processing was also inhibited; however, edited pri-miR-142 did not accumulate to high levels due to specific degradation of inosine-containing pri-miR-142 by Tudor-SN. Therefore, editing of pri-miR-142 serves to reduce the level of mature miR-142 by targeting the edited miRNA for degradation. Higher levels of endogenous miR-142 were found in ADAR-null mice confirming that this mechanism is used *in vivo* to regulate miR-142 levels (Yang *et al.*, 2006). Therefore, ADARs can reduce the abundance of a particular miRNA by inhibiting processing either by binding to or by editing miRNAs. As often different miRNAs can target the same transcript, it is not certain that reducing the abundance of one particular miRNA will affect the overall expression of the target transcript.

As mentioned previously, UTRs are highly edited (Li *et al.*, 2009) and editing of miRNA complementary seed sequences has been found (Borchert *et al.*, 2009). This is a more specific way of altering the RNAi response to a particular transcript as editing the 3'UTR is specific for that transcript and there is no miRNA redundancy problem.

Caenorhabditis elegans contains two *adar* genes and strains lacking both *adr-1* and *adr-2* were used to investigate antagonism between ADAR and RNAi (Knight and Bass, 2002). The *adr-1/adr-2* mutant worms had strong somatic transgene-induced RNAi which was absent from wild-type worms. This is due to editing of the dsRNA in wild-type worms which prevents the dsRNA from being cleavage by Dicer and entry into the RNAi pathway. However, the RNAi pathway is different in C. *elegans* so how it antagonize ADAR may be unique to nematodes.

XII. ADAR AND DISEASE

A link has been established between decreased levels of ADAR activity and tumor progression (Cenci *et al.*, 2008; Maas *et al.*, 2001; Paz *et al.*, 2007). Decreased editing of the GluR-B and serotonin receptor transcripts was observed in malignant gliomas and a correlation was shown between tumor stage and editing level indicating a progressive loss of editing (Maas *et al.*, 2001). However, the levels of ADAR proteins remained unchanged suggesting that the change in activity was due to posttranslational regulation of ADAR.

Analysis of editing of *Alu* repeat sequences revealed significant global hypoediting in tumors from brain, prostate, lung, kidney, and testis (Paz *et al.*, 2007). In addition, they observed a reduction in RNA levels of ADAR-1, -2, and -3 and overexpression of either ADAR1 or ADAR2 in a glioblastoma cell line resulted in decreased proliferation. The authors concluded that the loss of editing enzymes was involved in the pathogenesis of cancer.

ADAR activity was also found to be reduced in pediatric astrocytomas (Cenci *et al.*, 2008). The editing level also correlated with the grade of malignancy such that high-grade tumors had the lowest levels of editing. In tumor-derived astrocytoma cell lines, the low level of editing was restored by overexpression of ADAR2, and this inhibited proliferation in these cell lines. Although there was no difference in ADAR2 expression between tumor and control samples, both ADAR1 and ADAR3 were significantly over-expressed in the astrocytomas. This unbalanced expression could down-regulate ADAR2 activity as when the expression of ADAR1 was increased this led to decreased editing of ADAR2 substrates in HEK293 cell line. ADAR1 and ADAR2 were shown to co-immunoprecipitate indicating they could form heterodimers and the authors propose that a correct balance between ADAR1 and ADAR2 is required for their specific editing activity (Cenci *et al.*, 2008).

XIII. ADAR2 PROTECTS AGAINST NEURONAL DEGENERATION FOLLOWING ISCHEMIC INSULT

Following transient global ischemia, hippocampal CA1 pyramidal neurons degenerate, while neurons in other regions of the hippocampus remain undamaged. It was found that the CA1 pyramidal neurons have an 18-fold increased permeability to calcium following ischemic insult (Liu *et al.*, 2004). It is unclear what causes the increased calcium permeability and the specific vulnerability of these neuronal subset and not others. ADAR2 has recently been implicated in recovery from forebrain ischemia. It was found that expression of *GluR-B(R)* encoding the edited isoform that is calcium impermeable in the hippocampus of adult rats *in vivo* rescued vulnerable CA1 pyramidal neurons from forebrain ischemic injury. However, expressing the unedited calcium permeable isoform GluR-B(Q) resulted in postischemic degeneration of hippocampal granule neurons that are normally insensitive to ischemia. It was hypothesized that ADAR2 editing of *GluR-B Q/R* is disrupted in CA1 pyramidal neurons following ischemic insult, and this leads to an influx of calcium which results in neuronal degeneration. In support of this hypothesis, Peng and colleagues showed that siRNA knockdown of *Adar2* in the dentate gyrus increased the sensitivity of neurons to ischemic insult (Peng *et al.*, 2006). This led to decreased editing at the *GluR-B Q/R* site, which increased AMPA receptor calcium permeability and resulted in neuronal death. This neuronal death associated with knockdown of Adar2 was rescued by expression of the edited GluR-B (R) transcript in neuronal cells which is similar to the rescue of the ADAR2$^{-/-}$ mouse (Higuchi et al., 2000). In addition, ectopic expression of ADAR2 was shown to have a protective effect. The authors also demonstrated that expression of ADAR2 is regulated by cyclic AMP response element binding protein (CREB) that is decreased in CA1 pyramidal neurons following ischemic insult (Peng et al., 2006). Taken together, these results demonstrate that editing of GluR-B transcripts at the Q/R site by ADAR2 is critical in neuronal cell death following ischemia. However, other factors such as the abundance of calcium binding proteins and calcium pumps can influence the outcome.

XIV. ADAR2 AS A "NEURONAL GATEKEEPER"

ALS is a progressive, degenerative disease characterized by a selective loss of motor neurons that leads to paralysis and death, and currently there is no effective treatment. Editing of the GluR-B transcript was investigated as motor neurons are vulnerable to AMPA receptor-mediated neurotoxicity similar to CA1 pyramidal neurons (reviewed in Kwak and Kawahara, 2005). Editing at the Q/R position of the GluR-B transcript was analyzed in single motor neurons isolated by laser-capture microdissection from the spinal cord of patients with ALS and controls. Editing efficiency was 100% in control motor neurons; however, motor neurons form ALS patients showed a range of 0–100% editing, with an average of approximately 60% (Kawahara et al., 2004). The decrease in editing was specific to the motor neurons as editing in Purkinje cells from these patients was 100%, and it was also specific to sporadic ALS patients. However, loss of GluR-B Q/R site editing is not observed in motor neurons from patients with familial ALS (fALS) or rat models of fALS that contain mutations in the SOD1 gene, indicating despite presenting with similar symptoms there are multiple causes for the pathogenesis of ALS (Kawahara et al., 2006). This work on ALS has led to the hypothesis that GluR-B acts as a "neuronal gatekeeper" in the same manner as the model of ischemic injury, such that decreased editing of the GluR-B Q/R site renders the neuron permeable to calcium which leads to cell death (Buckingham et al., 2008).

XV. ADAR AND ALTERNATIVE SPLICING

Many mRNAs are alternatively spliced to generate multiple mature mRNAs which can have distinct functions from one pre-mRNA (reviewed in Black, 2003). As editing occurs prior to splicing and the editing sites are often close to splice sites, one obvious question is whether editing influences splicing and in particular alternative splicing. As described earlier, ADAR2 edits its own transcript within an intron creating an alternative splice acceptor site which leads to a truncated protein lacking both dsRBDs and the deaminase domain (Rueter et al., 1999). This generates a negative feedback loop to regulate the amount of active ADAR2, and alternative splicing directly correlates with the activity of the ADAR2 enzyme (Maas et al., 2001).

The Adar2$^{-/-}$ mice have an elevated level of incorrectly processed GluR-B transcripts that retain intron 11 (Higuchi et al., 2000). Intron 11 contains the ECS region that is required for editing at the GluR-B Q/R site, leading to the hypothesis that the processes of editing and splicing are linked. In addition to editing at the Q/R site, there are "hotspot" regions within intron 11 at the +60 site (hotspot 1) and the +262/263/264 sites (hotspot 2) that are also edited (Higuchi et al., 1993). Analysis of editing and splicing of this region of GluR-B transcript in a cell culture assay revealed that editing at both the Q/R site and intronic hotspot 2 is required for efficient splicing of intron 11 (Schoft et al., 2007). This may account for the observed 100% editing of mature GluR-B transcript in vivo, as unedited transcript is not processed correctly. The intronic hotspot 2 editing event could potentially alter a splice-repressor signal as mutations which disrupted the duplex structure of this region did not increase splicing efficiency (Schoft et al., 2007).

The GluR-B R/G editing site is 2 bp from the 5′ splice donor site of the mutually exclusive FLIP/FLOP exons. Unlike the Q/R site, editing at the R/G site reduced the efficiency of splicing downstream of the editing event; however, the decreased splicing efficiency correlated with an increase in splicing fidelity (Schoft et al., 2007). When an "uneditable" construct with a mutated ECS was used, there was skipping of both downstream FLIP/FLOP exons and this generated a transcript containing a premature stop codon (Schoft et al., 2007). These results demonstrate that editing at the R/G site facilitates correct splicing of the downstream FLIP/FLOP exons. Inhibition of splicing by ADAR2 was also observed with in vitro splicing reactions on a GluR-B transcript containing the R/G site, although the inhibition was partially alleviated by addition of RNA helicase A indicating that inhibition of splicing was in part due to the duplex structure (Bratt and Ohman, 2003). A construct containing an inosine residue at the R/G site that mimics constitutive editing also showed inhibition of downstream splicing. Therefore, the inhibitory effect is also due to the presence of inosine rather than solely due to ADAR2 binding and this may be due to the unedited sequence being closer to the consensus splice site (Schoft et al., 2007). Also, the edited R/G site shows greater similarity to the consensus binding site for hnRNP A1 and this could act as a splicing silencer; however, further work is required to prove this. In adult mouse brain, no correlation was found between editing and choice of the mutually exclusive FLIP/FLOP exons so editing and alternative splicing are not linked at this site (Schoft et al., 2007).

There are five known editing sites in the transcript encoding the human serotonin (5-HT_{2C}) receptor which are situated within a dsRNA stem–loop formed between exon 5 and an ECS in intron 5 (Burns et al., 1997). This region also undergoes alternative splicing with one canonical and two alternative $5'$ splice sites. Examination of editing and splicing in this region of the serotonin receptor transcript revealed a link between increased numbers of sites edited and use of the canonical $5'$ splice site (Flomen et al., 2004). As the alternative $5'$ splice sites produce truncated inactive protein, this suggests a link between editing and production of functional serotonin receptor protein. This effect is likely due to the disruption of the stem–loop structure which allows access of splicing factors to the canonical $5'$ splice site; however, editing could also possibly change splicing regulatory signals. A study of editing in malignant human gliomas which have hypoediting revealed an increase in the serotonin 5HT_{2C} receptor transcripts that utilize the alternative $5'$ splice site and therefore produce truncated inactive protein (Maas et al., 2001).

Characterization of the human protein tyrosine phosphatase *PTPN6* gene and its role in acute myeloid leukemia revealed increased expression of an isoform that retained a 251-bp intron (Beghini et al., 2000). Sequence analysis of this isoform demonstrated that multiple A-to-G mutations were present in the retained intron: one of these occurred 27 bp upstream of the $3'$ splice site and converted the branch point adenosine to an inosine residue. Loss of the branch point adenosine led to retention of the intron, which was confirmed with an *in vitro* splicing reaction. Furthermore, the retained intron contained an in-frame stop codon which would generate a truncated protein lacking the phosphatase domain. The level of the abnormally spliced isoform was significantly higher in patients at the time of diagnosis than when they had entered remission suggesting a correlation with the disease state.

XVI. COORDINATION OF EDITING AND ALTERNATIVE SPLICING

The carboxy-terminal domain (CTD) of RNAP II is involved in co-coordinating transcription with RNA processing events such as splicing, $3'$-end formation and $5'$-capping (McCracken et al., 1997a,b). As editing has to occur prior to splicing studies were performed to determine if the CTD of RNAP II could also coordinate splicing and editing. A truncated RNA pol II lacking the CTD region revealed that the CTD is required for efficient editing of the rat *ADAR2* transcript at the autoediting site (Laurencikiene et al., 2006). Further studies on the *GluR-B* transcript revealed that the RNA pol II CTD enhanced editing by preventing premature splicing that would remove the intron required for editing (Ryman et al., 2007). At the Q/R site which is located 24 nucleotides upstream of a splice donor site the effect was similar in that editing enhanced efficient splicing. These results indicate that the CTD of RNA pol II is required to coordinate editing and splicing, possibly through delaying the splicing event until editing has occurred thereby allowing these processes to occur sequentially.

XVII. CONCLUSION

Despite it being approximately 20 years since ADARs were first identified, many questions still remain unanswered. One of the main challenges is to identify the function of ADAR1. Is it an RNA-editing enzyme and if so what are its substrates? Recent publications suggest that instead it may play an important role in the interferon response particularly during embryogenesis in the liver. What is the function of ADAR3 and TENR both of which are evolutionarily conserved but catalytically inactive? Another question is what is the relationship between ADARs and RNA interference? Is there an important miRNA, similar to the *GluR-B* Q/R site that requires editing or do ADARs instead edit the miRNA target sequences in the 5'- and 3'-UTRs? The question that still remains a challenge to answer is why is there editing of the Q/R site in *GluR-B*? Why is it not genomically encoded if editing at this position is critical for the functioning of the AMPA receptor and failure of editing can lead to neuronal cell death? These and other mysteries remain to be addressed.

Acknowledgments

Some of the material for this chapter was taken from the PhD thesis of Marion Hogg. We would like to thank Craig Nicols for his help with figures. M. A. O. C. is funded by the Medical Research Council Grant U.1275.01.005.00001.01.

References

Anderson, P., and Kedersha, N. (2009). RNA granules: Post-transcriptional and epigenetic modulators of gene expression. *Nat. Rev. Mol. Cell Biol.* **10**(6), 430–436.

Athanasiadis, A., Rich, A., and Maas, S. (2004). Widespread A-to-I RNA editing of Alu-containing mRNAs in the human transcriptome. *PLoS Biol.* **2**(12), e391.

Bass, B. L., and Weintraub, H. (1987). A developmental regulated activity that unwinds RNA duplexes. *Cell* **48,** 607–613.

Bass, B. L., and Weintraub, H. (1988). An unwinding activity that covalently modifies its double-strand RNA substrate. *Cell* **55,** 1089–1098.

Beghini, A., Ripamonti, C. B., Peterlongo, P., Roversi, G., Cairoli, R., Morra, E., and Larizza, L. (2000). RNA hyperediting and alternative splicing of hematopoietic cell phosphatase (PTPN6) gene in acute myeloid leukemia. *Hum. Mol. Genet.* **9**(15), 2297–2304.

Benne, R., Van den Burg, J., Brakenhoff, J., Sloof, P., Van Boom, J. H., and Tromp, M. C. (1986). Major transcript of the frameshifted coxII gene from trypanosome mitochondria contains four nucleotides that are not encoded in the DNA. *Cell* **46**(6), 819–826.

Bernard, A., and Khrestchatisky, M. (1994). Assessing the extent of RNA editing in the TMII region of GluR5 and GluR6 kainate receptors during rat brain development. *J. Neurochem.* **62,** 2057–2060.

Betts, L., Xiang, S., Short, S. A., Wolfenden, R., and Carter, C. W., Jr. (1994). Cytidine deaminase. The 2.3 Å crystal structure of an enzyme: transition-state analog complex. *J. Mol. Biol.* **235,** 635–656.

Bhalla, T., Rosenthal, J. J., Holmgren, M., and Reenan, R. (2004). Control of human potassium channel inactivation by editing of a small mRNA hairpin. *Nat. Struct. Mol. Biol.* **11**(10), 950–956.

Black, D. L. (2003). Mechanisms of alternative pre-messenger RNA splicing. *Annu. Rev. Biochem.* **72,** 291–336.

Blow, M., Futreal, P. A., Wooster, R., and Stratton, M. R. (2004). A survey of RNA editing in human brain. *Genome Res.* **14**(12), 2379–2387.

Blow, M. J., Grocock, R. J., van Dongen, S., Enright, A. J., Dicks, E., Futreal, P. A., Wooster, R., and Stratton, M. R. (2006). RNA editing of human microRNAs. *Genome Biol.* **7**(4), R27.

Borchert, G. M., Gilmore, B. L., Spengler, R. M., Xing, Y., Lanier, W., Bhattacharya, D., and Davidson, B. L. (2009). Adenosine deamination in human transcripts generates novel microRNA binding sites. *Hum. Mol. Genet.* **18**(24), 4801–4807.

Buckingham, S. D., Kwak, S., Jones, A. K., Blackshaw, S. E., and Sattelle, D. B. (2008). Edited GluR2, a gatekeeper for motor neuron survival? *Bioessays* **30**(11–12), 1185–1192.

Burns, C. M., Chu, H., Rueter, S. M., Hutchinson, L. K., Canton, H., Sanders-Bush, E., and Emeson, R. B. (1997). Regulation of serotonin-2C receptor G-protein coupling by RNA editing. *Nature* **387,** 303–308.

Bratt, E., and Ohman, M. (2003). Coordination of editing and splicing of glutamate receptor pre-mRNA. *Rna* **9**(3), 309–318.

Carthew, R. W., and Sontheimer, E. J. (2009). Origins and Mechanisms of miRNAs and siRNAs. *Cell* **136**(4), 642–655.

Cenci, C., Barzotti, R., Galeano, F., Corbelli, S., Rota, R., Massimi, L., Di Rocco, C., O'Connell, M. A., and Gallo, A. (2008). Down-regulation of RNA editing in pediatric astrocytomas: ADAR2 editing activity inhibits cell migration and proliferation. *J. Biol. Chem.* **283** (11), 7251–7260.

Chen, C. X., Cho, D. S., Wang, Q., Lai, F., Carter, K. C., and Nishikura, K. (2000). A third member of the RNA-specific adenosine deaminase gene family, ADAR3, contains both single- and double-stranded RNA binding domains. *RNA* **6**(5), 755–767.

Chilibeck, K. A., Wu, T., Liang, C., Schellenberg, M. J., Gesner, E. M., Lynch, J. M., and MacMillan, A. M. (2006). FRET analysis of in vivo dimerization by RNA-editing enzymes. *J. Biol. Chem.* **281**(24), 16530–16535.

Cho, D. S., Yang, W., Lee, J. T., Shiekhattar, R., Murray, J. M., and Nishikura, K. (2003). Requirement of dimerization for RNA editing activity of adenosine deaminases acting on RNA. *J. Biol. Chem.* **278**(19), 17093–17102.

Clutterbuck, D. R., Leroy, A., O'Connell, M. A., and Semple, C. A. (2005). A bioinformatic screen for novel A-I RNA editing sites reveals recoding editing in BC10. *Bioinformatics* **21**(11), 2590–2595.

Connolly, C. M., Dearth, A. T., and Braun, R. E. (2005). Disruption of murine Tenr results in teratospermia and male infertility. *Dev. Biol.* **278**(1), 13–21.

Dawson, T. R., Sansam, C. L., and Emeson, R. B. (2004). Structure and sequence determinants required for the editing of ADAR2 substrates. *J. Biol. Chem.* **279**(6), 4941–4951.

Desterro, J. M., Keegan, L. P., Lafarga, M., Berciano, M. T., O'Connell, M., and Carmo-Fonseca, M. (2003). Dynamic association of RNA-editing enzymes with the nucleolus. *J. Cell Sci.* **116**(Pt. 9), 1805–1818.

Fierro-Monti, I., and Mathews, M. B. (2000). Proteins binding to duplexed RNA: One motif, multiple functions. *Trends Biochem. Sci.* **25**(5), 241–246.

Flomen, R., Knight, J., Sham, P., Kerwin, R., and Makoff, A. (2004). Evidence that RNA editing modulates splice site selection in the 5-HT2C receptor gene. *Nucleic Acids Res.* **32**(7), 2113–2122.

Fritz, J., Strehblow, A., Taschner, A., Schopoff, S., Pasierbek, P., and Jantsch, M. F. (2009). RNA-regulated interaction of transportin-1 and exportin-5 with the double-stranded RNA-binding domain regulates nucleocytoplasmic shuttling of ADAR1. *Mol. Cell. Biol.* **29**(6), 1487–1497.

Gallo, A., Keegan, L. P., Ring, G. M., and O'Connell, M. A. (2003). An ADAR that edits transcripts encoding ion channel subunits functions as a dimer. *EMBO J.* **22**(13), 3421–3430.

Gerber, A. P., and Keller, W. (1999). An adenosine deaminase that generates inosine at the wobble position of tRNAs. *Science* **286**(5442), 1146–1149.

Gerber, A. P., and Keller, W. (2001). RNA editing by base deamination: More enzymes, more targets, new mysteries. *Trends Biochem. Sci.* **26**(6), 376–384.

Gerber, A., O'Connell, M. A., and Keller, W. (1997). Two forms of human double-stranded RNA-specific editase 1 (hRED1) generated by the insertion of an Alu cassette. *RNA* **3**, 453–463.

Gommans, W. M., Tatalias, N. E., Sie, C. P., Dupuis, D., Vendetti, N., Smith, L., Kaushal, R., and Maas, S. (2008). Screening of human SNP database identifies recoding sites of A-to-I RNA editing. *Rna* **14**(10), 2074–2085.

Greger, I. H., Khatri, L., and Ziff, E. B. (2002). RNA editing at arg607 controls AMPA receptor exit from the endoplasmic reticulum. *Neuron* **34**(5), 759–772.

Greger, I. H., Khatri, L., Kong, X., and Ziff, E. B. (2003). AMPA receptor tetramerization is mediated by q/r editing. *Neuron* **40**(4), 763–774.

Gregory, R. I., Chendrimada, T. P., Cooch, N., and Shiekhattar, R. (2005). Human RISC couples microRNA biogenesis and posttranscriptional gene silencing. *Cell* **123**(4), 631–640.

Gurney, M. E., Pu, H., Chiu, A. Y., Dal Canto, M. C., Polchow, C. Y., Alexander, D. D., Caliendo, J., Hentati, A., Kwon, Y. W., Deng, H. X., *et al.* (1994). Motor neuron degeneration in mice that express a human Cu, Zn superoxide dismutase mutation. *Science* **264**(5166), 1772–1775.

Hartner, J. C., Schmittwolf, C., Kispert, A., Muller, A. M., Higuchi, M., and Seeburg, P. H. (2004). Liver disintegration in the mouse embryo caused by deficiency in the RNA-editing enzyme ADAR1. *J. Biol. Chem.* **279**(6), 4894–4902.

Hartner, J. C., Walkley, C. R., Lu, J., and Orkin, S. H. (2009). ADAR1 is essential for the maintenance of hematopoiesis and suppression of interferon signaling. *Nat. Immunol.* **10**(1), 109–115.

Heale, B. S., Keegan, L. P., McGurk, L., Michlewski, G., Brindle, J., Stanton, C. M., Caceres, J. F., and O'Connell, M. A. (2009a). Editing independent effects of ADARs on the miRNA/siRNA pathways. *EMBO J.* **28**(20), 3145–3156.

Heale, B. S., Keegan, L. P., and O'Connell, M. A. (2009b). ADARs have effects beyond RNA editing. *Cell Cycle* **8**(24), 4011–4012.

Herbert, A., and Rich, A. (2001). The role of binding domains for dsRNA and Z-DNA in the in vivo editing of minimal substrates by ADAR1. *Proc. Natl. Acad. Sci. USA* **98**(21), 12132–12137.

Higuchi, M., Single, F. N., Köhler, M., Sommer, B., Sprengel, R., and Seeburg, P. H. (1993). RNA editing of AMPA receptor subunit GluR-B: A base-paired intron-exon structure determines position and efficiency. *Cell* **75**, 1361–1370.

Higuchi, M., Maas, S., Single, F. N., Hartner, J., Rozov, A., Burnashev, N., Feldmeyer, D., Sprengel, R., and Seeburg, P. H. (2000). Point mutation in an AMPA receptor gene rescues lethality in mice deficient in the RNA-editing enzyme ADAR2. *Nature* **406**(6791), 78–81.

Honjo, T., Kinoshita, K., and Muramatsu, M. (2002). Molecular mechanism of class switch recombination: Linkage with somatic hypermutation. *Annu. Rev. Immunol.* **20**, 165–196.

Hough, R. F., and Bass, B. L. (1994). Purification of the *Xenopus laevis* dsRNA adenosine deaminase. *J. Biol. Chem.* **269**, 9933–9939.

Jaikaran, D. C., Collins, C. H., and MacMillan, A. M. (2002). Adenosine to inosine editing by ADAR2 requires formation of a ternary complex on the GluR-B R/G site. *J. Biol. Chem.* **277**(40), 37624–37629.

Jiang, Q. X., Wang, D. N., and MacKinnon, R. (2004). Electron microscopic analysis of KvAP voltage-dependent K+ channels in an open conformation. *Nature* **430**(7001), 806–810.

Kawahara, Y., Ito, K., Sun, H., Aizawa, H., Kanazawa, I., and Kwak, S. (2004). Glutamate receptors: RNA editing and death of motor neurons. *Nature* **427**(6977), 801.

Kawahara, Y., Sun, H., Ito, K., Hideyama, T., Aoki, M., Sobue, G., Tsuji, S., and Kwak, S. (2006). Underediting of GluR2 mRNA, a neuronal death inducing molecular change in sporadic ALS, does not occur in motor neurons in ALS1 or SBMA. *Neurosci. Res.* **54**(1), 11–14.

Kawahara, Y., Zinshteyn, B., Chendrimada, T. P., Shiekhattar, R., and Nishikura, K. (2007a). RNA editing of the microRNA-151 precursor blocks cleavage by the Dicer-TRBP complex. *EMBO Rep.* **8**(8), 763–769.

Kawahara, Y., Zinshteyn, B., Sethupathy, P., Iizasa, H., Hatzigeorgiou, A. G., and Nishikura, K. (2007b). Redirection of silencing targets by adenosine-to-inosine editing of miRNAs. *Science* **315** (5815), 1137–1140.

Kawahara, Y., Megraw, M., Kreider, E., Iizasa, H., Valente, L., Hatzigeorgiou, A. G., and Nishikura, K. (2008). Frequency and fate of microRNA editing in human brain. *Nucleic Acids Res.* **36**(16), 5270–5280.

Kawakubo, K., and Samuel, C. E. (2000). Human RNA-specific adenosine deaminase (ADAR1) gene specifies transcripts that initiate from a constitutively active alternative promoter. *Gene* **258** (1–2), 165–172.

Keegan, L. P., Leroy, A., Sproul, D., and O'Connell, M. A. (2004). Adenosine deaminases acting on RNA (ADARs): RNA-editing enzymes. *Genome Biol.* **5**(2), 209.

Kim, U., Garner, T. L., Sanford, T., Speicher, D., Murray, J. M., and Nishikura, K. (1994). Purification and characterization of double-stranded RNA adenosine deaminase from bovine nuclear extracts. *J. Biol. Chem.* **269**, 13480–13489.

Kim, D. D., Kim, T. T., Walsh, T., Kobayashi, Y., Matise, T. C., Buyske, S., and Gabriel, A. (2004). Widespread RNA editing of embedded alu elements in the human transcriptome. *Genome Res.* **14** (9), 1719–1725.

Knight, S. W., and Bass, B. L. (2002). The role of RNA editing by ADARs in RNAi. *Mol. Cell* **10**(4), 809–817.

Köhler, M., Burnashev, N., Sakmann, B., and Seeburg, P. H. (1993). Determinants of Ca^{2+} permeability in both TM1 and TM2 of high affinity kainate receptor channels: Diversity by RNA editing. *Neuron* **10**, 491–500.

Kondo, T., Suzuki, T., Ito, S., Kono, M., Negoro, T., and Tomita, Y. (2008). Dyschromatosis symmetrica hereditaria associated with neurological disorders. *J. Dermatol.* **35**(10), 662–666.

Kuner, R., Groom, A. J., Bresink, I., Kornau, H. C., Stefovska, V., Muller, G., Hartmann, B., Tschauner, K., Waibel, S., Ludolph, A. C., Ikonomidou, C., Seeburg, P. H., *et al.* (2005). Late-onset motoneuron disease caused by a functionally modified AMPA receptor subunit. *Proc. Natl. Acad. Sci. USA* **102**(16), 5826–5831.

Kwak, S., and Kawahara, Y. (2005). Deficient RNA editing of GluR2 and neuronal death in amyotropic lateral sclerosis. *J. Mol. Med.* **83**(2), 110–120.

Lai, F., Chen, C.-X., Carter, K. C., and Nishikura, K. (1997). Editing of glutamate receptor B subunit ion channel RNAs by four alternatively spliced DRADA2 double-stranded RNA adenosine deaminases. *Mol. Cell. Biol.* **17**, 2413–2424.

Laurencikiene, J., Kallman, A. M., Fong, N., Bentley, D. L., and Ohman, M. (2006). RNA editing and alternative splicing: The importance of co-transcriptional coordination. *EMBO Rep.* **7**(3), 303–307.

Lee, Y., Jeon, K., Lee, J. T., Kim, S., and Kim, V. N. (2002). microRNA maturation: stepwise processing and subcellular localization. *EMBO J.* **21**(17), 4663–4670.

Lehmann, K. A., and Bass, B. L. (2000). Double-stranded RNA adenosine deaminases ADAR1 and ADAR2 have overlapping specificities. *Biochemistry* **39**(42), 12875–12884.

Levanon, E. Y., Eisenberg, E., Yelin, R., Nemzer, S., Hallegger, M., Shemesh, R., Fligelman, Z. Y., Shoshan, A., Pollock, S. R., Sztybel, D., Olshansky, M., Rechavi, G., et al. (2004). Systematic identification of abundant A-to-I editing sites in the human transcriptome. *Nat. Biotechnol.* **22**(8), 1001–1005.

Levanon, E. Y., Hallegger, M., Kinar, Y., Shemesh, R., Djinovic-Carugo, K., Rechavi, G., Jantsch, M. F., and Eisenberg, E. (2005). Evolutionarily conserved human targets of adenosine to inosine RNA editing. *Nucleic Acids Res.* **33**(4), 1162–1168.

Li, J. B., Levanon, E. Y., Yoon, J. K., Aach, J., Xie, B., Leproust, E., Zhang, K., Gao, Y., and Church, G. M. (2009). Genome-wide identification of human RNA editing sites by parallel DNA capturing and sequencing. *Science* **324**(5931), 1210–1213.

Liu, Y., Lei, M., and Samuel, C. E. (2000). Chimeric double-stranded RNA-specific adenosine deaminase ADAR1 proteins reveal functional selectivity of double-stranded RNA-binding domains from ADAR1 and protein kinase PKR. *Proc. Natl. Acad. Sci. USA* **97**(23), 12541–12546.

Liu, S., Lau, L., Wei, J., Zhu, D., Zou, S., Sun, H. S., Fu, Y., Liu, F., and Lu, Y. (2004). Expression of Ca(2+)-permeable AMPA receptor channels primes cell death in transient forebrain ischemia. *Neuron* **43**(1), 43–55.

Liu, Q., Jiang, L., Liu, W. L., Kang, X. J., Ao, Y., Sun, M., Luo, Y., Song, Y., Lo, W. H., and Zhang, X. (2006). Two novel mutations and evidence for haploinsufficiency of the ADAR gene in dyschromatosis symmetrica hereditaria. *Br. J. Dermatol.* **154**(4), 636–642.

Lomeli, H., Mosbacher, J., Melcher, T., Höger, T., Geiger, J. R., Kuner, T., Monyer, H., Higuchi, M., Bach, A., and Seeburg, P. H. (1994). Control of kinetic properties of AMPA receptor channels by nuclear RNA editing. *Science* **266,** 1709–1713.

Maas, S., Patt, S., Schrey, M., and Rich, A. (2001). Underediting of glutamate receptor GluR-B mRNA in malignant gliomas. *Proc. Natl. Acad. Sci. USA* **98**(25), 14687–14692.

Macbeth, M. R., Lingam, A. T., and Bass, B. L. (2004). Evidence for auto-inhibition by the N terminus of hADAR2 and activation by dsRNA binding. *RNA* **10**(10), 1563–1571.

Macbeth, M. R., Schubert, H. L., Vandemark, A. P., Lingam, A. T., Hill, C. P., and Bass, B. L. (2005). Inositol hexakisphosphate is bound in the ADAR2 core and required for RNA editing. *Science* **309**(5740), 1534–1539.

McCracken, S., Fong, N., Rosonina, E., Yankulov, K., Brothers, G., Siderovski, D., Hessel, A., Foster, S., Shuman, S., and Bentley, D. L. (1997a). 5′-Capping enzymes are targeted to pre-mRNA by binding to the phosphorylated carboxy-terminal domain of RNA polymerase II. *Genes Dev.* **11**(24), 3306–3318.

McCracken, S., Fong, N., Yankulov, K., Ballantyne, S., Pan, G., Greenblatt, J., Patterson, S. D., Wickens, M., and Bentley, D. L. (1997b). The C-terminal domain of RNA polymerase II couples mRNA processing to transcription. *Nature* **385**(6614), 357–361.

Melcher, T., Maas, S., Herb, A., Sprengel, R., Higuchi, M., and Seeburg, P. H. (1996). RED2, a brain specific member of the RNA-specific adenosine deaminase family. *J. Biol. Chem.* **271**(50), 31795–31798.

Miyamura, Y., Suzuki, T., Kono, M., Inagaki, K., Ito, S., Suzuki, N., and Tomita, Y. (2003). Mutations of the RNA-specific adenosine deaminase gene (DSRAD) are involved in dyschromatosis symmetrica hereditaria. *Am. J. Hum. Genet.* **73**(3), 693–699.

Nutt, S. L., and Kamboj, R. K. (1994). RNA editing of human kainate receptor subunits. *Neuroreport* **5**(18), 2625–2629.

O'Connell, M. A., and Keller, W. (1994). Purification and properties of double-stranded RNA-specific adenosine deaminase from calf thymus. *Proc. Natl. Acad. Sci. USA* **91,** 10596–10600.

Ohlson, J., Pedersen, J. S., Haussler, D., and Ohman, M. (2007). Editing modifies the GABA(A) receptor subunit alpha3. *RNA* **13**(5), 698–703.

Palavicini, J. P., O'Connell, M. A., and Rosenthal, J. J. (2009). An extra double-stranded RNA binding domain confers high activity to a squid RNA editing enzyme. *RNA* **15**(6), 1208–1218.

Palladino, M. J., Keegan, L. P., O'Connell, M. A., and Reenan, R. A. (2000). dADAR, a *Drosophila* double-stranded RNA-specific adenosine deaminase is highly developmentally regulated and is itself a target for RNA editing. *RNA* **6**, 1004–1018.

Paschen, W., and Djuricic, B. (1994). Extent of RNA editing of glutamate receptor subunit GluR5 in different brain regions of the rat. *Cell. Mol. Neurobiol.* **14**(3), 259–270.

Patterson, J. B., and Samuel, C. E. (1995). Expression and regulation by interferon of a double-stranded-RNA-specific adenosine deaminase from human cells: Evidence for two forms of the deaminase. *Mol. Cell. Biol.* **15**(10), 5376–5388.

Patton, D. E., Silva, T., and Bezanilla, F. (1997). RNA editing generates a diverse array of transcripts encoding squid Kv2 K+ channels with altered functional properties. *Neuron* **19**(3), 711–722.

Paul, M., and Bass, B. L. (1998). Inosine exists in mRNA at tissue-specific levels and is most abundant in brain mRNA. *EMBO J.* **17**, 1120–1127.

Paz, N., Levanon, E. Y., Amariglio, N., Heimberger, A. B., Ram, Z., Constantini, S., Barbash, Z. S., Adamsky, K., Safran, M., Hirschberg, A., Krupsky, M., Ben-Dov, I., et al. (2007). Altered adenosine-to-inosine RNA editing in human cancer. *Genome Res.* **17**(11), 1586–1595.

Peng, P. L., Zhong, X., Tu, W., Soundarapandian, M. M., Molner, P., Zhu, D., Lau, L., Liu, S., Liu, F., and Lu, Y. (2006). ADAR2-dependent RNA editing of AMPA receptor subunit GluR2 determines vulnerability of neurons in forebrain ischemia. *Neuron* **49**(5), 719–733.

Polson, A. G., Bass, B. L., and Casey, J. L. (1996). RNA editing of hepatitis delta antigenome by dsRNA-adenosine deaminase. *Nature* **380**, 454–456.

Poulsen, H., Nilsson, J., Damgaard, C. K., Egebjerg, J., and Kjems, J. (2001). CRM1 mediates the export of ADAR1 through a nuclear export signal within the Z-DNA binding domain. *Mol. Cell. Biol.* **21**(22), 7862–7871.

Poulsen, H., Jorgensen, R., Heding, A., Nielsen, F. C., Bonven, B., and Egebjerg, J. (2006). Dimerization of ADAR2 is mediated by the double-stranded RNA binding domain. *RNA* **12**(7), 1350–1360.

Ramos, A., Grunert, S., Adams, J., Micklem, D. R., Proctor, M. R., Freund, S., Bycroft, M., St. Johnston, D., and Varani, G. (2000). RNA recognition by a Staufen double-stranded RNA-binding domain. *EMBO J.* **19**(5), 997–1009.

Roberts, R. J., and Cheng, X. (1998). Base flipping. *Annu. Rev. Biochem.* **67**, 181–198.

Rueter, S. M., Dawson, T. R., and Emeson, R. B. (1999). Regulation of alternative splicing by RNA editing. *Nature* **399**(6731), 75–80.

Ryman, K., Fong, N., Bratt, E., Bentley, D. L., and Ohman, M. (2007). The C-terminal domain of RNA Pol II helps ensure that editing precedes splicing of the GluR-B transcript. *RNA* **13**(7), 1071–1078.

Sansam, C. L., Wells, K. S., and Emeson, R. B. (2003). Modulation of RNA editing by functional nucleolar sequestration of ADAR2. *Proc. Natl. Acad. Sci. USA* **100**(24), 14018–14023.

Scadden, A. D. (2005). The RISC subunit Tudor-SN binds to hyper-edited double-stranded RNA and promotes its cleavage. *Nat. Struct. Mol. Biol.* **12**(6), 489–496.

Scadden, A. D. (2007). Inosine-containing dsRNA binds a stress-granule-like complex and down-regulates gene expression in trans. *Mol. Cell* **28**(3), 491–500.

Scadden, A. D., and O'Connell, M. A. (2005). Cleavage of dsRNAs hyper-edited by ADARs occurs at preferred editing sites. *Nucleic Acids Res.* **33**(18), 5954–5964.

Schoft, V. K., Schopoff, S., and Jantsch, M. F. (2007). Regulation of glutamate receptor B pre-mRNA splicing by RNA editing. *Nucleic Acids Res.* **35**(11), 3723–3732.

Schumacher, J. M., Lee, K., Edelhoff, S., and Braun, R. E. (1995). Distribution of Tenr, an RNA-binding protein, in a lattice-like network within the spermatid nucleus in the mouse. *Biol. Reprod.* **52**(6), 1274–1283.

Serra, M. J., Smolter, P. E., and Westhof, E. (2004). Pronounced instability of tandem IU base pairs in RNA. *Nucleic Acids Res.* **32**(5), 1824–1828.

Sharma, P. M., Bowman, M., Madden, S. L., Rauscher, F. J., and Sukumar, S. (1994). RNA editing in the Wilm's tumor susceptibility gene WT1. *Genes Dev.* **8**, 720–731.

Singh, M., Kesterson, R. A., Jacobs, M. M., Joers, J. M., Gore, J. C., and Emeson, R. B. (2007). Hyperphagia-mediated obesity in transgenic mice misexpressing the RNA-editing enzyme ADAR2. *J. Biol. Chem.* **282**(31), 22448–22459.

Sommer, B., Kohler, M., Sprengel, R., and Seeburg, P. H. (1991). RNA editing in brain controls a determinant of ion flow in glutamate-gated channels. *Cell* **67**(1), 11–19.

Stefl, R., Xu, M., Skrisovska, L., Emeson, R. B., and Allain, F. H. (2006). Structure and specific RNA binding of ADAR2 double-stranded RNA binding motifs. *Structure* **14**(2), 345–355.

Stephens, O. M., Haudenschild, B. L., and Beal, P. A. (2004). The binding selectivity of ADAR2's dsRBMs contributes to RNA-editing selectivity. *Chem. Biol.* **11**(9), 1239–1250.

Strehblow, A., Hallegger, M., and Jantsch, M. F. (2002). Nucleocytoplasmic distribution of human RNA-editing enzyme ADAR1 is modulated by double-stranded RNA-binding domains, a leucine-rich export signal, and a putative dimerization domain. *Mol. Biol. Cell* **13**(11), 3822–3835.

Sun, W., Ferrer-Montiel, A. V., and Montal, M. (1994). Primary structure and functional expression of the AMPA/kainate receptor subunit 2 from human brain. *Neuroreport* **5**(4), 441–444.

Tojo, K., Sekijima, Y., Suzuki, T., Suzuki, N., Tomita, Y., Yoshida, K., Hashimoto, T., and Ikeda, S. (2006). Dystonia, mental deterioration, and dyschromatosis symmetrica hereditaria in a family with ADAR1 mutation. *Mov. Disord.* **21**(9), 1510–1513.

Valente, L., and Nishikura, K. (2007). RNA binding-independent dimerization of adenosine deaminases acting on RNA and dominant negative effects of nonfunctional subunits on dimer functions. *J. Biol. Chem.* **282**(22), 16054–16061.

Vitali, P., and Scadden, A. D. (2010). Double-stranded RNAs containing multiple IU pairs are sufficient to suppress interferon induction and apoptosis. *Nat. Struct. Mol. Biol.* **17**(9), 1043–1050.

Wang, Q., Miyakoda, M., Yang, W., Khillan, J., Stachura, D. L., Weiss, M. J., and Nishikura, K. (2004). Stress-induced apoptosis associated with null mutation of ADAR1 RNA editing deaminase gene. *J. Biol. Chem.* **279**(6), 4952–4961.

Wilding, T. J., Zhou, Y., and Huettner, J. E. (2005). Q/R site editing controls kainate receptor inhibition by membrane fatty acids. *J. Neurosci.* **25**(41), 9470–9478.

Wilson, D. K., Rudolph, F. B., and Quiocho, F. A. (1991). Atomic structure of adenosine deaminase complexed with a transition-state analog: Understanding catalysis and immunodeficiency mutations. *Science* **252**, 1278–1284.

Wong, S. K., Sato, S., and Lazinski, D. W. (2001). Substrate recognition by ADAR1 and ADAR2. *RNA* **7**(6), 846–858.

XuFeng, R., Boyer, M. J., Shen, H., Li, Y., Yu, H., Gao, Y., Yang, Q., Wang, Q., and Cheng, T. (2009). ADAR1 is required for hematopoietic progenitor cell survival via RNA editing. *Proc. Natl. Acad. Sci. USA* **106**(42), 17763–17768.

Yang, W., Chendrimada, T. P., Wang, Q., Higuchi, M., Seeburg, P. H., Shiekhattar, R., and Nishikura, K. (2006). Modulation of microRNA processing and expression through RNA editing by ADAR deaminases. *Nat. Struct. Mol. Biol.* **13**(1), 13–21.

Yi-Brunozzi, H. Y., Stephens, O. M., and Beal, P. A. (2001). Conformational changes that occur during an RNA editing adenosine deamination reaction. *J. Biol. Chem.* **7**, 846–858.

Zhang, X. J., He, P. P., Li, M., He, C. D., Yan, K. L., Cui, Y., Yang, S., Zhang, K. Y., Gao, M., Chen, J. J., Li, C. R., Jin, L., *et al.* (2004). Seven novel mutations of the ADAR gene in Chinese families and sporadic patients with dyschromatosis symmetrica hereditaria (DSH). *Hum. Mutat.* **23**(6), 629–630.

Zhang, F., Liu, H., Jiang, D., Tian, H., Wang, C., and Yu, L. (2008). Six novel mutations of the ADAR1 gene in Chinese patients with dyschromatosis symmetrica hereditaria. *J. Dermatol. Sci.* **50**(2), 109–114.

4

Cell Entry of Enveloped Viruses

François-Loic Cosset[*,†,‡] and Dimitri Lavillette[*,†,‡]

*Université de Lyon, UCB-Lyon1, IFR128, Lyon, France
†INSERM, U758, Lyon, France
‡Ecole Normale Supérieure de Lyon, Lyon, France

I. Introduction
II. Architecture and Structure of the Viral Fusion Machinery
    A. Membrane fusion according to the stalk-pore model
    B. pH-dependent and -independent molecular switches
    C. The entry process can be achieved by different number of EnvGP
    D. Structural classification: Class I, class II, and class III fusion proteins
    E. Common refolding process
    F. Domain organizations/fusion domains
III. How Viruses Subvert Different Cell Proteins for Entry?
    A. Definition of receptors, adsorption molecules, and cofactors
    B. Use of multiple receptors—Receptor switch
    C. Separation of the binding and fusion functions
IV. Basic and Translational Research Exploiting Entry Properties of Viruses
    A. Tropism properties and use of pseudoparticles in gene therapy
    B. Identification of viral cell entry receptors using pseudoparticles
    C. Determining the endocytosis pathway of entry and the different cell proteins involved in entry by RNA interference screens

Advances in Genetics, Vol. 73
0065-2660/11 $35.00
DOI: 10.1016/B978-0-12-380860-8.00004-5

ABSTRACT

Enveloped viruses penetrate their cell targets following the merging of their membrane with that of the cell. This fusion process is catalyzed by one or several viral glycoproteins incorporated on the membrane of the virus. These envelope glycoproteins (EnvGP) evolved in order to combine two features. First, they acquired a domain to bind to a specific cellular protein, named "receptor." Second, they developed, with the help of cellular proteins, a function of finely controlled fusion to optimize the replication and preserve the integrity of the cell, specific to the genus of the virus. Following the activation of the EnvGP either by binding to their receptors and/or sometimes the acid pH of the endosomes, many changes of conformation permit ultimately the action of a specific hydrophobic domain, the fusion peptide, which destabilizes the cell membrane and leads to the opening of the lipidic membrane. The comprehension of these mechanisms is essential to develop medicines of the therapeutic class of entry inhibitor like enfuvirtide (Fuzeon) against human immunodeficiency virus (HIV). In this chapter, we will summarize the different envelope glycoprotein structures that viruses develop to achieve membrane fusion and the entry of the virus. We will describe the different entry pathways and cellular proteins that viruses have subverted to allow infection of the cell and the receptors that are used. Finally, we will illustrate more precisely the recent discoveries that have been made within the field of the entry process, with a focus on the use of pseudoparticles. These pseudoparticles are suitable for high-throughput screenings that help in the development of natural or artificial inhibitors as new therapeutics of the class of entry inhibitors. © 2011, Elsevier Inc.

I. INTRODUCTION

Enveloped viruses have a core incorporating the genetic material of the virus surrounded by a lipidic membrane acquired from the cells they bud from. Their genetic material enters target cells following the merging of their membrane with the membrane of the cells at either the plasma membrane or the membrane of another internal compartment. The process of membrane merging is called membrane fusion, which preserves the integrity of the cell membrane. The fusion of two separate lipid bilayers in a nonaqueous environment first requires that they come into close contact. This is followed by an intermediate stage characterized by the merger of only the closest contacting monolayer, a process called hemifusion. Third, the fully completed fusion results in whole bilayer merging, followed by the opening of the pore. To achieve this process, the envelope glycoproteins (EnvGP) have evolved in order to combine two features. On the one hand, they acquired a domain to bind to a specific cellular protein, named "receptor." On the other hand, they developed in a different manner, according to the genus of the virus, a function of fusion that allows the destabilization of the membrane and the opening of a pore through which the genetic material will enter the cell.

Despite three different classes of fusion protein having been described so far, three common main steps are described for achieving the pre- to postconformational changes. The first one, after EnvGP activation upon receptor binding or acidification of the endosomal compartment, exposes the fusion peptide that is projected toward the top of the glycoprotein, allowing the initial interaction with the target membrane. The second one is the folding back of the C-terminal region onto a trimeric N-terminal region that leads to the formation of a postfusion protein structure. The final and third step also requires further refolding of the membrane proximal and transmembrane regions in order to obtain a full-length postfusion structure where both membrane anchors (fusion peptide and tm domains) are present in the same membrane.

The viral glycoprotein-induced fusion must be controlled to allow the virus to leave the cell, to prevent the aggregation of the viruses, and to release genetic material next to a compartment that will permit the continuity of the infectious cycle. This regulation is executed principally by the inactivation or the masking of the fusion machinery that directly disorganizes the lipid bilayer. On one hand, most of these EnvGP are synthesized as a precursor that requires a cleavage by a cellular protease (like furin) and prepares the molecules for the subsequent necessary changes of conformation for the fusion process. On the other hand, the activation of the conformational changes is induced by interaction with the receptor and/or the action of the pH (that protonates some amino acids), modifying the interactions of the EnvGP and their structure. In addition, some proteases or other enzymes are necessary to achieve some complementary priming of the EnvGP to make it competent for fusion (such as cathepsin B and L).

Two types of fusion mechanisms can occur, namely, pH independent and pH dependent. In the first case, the recognition between virus and receptor directly triggers conformational changes in the EnvGP that leads to the direct fusion between the two membranes (viral and plasma) and to the liberation of the viral genetic material. This activation of EnvGP at neutral pH allows the fusion *in vitro* and *in vivo* of EnvGP-expressing cells with receptor-expressing cells. This fusion leads to the merging of cell cytoplasms and to the generation of multinucleated cells called syncytia. In the second case, for pH-dependent fusion, the interaction between the EnvGP and the receptor leads to the obligatory endocytosis of the virus–receptor complex, and the acidification of endosomes triggers conformational changes in the EnvGP. For the pH-dependent virus, such a fusion can be reproduced in cell culture *in vitro* or in a liposome–virus fusion assay in the tube after decreasing the pH, but cannot occur *in vivo*.

Research during the last few years has greatly advanced our understanding of the cell surface receptors for viruses and has provided many surprising insights. These advances were achieved largely by identification and molecular cloning of the cell surface or cytoplasmic proteins that have been subverted for use as viral receptors or cofactor, and by parallel advances in studies of the viral EnvGP that bind to the receptors.

Another key area of new insights concerns the physical–chemical process of viral adsorption and of pulling the virus closely onto the cellular membrane. Indeed, adsorption is a severely limiting step in infections of cultured cells, and the initial attachment often does not involve the receptors that ultimately mediate infections (Andreadis *et al.*, 2000; Guibinga *et al.*, 2002; Pizzato *et al.*, 1999, 2001; Ugolini *et al.*, 1999). Thus, we need to distinguish cell surface molecules such as heparan sulfate proteoglycans, DC-SIGN, or integrins that can enhance infections by concentrating retroviruses onto cells (Bounou *et al.*, 2002; Geijtenbeek *et al.*, 2000; Jinno-Oue *et al.*, 2001; Mondor *et al.*, 1998; Pohlmann *et al.*, 2001; Saphire *et al.*, 2001) from authentic receptors that induce conformational changes in EnvGP that are a prerequisite for fusion of the viral and cellular membranes. In contrast to other cell surface components such as lectins or proteoglycans that influence infections indirectly by enhancing virus adsorption onto specific cells, the true receptors induce conformational changes in the viral EnvGP that are essential for infection. However, it appears that more and more intracellular proteins have roles in controlling viral host ranges, and the proteins involved in traffic of intracellular vesicles like endosomes, play a critical role in entry. Therefore, proteins from pathways specifically characterized might be considered as a cofactor as their role is not to mediate direct contact with EnvGP but is crucial for entry. In addition, some viruses are able to use more than one endocytic pathway and the cellular proteins involved thus direct the virus to the entry door beneficial for the virus under a particular condition.

Pseudoparticles are retroviridae cores which incorporate heterogeneous envelope protein from a different virus, possibly of a different family. Their entry process mimics precisely that of the wild-type viruses from which the envelope glycoprotein was derived. They represent a useful tool to study molecular processes of envelope viruses as they are very flexible, allowing the analysis of numerous mutants. Moreover, they allow a precise measurement of the infectivity that depends on the envelope glycoprotein for the entry step and they allow the establishment of virus–liposome fusion assays.

We will first introduce envelope architecture and structure of the viral fusion machineries that have been developed to achieve the fusion and entry steps. Despite many differences of structure, they share a common refolding process that activates different fusion domains found in most fusion proteins. We will then illustrate the different fusion regulation processes that have been developed by viruses to lead to functional virions, both in terms of cleavage activation of fusion proteins during exit and in terms of activation during binding and endocytosis. Despite some differences between distinct players, some common principles can be proposed for all fusion processes. We will illustrate the molecular details characterizing the maturation of the different fusion proteins, defined by the following three characteristics: the cleavage of an envelope protein precursor, the presence and triggering of the exposition of a fusion peptide, and an association as a trimeric complex association in its active fusion conformation. The progression of these structural rearrangements slows down the kinetic barrier between hemifusion and fusion-pore formation. In a second part, we will present the different entry pathways and cell proteins that are used by viruses to infect cells. Viruses have emerged as valuable tools for the study of endocytic mechanisms. These properties have been crucial for the development of pseudoparticles for their use in terms of vector for gene therapy and for their use in terms of tool for receptor cloning, transgenesis, or transduction. Finally, we will give examples of strategies that have been developed *in vivo* and *in vitro* to inhibit the entry step of the enveloped viruses which have led, in the case of HIV, to the development of inhibitors that are used in the clinic.

II. ARCHITECTURE AND STRUCTURE OF THE VIRAL FUSION MACHINERY

A. Membrane fusion according to the stalk-pore model

EnvGP are responsible for bringing the membranes closer, triggering the link and the destabilization of the outer leaflet, the merging of the whole membrane, and the opening of a pore through the cell and virus membrane to allow the core to enter the cytoplasm of the cell. The hypothesis of the pore model in viral membrane fusion mechanism is supported by experimental results. The first evidence for a hemifusion intermediate was achieved by studying influenza virus entry that occurs after the hemagglutinin glycoprotein binding to the host cell. The substitution of the hemagglutinin transmembrane domain by a glycosylphosphatidylinositol (GPI) revealed the importance of the transmembrane region for the fusion pore opening and expansion. Hemifusion structures are connections between outer leaflets of apposed membranes, whereas the inner leaflets remain distinct. This is a transient structure that either dissociates or gives rise to the fusion pore (Chernomordik and Kozlov, 2008). Interestingly, the helix breaker residues within the tm domain are critical for the fusogenicity of different retroviral Env, such as HIV (Owens et al., 1994) and Mo-MLV (Taylor and Sanders, 1999), being important for both the hemifusion and pore opening step of the fusion process. In addition, hemifusion intermediate has been detected in the case of HIV Env-mediated fusion (Munoz-Barroso et al., 1998) by using peptide inhibitors that target a prefusion or prehairpin structure such as HIV-1 gp41 T-20. Once the pore is formed, it allows a connection between two compartments initially separated by the apposed membranes.

The ability of the membrane to hemifuse and develop a fusion pore has been found to depend on the lipid microdomain composition, for example, cholesterol (Chernomordik and Kozlov, 2003). Indeed, a potential lipid dependence of virus entry processes was first deduced from experiments on influenza virus, implying a role for detergent-resistant lipidic microdomain (Takeda et al., 2003). For retroviruses, the tm palmitoylations which contribute to the Env localization in detergent-resistant lipidic microdomain domains (Li et al., 2002) indirectly influence the fusion process (Ochsenbauer-Jambor et al., 2001). As an alternative to the lipidic pore hypothesis, a direct fusion mechanism has also been proposed. The fusion pore is a full proteic channel-like structure dependent only on the transmembrane domains of the glycoproteins. In this model, the pore is opened by the joining of two hemipores located on each membrane (Chernomordik and Kozlov, 2005, 2008).

After fusion pore opening and enlargement (Melikyan et al., 2005), the genetic material enters the cytoplasm of the cell, instigating the virus cell cycle.

B. pH-dependent and -independent molecular switches

The fusion of the viruses that enter directly at membrane plasmic level (as with the paramyxoviruses and most retroviruses) is triggered by the activation of the viral envelope protein at neutral pH. In this case, only the binding of the virus to its receptor activates the fusogenic potential of the complexes of EnvGP. This mechanism is pH-independent. In the case of retroviruses, the binding of the surface subunit (SU) to its receptor induces conformational changes not only in itself but also in the TM with which it interacts, thus inducing fusion. This activation at neutral pH permits the envelope glycoprotein of pH-independent viruses, in certain experimental conditions or in vivo, to induce the fusion between the cells expressing the envelope glycoprotein and the cells expressing the receptor (Harrison, 2008; Kielian and Rey, 2006; Weissenhorn et al., 2007). This intercellular fusion, by merging their plasma membranes, places the cell cytoplasms in continuity, and one or more cells become giant multinucleated cells named syncytia.

In contrast, the fusion of Fig. 4.1 most other viruses depends strictly on their internalization into one of the numerous endocytic pathways such as the clathrin-dependent, clathrin-independent, and caveolae-independent, as well as the macropinocytosis (Vaccinia virus) (Mercer and Helenius, 2008) or the phagocytose (as recently described for Equine herpesvirus 1 virus (Frampton *et al.*, 2007) and the HIV virus (Trujillo *et al.*, 2007), although in this case not resulting in a productive infection; Marsh and Helenius, 2006; Sieczkarski and Whittaker, 2002; Figure 4.1). As described so far, the enveloped viruses that use these itineraries have fusion reactions that require exposure to a moderately acid pH in the different endocytosis vesicles (pH-dependent viruses). Classical examples of viruses that fuse at low pH with the membrane of endosomes or artificial membranes in the test tube include the influenza orthomyxovirus, the dengue or the tick-borne encephalitis (TBEV) flaviviruses, and the Semliki forest alphavirus (SFV) (Harrison, 2008; Kielian and Rey, 2006; Roche *et al.*, 2008; Weissenhorn *et al.*, 2007).

Interestingly, an intermediate mechanism has been described for two retroviruses, ASLS and JSRV. The proposed mechanism of ASLV virion entry occurs in two steps involving a receptor-priming step that induces Env conformational changes, thus allowing the Env to become sensitive to the action of acid pH (Mothes *et al.*, 2000). This hybrid mechanism does not lead to cell–cell fusion *in vivo*. JSRV also uses receptor priming for fusion activation of JSRV Env at a low pH, but the mechanism differs slightly to ASLV, requiring dynamin-associated endocytosis (Bertrand *et al.*, 2008).

So far, many different endocytosis pathways have been described (Marsh and Helenius, 2006; Mercer and Helenius, 2009; Mercer *et al.*, 2010b) as being used by both pH-dependent and pH-independent viruses. However, reinvestigations of these entry pathways are clearly needed for many pH-independent viruses that were originally thought not to rely on endocytosis. For example, Nipah paramyxoviruses induce fusion between cells at neutral pH and were considered as a pH-independent virus not reliant on endocytosis for entry. However, recently, it was proposed that Nipah viruses (NiVs) use macropinocytosis for entry (Pernet *et al.*, 2009). It should be noted that viruses that use a pH-independent mechanism of activation of EnvGP may still enter the cell by endocytosis without any imperative requirement for acidification activation of EnvGP in the endosomes.

Clathrin-dependent endocytosis is the best-characterized pathway of entry and is the major itinerary of entry for pH-dependent viruses. This process is initiated by the formation of the characteristic invaginations of membrane, known as clathrin-coated pits (CCP) (Fig. 4.1). The CCP assembly takes place on the internal face of the plasma membrane following a signal induced by the activation of a receptor. This clathrin-dependent method of internalization is used by the Semliki forest and Sindbis alphaviruses, the rubella rubivirus, and the

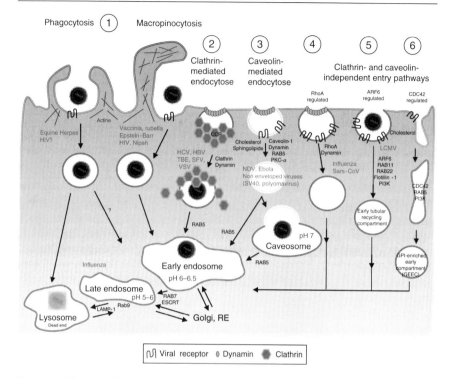

Figure 4.1. Viruses exploit different endocytosis pathways to enter host cells. Multiple mechanisms have been defined as pinocytic, that is, they are involved in the uptake of fluids, solutes, and small particles. These include clathrin-mediated, macropinocytosis, caveolar/raft-mediated mechanisms, as well as several novel mechanisms. Large particles are taken up by phagocytosis, a process restricted to a few cell types. Though poorly documented, the entry of some viruses can be mediated by numerous cargoes which can be endocytosed by mechanisms that are independent of the clathrin coat protein and the fission GTPase, dynamin. These pathways include RhoA- (IL-2 pathway), ARF6-, and CDC42-regulated pathway (GEEC pathway). After binding of the particles to the cell surface entry receptor, genetic material is delivered into the cytoplasm of the cell via a specific endocytosis pathway. Recently, macropinocytosis has emerged as an important entry pathway for viruses (1). The entry of most pH-dependent viruses is mediated by the use of clathrine-coated endocytic vesicles (2). Other viruses penetrate host cells via the formation of vesicles covered by caveola molecule (3). Using this pathway, the viral particles are targeted to neutral pH caveosomes or to early endosomes with moderately acid pH. Alternately, the virus can use non-clathrin, non-caveolin-dependent pathways, both dynamin-dependent or independent (4, 5, 6). These pathways are differentiated by the implication of different GTPases (RhoA, CDC42, or ARF6) and their different compositions (cholesterol, flotillin, TEM, etc). (See Color Insert.)

Hantaan hantavirus, for example (DeTulleo and Kirchhausen, 1998; Helenius *et al.*, 1980; Jin *et al.*, 2002; Kee *et al.*, 2004). Macropinocytosis is another endocytosis pathway utilized by viruses belonging to vaccinia, adeno, picorna, and other virus families. Macropinocytosis is an endocytic mechanism normally involved in fluid uptake induced by growth factors, phorbol ester. However, the binding of virus particles can also activate signaling pathways that trigger actin-mediated membrane ruffling and blebbing. This is followed by the formation of large vacuoles (macropinosomes) at the plasma membrane, internalization of virus particles, and penetration by the viruses or their capsids into the cytosol through the limiting membrane of the macropinosomes (Mayor and Pagano, 2007; Mercer and Helenius, 2009). A variety of evidences suggest that macro-pinocytosis can be a mechanism for HIV-1 and Epstein-Barr herpes virus entry in some primary cells or cell lines (Marechal *et al.*, 2001; Miller and Hutt-Fletcher, 1992). Based on the more defined criteria of macropinocytosis (Mercer and Helenius, 2009), these conclusions are probably premature and may depend on the experimental conditions that use very concentrated viruses which may influence the utilization of less specific pathways. It should be noted that different strains of the same virus can elicit dramatically different responses in host cells during entry, and different macropinocytic mechanisms are possible in the same cell line through subtle differences in the activating ligand (Mercer *et al.*, 2010a).

An additional endocytosis pathway, whilst badly defined, is known as a clathrin- and caveola-independent pathway. The dependence on different GTPase and cargo proteins (dynamin, Rho, CDC42, ARF6, etc.) defines this endocytosis pathway. Several viruses are using this clathrin- and caveola-independent pathway, including the influenza orthomyxovirus (as a second possible pathway), the SARS-CoV coronavirus, the lymphocytic choriomeningitis arenavirus (LCMV), and picornaviruses (Madshus *et al.*, 1987; Matlin *et al.*, 1981; Wang *et al.*, 2008). Finally, other vesicles (characterized by caveolin or caveosome, with a high concentration in cholesterol and sphingolipids) are clathrin-independent but dynamin-dependent and do not have acid compartments before their merging with early endosomes (Mayor and Pagano, 2007). The nonenveloped SV40 virus and some human enteroviruses are the prototypes of the pH-independent viruses that use this endocytosis pathway. So far, only two enveloped viruses have been described to use this pathway, the Newcastle disease paramyxovirus (NDV) (Cantin *et al.*, 2007) and Ebola virus (Empig and Goldsmith, 2002; Sanchez, 2007). Most viruses have been shown to use only one entry pathway; however, there are more and more recent examples that indicate that some viruses use multiple endocytosis pathways to enter their target cells (influenza, HIV). For example, Ebola viruses are known to enter cells by clathrin-mediated endocytosis (Sanchez, 2007), but lipid raft-associated, caveo-lin-mediated endocytosis has also been proposed as a mechanism of Ebola virus uptake (Empig and Goldsmith, 2002).

In terms of the activation process of the membrane fusion and the need for particular microdomains, it is not always clear why the viruses are using so many different pathways. It is possible that the viruses that evolved to fuse intracellularly have a selective advantage to release their genome to specific intracellular sites that will allow the rapid and efficient establishment of the infectious cycle. One selective advantage could also be the requirement for particular lipid microdomains, like those enriched in cholesterol. Indeed, it was shown that contrary to the class II flavivirus dengue virus and the yellow fever virus, the E1 fusion protein from Semliki forest virus (SFV) binds cholesterol which explains its dependence for cholesterol and compartment containing cholesterol (Umashankar et al., 2008). For other viruses, the dependence on cholesterol is linked to bulk effects on membrane fluidity and the maintenance of particular microdomains where receptors are located and where they have some particular diffusion.

C. The entry process can be achieved by different number of EnvGP

In addition to these dichotomies of pH-independent or pH-dependent fusion process, of plasma or endosomes membrane fusion, the mechanisms of activation of the fusion proteins of the different enveloped viruses are very diverse. Some reactions of fusion are triggered by the interaction of one envelope glycoprotein with a unique receptor. One example is the interaction of the envelope glyco-protein SU–TM of the γ-retrovirus with the multitransmembrane receptor (Table 4.1). In another case, one unique envelope glycoprotein can interact with several receptors. For example, the fusion of HIV-1 is triggered by the sequential interaction of the SU gp120 with the CD4 receptor, followed by interaction with a coreceptor such as CCR5 or CXCR4, chimiokine receptors of seven transmembrane domains (Hunter, 1997). Previously, it was believed that binding to receptors directly triggered a series of conformational changes in the viral EnvGP culminating in fusion of the viral and cellular membranes. However, new evidence suggests that γ-retroviral association with receptors triggers an obligatory cooperative interaction or cross-talk between EnvGP on the viral surface for HIV and MLV (Tailor et al., 2003). If this intermediate step is prevented, infection fails. Conversely, in several circumstances, this cross-talk can be induced in the absence of a cell surface receptor for the virus, in which case, infection can proceed efficiently. This new evidence strongly implies that the role of cell surface receptors in infections of γ-retroviruses (and perhaps of other enveloped animal viruses) is more complex and interesting than was previously imagined.

In all these cases, the EnvGP have a double function attachment to the receptor and an exclusive role in fusion in the same protein. However, for other viruses, these two functions are filled by different proteins. The E2 protein of the

Table 4.1. Example of Different Viruses, Their Envelope Glycoproteins, and Their Receptors

Virus Family	Virus	Envelope Glycoprotein	Receptor	Function
pH-dependant				
Alpha	Semliki forest virus (SFV)	E3-E2-6K-E1	MHC-I	Immune recognition (1 tmd)
	Chikungunya	E3-E2-6K-E1		
Flavi	Tick-borne encephalitis virus (TBEV)	PrM-E	HS	Glycoaminoglycan (1 tmd)
	Dengue	PrM-E	MR (CD206)	Mannose receptor
			HS, DC-SIGN?	Glycoaminoglycan, lectin
Hepaci	Hepatitis C virus (HCV)	E1-E2	CD81,	Tetraspanin (adhesion, activation...) (4 tmd)
				HDL Receptor (2 tmd)
			SRB1,	Tight junction (4tmd)
			Claudin,	Tight junction (4tmd)
			Occludin	
Orthomyxo	Influenza A	HA1–HA2	Sialic acid	Carbohydrate
Rhabdo	Vesicular stomatitis virus (VSV)	G	PS?	Phosphatidylserine
Bunya	Haanta virus	G1–G2	beta 3 integrin	Integrin (1tmd)
			DAF(CD55)	Complement system
Filo	Ebolavirus	GP1–GP2	L-DC-SIGN,	C-type lectin (1tmd)
			hMGL	C-type lectin (1tmd)
			Cathepsine L, B	Endosomal protease
			FRα	Folate receptor α (1tmd)
Rubella	Rubellavirus	E2-E1	PS, PI, PE, PC	Phospholipids
			ganglioside	Glycolipids
Retro	Avian leukosis and sarcoma virus (ALSV)	Gp85–gp37	Tva	LDL-R homology (1tmd)
Arena	Lymphocytic choriomeningitis virus (L-CMV)	GP1–GP2	α-Dystroglycan	Laminin receptor extracellular matrix
	Lassa virus	GP1–GP2	α-Dystroglycan	Laminin receptor extracellular matrix

(Continues)

Table 4.1. (*Continued*)

Virus Family	Virus	Envelope Glycoprotein	Receptor	Function
pH-Independent				
Retro	Human immunodeficiency virus 1 (HIV-1)	Gp120–gp41	CD4, CCR5, CXCR4	Immune recognition (1 tmd) Chimiokine receptor (7 tmd)
	Amphotropic Murine leukemia virus (A-MLV)	Gp70–p15	PiT-2	Na-Pi cotransporter (10 tmd)
	Feline leukemia virus (FeLV)	Gp70–p15	PiT-1	Na-Pi cotransporter (10 tmd)
	Gibbon ape leukemia virus (GaLV)	Gp70–p15	PiT-1	Na-Pi cotransporter (10 tmd)
	Pig endogenous retrovirus A (PERV-A)	Gp70–p15	HuPAR2, GHBh1	G-protein coupled receptor (10 tmd)
	Feline endogenous retrovirus (RD114)	GP70–p15	ASCT-2, SLC1A5	Cotransport Na-a.a. neutres (7 tmd)
	Human T-cell lymphotropic virus type 1 (HTLV-1)	Gp46–gp21	GluT-1 Neuropilin 1 HS	Transport glucose (12tmd) VEGF receptor, synapse(1tmd) Glycoaminoglycan (1 tmd)
Paramyxo	Measle virus	H, F	CD46 SLAM (CD150)	Complement regulator Ig-like, CD4 regulation
	Newcastle disease virus	HN, F	Sialic acid	On gangliosides and glycoproteins
	Human parainfluenza virus	H, F2-F1		
Alpha-herpes	Herpes simplex virus 1 (HSV-1)	gB, gD, gL, gH	HVEM, nectin 1α et β, HS	Co-stimulation factor (1 tmd) Ig-like; adherens-junction, synapse (1 tmd) Glycoaminoglycan (1 tmd)
Beta-herpes	Epstein-Barr virus (EBV)	gB, gL, gH	CD21	Complement cascade component
Pox	Vaccinia virus	L1, A27L, D8L, A33, H3L, B5, etc.	HS EGF receptor	Glycoaminoglycan (1 tmd) Signaling receptor
Corona	Murine hepatitis virus 4 (MHV-4)	S	Bgp (biliary gp)	Ig-like (1 tmd)
	SARS-CoV	S	ACE-2 Cathepsine L, B	Metallocarboxypeptidase Endosomal protease

SFV alphavirus binds the receptor that permits the endocytose of the virus in a compartment where the acid pH will activate the E1 protein that will induce the fusion (Table 4.1). For most paramyxoviruses (including Newcastle parainfluenza virus, human parainfluenza virus type 3 (HPIV-3), mumps virus, bovine RSV (BRSV), ovine RSV (ORSV), and human RSV (HRSV)), the hypothesis is that the binding of HN or G/SH to their receptor induced not only conformational changes in the receptor but also in the F protein with which it interacts. These changes make F competent for the membrane fusion, with the exception of the F protein of the simian parainfluenza virus 5 (SV5), the BRSV, ORSV, and HRSV that do not strictly require HN or G/SH to induce the fusion (though the fusion is greatly increased by HN or G/SH, respectively). This complexity in the distribution of the functions is still more evident in, for example, the alphaherpesvirus (HSV-1, HSV-2) (Rey, 2006). In this case, the binding of the gD glycoprotein to one of its three receptors (HVEM for herpesvirus entry molecule mediator, or nectin 1, or a specific heparan sulfate) activates a complex of gB trimers associated to gL and gH proteins, all the three being essential to the fusion (Gianni et al., 2006; Table 4.1). In the same way as for the hepatitis C virus, the E1E2 heterodimer induces entry following the recognition of at least four receptors, CD81, SRB1, claudin, and ocludine (Zeisel et al., 2009).

Another interesting variation of activation of fusion is the artificial reverse process. In some cases, it was described that some exosomes or pseudo-particle-incorporating receptors were able to enter cells expressing envelope. For example, the multitransmembrane receptor mCAT-1 from ectropic murine leukemia virus (MLV) or the Tva receptor for avian sarcoma-leukosis virus (ASLV-A) can be used to infect cells that express their respective retroviral EnvGP (Balliet and Bates, 1998). Similarly, the incorporation of both CD4 and one other coreceptor of human immunodeficiency virus (HIV), CCR5 (Endres et al., 1997) or CXCR4 (Endres et al., 1997; Mebatsion et al., 1997; Schnell et al., 1997), allows the production of viral particles infecting cells that express simian immunodeficiency virus (SIV) or HIV envelope glycoprotein. Interestingly, it was shown that Nef, an accessory protein of human immunodeficiency virus type 1 (HIV-1) that enhances the infectivity of progeny virions when expressed in virus-producing cells, significantly enhanced the infectivity of CD4-chemokine receptor pseudotypes for cells expressing HIV-1 Env. Surprisingly, Nef also increased the infectivity of HIV-1 particles pseudotyped with Tva, even though Nef had no effect if the pH-dependent Env protein of ASLV-A was used for pseudotyping (Pizzato et al., 2008). This process indicates that the difference of superficial tension between the cell (weak because the radius of the cell is big) and the virus (strong as the radius of the virus is small) is not crucial for the development of an oriented fusion process with EnvGP on the one side and receptor on the other. Having said that, the cell is not an empty bubble, and this efficacy of the fusion process regardless of which membrane harbors the receptor or the EnvGP may reflect the adaptation of the virus to optimize the membrane merging independently of the superficial tension. Cells are not completely round, and superficial tension depends on the localization where the virus binds. Moreover, the plasma membrane has different lipid compositions and interacts with the cytoskeleton which will attenuate or increase the superficial tension at certain localizations.

D. Structural classification: Class I, class II, and class III fusion proteins

Glycoproteins from enveloped viruses have evolved to combine two main features. First, they have the capacity to bind with a specific cellular receptor and second, they include a fusion domain (peptide fusion and transmembrane domain) that can be activated to mediate the merging (fusion) of viral and cellular membranes.

Three different classes of viral fusion proteins have been identified to date based on key structural features at pre- and postfusion stages. Many studies have demonstrated that the structural transition from a pre- to a postfusion conformation leads to a stable hairpin conformation. This includes class I fusion proteins, characterized by trimers of hairpins containing a central alpha-helical coiled-coil structure, and class II fusion proteins, characterized by trimers of hairpins composed of beta structures. A third class of fusion proteins has been described recently, that also forms trimers of hairpins by combining the two structural elements alpha-helix and beta-sheet structures (Roche *et al.*, 2006, 2007).

The synthesis and the conformation of these classes I, II, or III fusion proteins are different. Class I viral fusion proteins of diverse virus families, including Retroviridae, Filoviridae, Orthomyxoviridae, Paramyxoviridae, and Coronaviridae, differ greatly in size and amino acid sequence, but their membrane-anchored domains share common structural features that are essential for membrane fusion, including two heptad repeats (called HR-1 and HR-2), preceded by a hydrophobic fusion peptide. Class I membrane-fusion reaction is mediated by the refolding of the fusion protein to a highly stable rod-like structure with a central trimeric α-helical coiled coil. Such coiled-coil structures are emblematic of class I proteins, and physical demonstration or computer prediction of such a structure is frequently used to help define a fusion protein as belonging to class I. The envelope glycoprotein of a retrovirus is generated by the cleavage of its precursor in one SU and one transmembrane subunit containing an anchoring sequence to the membrane (Hunter, 1997). This maturation, essential to the process of fusion, frees a hydrophobic sequence to the N terminal extremity of the TM, named fusion peptide, as it is supposed to insert into the target membrane and initiates the fusion. Initially, the fusion peptide is masked inside the trimer of EnvGP, a form competent for the fusion (Fig. 4.2A).

In general, the cleavage of the fusion proteins is mandatory for most class I proteins to render them competent for fusion (Earp *et al.*, 2005; Harrison, 2005). Exceptions are the envelope glycoprotein from Ebola or SARS-CoV viruses, classified as class I on the grounds of the three-dimensional structure of one of their fragments, which are not cleaved but yet are functional. Though an N-terminal fusion peptide is predicted in the potential transmembrane subunit (localized by analogy to homologous viruses), their functionality can be explained also by the presence of an internal fusion peptide (Ebola). Another explanation of their functionality is their requirement for the L or B cathepsins that cleave the EnvGP during endocytosis (see below). The number of complexes on the surface of viruses harboring class I fusion proteins is very variable. It seems that lentiviruses have only 10–20 SU–TM trimers, whereas the coronaviruses have hundreds of trimers coating the surface, giving them their characteristic morphology of "crown" which is the origin of the name of this family (Tables 4.1 and 4.2).

The process of assemblage and biosynthesis of class virus II is very different to the class I proteins (Kielian and Rey, 2006) (Fig. 4.2B). During biosynthesis, the alphavirus (E1) and flavivirus (E) fusion proteins fold cotranslationally with a companion or regulatory protein, termed p62 (or PE2) for alphaviruses and prM for flaviviruses (Garoff *et al.*, 2004). This heterodimeric interaction is important for the correct folding and transport of the fusion protein. Both p62 and prM are cleaved by the cellular protease furin late in the secretory pathway, in a maturation reaction that is a crucial regulatory step for subsequent virus fusion (Salminen *et al.*, 1992; Stadler *et al.*, 1997; Wengler, 1989; Zhang *et al.*, 2003a). Though this cleavage in the envelope glycoprotein complex is important, contrary to class I fusion protein, it is not crucial, since mutated alphavirus and flavivirus for which the regulating proteins are not cleaved have only a decreased infectivity (Salminen *et al.*, 1992; Zhang *et al.*, 2003b). In the case of the SFV alphavirus, the fusion process induced by the uncleaved fusion proteins can be a trigger after treatment of the virus at pH5 or less, rather than the normal fusion threshold pH 6 of the wild-type virus (Salminen *et al.*, 1992; Zhang *et al.*, 2003a). This variation in the threshold of pH of the fusion is necessary for the dissociation of the heterodimer. The class II fusion proteins are either homodimers or heterodimers (E homodimer for the flaviviruses or E2–E1 heterodimer for alphaviruses) that form an envelope netting the viral membrane (Lescar *et al.*, 2001; Rey *et al.*, 1995). The internal fusion peptide is masked at the interface of the dimers at the extremity of a long beta sheet. One important difference between these two groups of viruses is the budding site (Garoff *et al.*, 2004). In alphaviruses, the p62–E1 complex is transported to the plasma membrane, and the heterodimer interaction is maintained after p62 processing. New virions bud at the plasma membrane, in a process that is driven by lateral contacts between E2 and E1 heterodimers (E2 being the mature companion protein) to induce the required curvature of the lipid bilayer, in combination

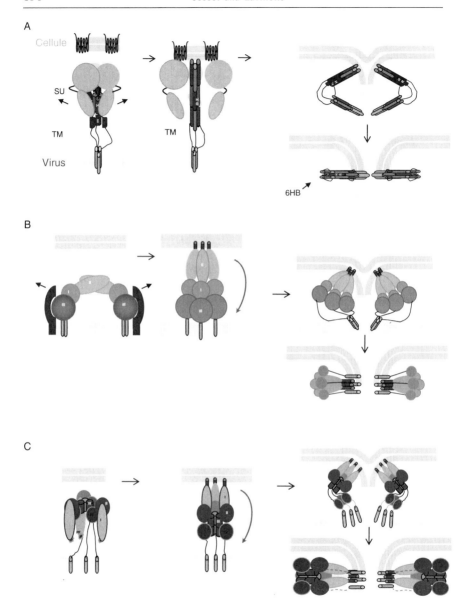

Figure 4.2. Possible fusion models have evolved for class I (A), class II (B), and class III (C) fusion proteins from their pre- to postfusion conformations. Ambiguities remain regarding certain events in membrane fusion promoted by a viral fusion protein. Despite the diversity in the structure of the fusion proteins, the major steps of the fusion process are similar. The first step, after EnvGP activation upon receptor binding or acidification

with interactions of the cytosolic tail of E2 with the nucleocapsid. Budding results in the formation of icosahedral-enveloped particles of triangulation $T = 4$, containing 80 trimeric E2/E1 spikes.

By contrast, flavivirus particles bud into the endoplasmic reticulum as immature virions formed by 60 trimers of prM–E. The immature particles have an organization similar to that of mature alphaviruses, with each trimer forming a

---

of the endosomal compartment, exposes the fusion peptide that is projected toward the top of the glycoprotein, allowing the initial interaction with the target membrane. The second step is the folding back of the C-terminal region onto a trimeric N-terminal region that leads to the formation of a postfusion protein structure. The third and final step also requires further refolding of the membrane proximal and transmembrane regions in order to obtain a full-length postfusion structure where both membrane anchors (fusion peptide and tm domains) are present in the same membrane. Three different classes of fusion have been identified so far based on common structural motives. (A) The class I fusion proteins are characterized by trimers of hairpins containing a central alpha-helical coiled-coil structure. For retroviruses, receptor binding induces the movement of the SU, allowing a loop-to-helix transition of a polypeptide segment of TM that was previously buried underneath the SU heads, projecting the fusion peptide ~ 100 Å toward the target membrane, where it inserts irreversibly. This occurs by a "spring-loaded" mechanism. The HR2 C-terminal end (green) of the long TM α-helix jackknifes back, reversing the direction of the viral-membrane-proximal segment of TM, which then interacts in an antiparallel fashion with the groove formed by the N-terminal HR1 (blue) trimeric coiled coil. The final postfusion conformation of TM is, therefore, a highly stable rod with the TM and fusion-peptide segments together at the same end of the molecule, a structure termed a "trimer of hairpins" or helix buddle (HB). (B) Class II fusion proteins are characterized by trimers of hairpins composed of beta structures. The red, yellow, and blue parts of each subunit correspond, respectively, to domains I, II, and III of the ectodomain. The fusion loop is at the tip of domain II. Monomeric transition between the prefusion dimer and the trimeric-extended intermediate is shown. After exposure to the low pH of the endosomes, domains I and II swing outward, while domain III and the stem remain oriented against the membrane roughly similar to the prefusion state. The fusion loop, at the top of the diagram, interacts with the target bilayer. Domains I and II associate into the trimeric core of the postfusion conformation, and domain III must then zip back along the trimer core, thus reorientating the domain III. (C) A third class of fusion proteins has been described recently, which also forms trimers of hairpins by combining the two structural elements alpha-helix and beta-sheet structures. Class III fusion proteins are composed of five domains that give rise to a molecular architecture very distinct from any reported class I or class II fusion proteins. Interestingly, the ectodomain of G has been crystallized in its pre and postfusion (low-pH) state. During the conformational change that occurs upon low pH exposure, the domains of G radically change their position and orientation as a result of rearrangements that occur in the linker regions. Domain I (yellow), carrying the fusion loops, and the transmembrane domain move 16 nm from one end of the molecule to the opposite (Backovic and Jardetzky, 2009). Only domain III (blue) undergoes significant refolding with extension of the central helix F. To complete the process, the C-terminal helices of domain IV (red) insert into crevices formed by two other protomers in the postfusion form, reminiscent of the structural changes observed during refolding events of class I fusion proteins with HB formation. (See Color Insert.)

Table 4.2. Classification of Fusion Proteins Based on Their Family, Class, and Activation Mechanism

Virus family	Virus species	Fusion proteins	Fusion pH for activation
Class I			
Orthomyxoviridae	Influenza A virus	HA2	Low
	Influenza C virus	HEF	Low
Paramyxoviridae	Simian parainfluenza virus 5	F (F2–F1)	Neutral
	Human parainfluenza virus	F	Neutral
	Newcastle disease virus	F	Neutral
	Respiratory syncytial	F	Neutral
Filoviridae	Ebola virus	Gp1-Gp2	Low (for cathepsin cleavage)
Retroviridae	Moloney Murine leukemia virus	TM (gp21)	Neutral
	Human immunodeficiency virus 1	gp41	Neutral
	Simian immunodeficiency virus	gp41	Neutral
	Human T cell leukemia virus 1	gp21	Neutral
	Human syncytin-2	TM	Neutral
	Visna virus	TM	Neutral
Coronaviridae	Mouse hepatitis virus	S2	Low (for cathepsin cleavage)
	Sars corona virus	E2	Low (for cathepsin cleavage)
Class II			
Flaviviridae	Tick-borne encephalitis virus	E	Low
	Dengue virus	E	Low
Togaviridae	Semliki forest virus	E1	Low
Class III			
Rhabdoviridae	Vesicular stomatitis virus	G	Low
Herpesviridae	Herpes simplex virus	gB	Neutral
	Eptsein Barr virus	gB	Neutral

There is no correlation between these criteria for class I and class III fusion proteins, but all the viruses harboring class II fusion proteins are pH-dependent.

spike in which prM covers the fusion protein E. The newly formed virions are then transported to the external milieu through the exocytic pathway. Processing of prM generates the mature M protein with a short ($\sim$40 residues) ectodomain (Kuhn *et al.*, 2002; Zhang *et al.*, 2003b). Presumably because of the removal of a large portion of the prM ectodomain, the flavivirus surface dramatically reorganizes after processing to give 90 E–E homodimers arranged with icosahedral symmetry. The mature flavivirus particles display a smooth, spikeless surface, with E dimers ordered in a characteristic "herring bone" pattern.

Finally, the four class III fusion proteins resolved (VSV-G, gB of HSV1 and EBV and baculovirus gp64) in spite of their homology of trimeric structure have some major functional differences arising from their difference of biosynthesis. G protein is the only class III fusion protein whose prefusion structure is known. The proteins of class fusion III have a combination of the two structural elements, alpha helix like class I and beta sheet like class II (Backovic et al., 2009; Heldwein et al., 2006; Kadlec et al., 2008; Roche et al., 2006, 2007) (Fig. 4.2C). The trimers are maintained by interaction between central alpha helices, but each domain of fusion exposes two buckles of internal fusion peptide placed at the extremity of a long beta sheet. The G protein is the only protein responsible for the binding and the entry of the VSV. The rhabdovirus VSV possesses 1200 molecules of the G protein on its surface and forms 400 trimers. In contrast, the HSV-1 virus incorporates 12 different EnvGP on its surface, of which 4 are essential for the entry step (gD, gB, and gH/gL) (Turner et al., 1998). During their biosynthesis, neither the G protein of VSV nor gB of HSV-1 are cleaved, and they present internal fusion peptides. However, whereas the G is functional by itself, the gB envelope glycoprotein alone is not sufficient to induce the entry of the virus or the membrane fusion. Nevertheless, currently, no precise interaction between gD, gB, and gH/gL has been identified, even though the gD ectodomain has been shown to allow the entry of engineered HSV-1 virus particles that lack gD (that is gD-null mutants; Cocchi et al., 2004). Gp64 is the major component of the viral envelope, and the sole fusogenic proteins that are triggered to induce the fusion in the low pH environment of endosomes for baculovirus. Interestingly, the distinguishing feature between G and gp64 and any other fusion protein is that they can undergo a reversible conformational change, unlike class I and most class II fusion proteins, for which the postfusion conformation is thermodynamically more stable at all pH values, and the conformational rearrangement is effectively irreversible. VSV or gp64 exposure to low pH inactivates the virus, but the fusion activity can be fully recovered when the pH is raised. It has been proposed that the reversibility of the conformational change allows G to avoid unspecific activation during transport through the acidic Golgi vesicles.

As seen in the previous sections, though there is much described variation in the manner of activation of the fusion proteins and additional mechanisms await discovery, only three classes of fusion proteins have been defined so far (Weissenhorn et al., 2007). As previously described, based on the main structural organization of their EnvGP, the viruses are now assigned to the class I, II, or III. In parallel, based on the activation mechanism of the fusion protein, viruses have been classified as pH-dependent or -independent. Interestingly, the relationship between the mechanism of activation of the fusion and the class of protein of fusion is not correlated. The HIV or Influenza EnvGP are both class I and yet they have a process of activation that is pH-independent and

-dependent, respectively, for the entry of the virus. Similarly, the class III fusion proteins enclose the Herpes virus simplex 1 (HSV-1) gB protein and the rhabdovirus VSV-G protein that are pH-independent and pH-dependent, respectively, for their activation. So far, all the class II fusion proteins have been pH-dependent for their activation.

E. Common refolding process

Although there are notable differences between the activation processes, the structural motives used, and the initial oligomeric states of the fusion proteins (the native trimeric conformation of class I and III proteins in opposition to the homo- or heterodimers of the class II fusion proteins), the common features of the final structures obtained after fusion seem to suggest some generic mechanism of conformational changes common to all EnvGP of enveloped viruses (see Fig. 4.2).

First, the activation of the fusion protein following the interaction with the cellular receptor(s), coupled or not to the exposure to the acid environment of the endosomes, exposes the fusion peptide that is projected toward the top of the glycoprotein, allowing the initial interaction with the cellular target membrane. For class I fusion proteins, the proposed model indicates that the transition of conformation requires the transformation of a part of the molecule in alpha helix and the association of this in three helixes bundle (named the "coiled coil"; Weissenhorn et al., 2007). For retroviruses, this movement allows a loop-to-helix transition of a polypeptide segment of TM that was previously buried underneath the SU heads, projecting the fusion peptide ~ 100 Å toward the target membrane, where it inserts irreversibly. In the case of class I fusion proteins like retroviruses, this occurs by a "spring-loaded" mechanism. This initial change is proposed to result in a "prehairpin intermediate," an extended structure that is anchored both in the target membrane by the fusion peptide and in the virus membrane by the TM transmembrane segment. For the class II fusion proteins, the projection of the fusion peptide requires the dissociation of the hetero- or homodimers and modifications in the "hinge region" unstructured before conformational changes (Stiasny and Heinz, 2006). For the class III (Roche et al., 2008), similarly to the fusion proteins of class I, the exhibition of the fusion peptide requires a rearrangement of the domains mediated by modification of the central helixes that do remain parallels.

Second, the folding back of the region including the fusion peptide onto a trimeric C-terminal region leads to the formation of a postfusion protein structure with the outer regions zipped up against an inner trimeric core. Interestingly, it has been described that all the class I, II, and III peptides can inhibit this step by competing with the interaction of EnvGP-specific domain for the formation of this structure.

Third, the final steps require further refolding of the juxtamembrane and transmembrane regions to obtain a stable postfusion structure with the fusion peptides and the transmembrane domains at the same extremity of a stable stem of a protein complex anchored in the target membrane. This structure brings the two membranes proximal and provides free energy to overcome the barrier of membrane merging (Melikyan, 2008). Membrane fusion occurs, which leads to pore formation and release of the viral genome into the cytoplasm.

The refolding of class II fusion proteins generates trimers from monomeric intermediate. The existence of monomeric intermediates for class I and III is not well known. However, if the steps that led to exposition of the fusion peptide and its interaction with the membrane targets maintain a three-order symmetry, the refolding of the C-terminal region requires the destruction of the trimer at least to the juxtamembrane region. Moreover, from the differences observed between the amino acids involved in the interface of the reconstituted resolved trimers in their pre- and postfusion conformations, it is probable that the conformational changes are going through a monomeric intermediate for the class III G envelope glycoprotein of VSV and the class I F protein of paramyxovirus. On the contrary, the interface of the class I HA2 subunit of the HA envelope glycoprotein trimer is very similar between the pre- and postfusion conformation, shedding doubt on the existence of monomeric intermediates. In contrast, when fusion is initiated at low pH, the dimers of the class II fusion proteins are broken, freeing monomers that reassociate in trimers.

Finally, precise structural information of the native metastable conformation (prefusion) and the final stable conformation (postfusion) is available only for a limited number of viruses (for envelope glycoprotein of influenza virus, SFV, TBEV, VSV, and the parainfluenza 5 F protein). The structural conversion of the native metastable conformation in a final stable conformation is not precisely known and is highly speculative for most viruses, and the envelope glycoprotein domains implied in these molecular rearrangements are little-referenced. Clearly, more studies are necessary to identify the intermediaries of envelope glycoprotein conformations. These intermediates would identify the domains that interact during the conformational changes which will highlight ways to generate inhibitory peptides.

F. Domain organizations/fusion domains

1. Acquisition of fusion competence: Priming by cleavage

During virus production, the host cell is basically preserved, since the expression of fusogenic competent glycoproteins is highly controlled for most viruses. However, for some viruses, the EnvGP induce a cytopathic effect that leads to the generation of multinucleated cells, called syncytia, which are induced by the fusion of cells that express the envelope glycoprotein and receptors. Abundant glycoprotein at the surface of the cell could induce cellular death by syncytia formation, toxicity via receptor interaction, or immune recognition. For these reasons, the localization and the amount of the oligomerized envelope glycoprotein at the host cellular surface are highly modulated by fine trafficking and sequestration mechanisms. The receptor interference mechanism (either by saturation of binding site on receptor by envelope glycoprotein or by internalization of receptor by different viral proteins, as it is achieved by some retroviruses; Hunter, 1997) can also limit the amount of receptors available for fusion between infected cells. The control of the cleavage of the EnvGP to free the fusion peptide is also a regulation process of the fusion protein fusogenicity (Labonte and Seidah, 2008). Finally, EnvGP fusion competency may be a late event that occurs during virus budding, as described for MuLV retroviruses (Rein *et al.*, 1994).

Proteolytic priming is a common method of controlling the activation of membrane fusion mediated by viral glycoproteins. The members of the proprotein convertase (PC) family play a central role in the processing and/or activation of various protein precursors involved in many physiological processes and various pathologies such as neurodegenerative pathology, cancer bacterial toxins activation, and viral infections. The proteolysis of these precursors that occurs at basic residues within the general motif (K/R)-(X)-(K/R) is mediated by the proprotein convertases PC1/3, PC2, Furin, PACE4, PC4, PC5 (also called PC6), and PC7 (also called PC8, LPC, or SPC7), whereas the proteolysis of precursors within hydrophobic residues performed by the convertase S1P/SKI-1 and the convertase NARC-1/PCSK9 seems to prefer to cleave at a LVFAQSIP motif (Lahlil *et al.*, 2009). The seven PCs have different, albeit partly, overlapping expression patterns and subcellular localization. They have conserved aminotermini with highest homology in the subtilisin-like catalytic domain. Data on various infectious viruses revealed that the cleavage of their envelope glycoprotein precursors by one or more PCs is a required step for the acquisition of the infectious capacity of viral particles. Indeed, various studies have demonstrated the capacity of the PCs to correctly cleave a variety of viral surface glycoproteins. These include the HIV-1 gp160 (Decroly *et al.*, 1996) and surface glycoproteins of Hong Kong, Ebola virus, the severe acute respiratory syndrome coronavirus and chikungunya virus (Basak *et al.*, 2001; Bergeron *et al.*, 2005; Ozden *et al.*, 2008). In parallel, other studies revealed that the inhibition of processing of these viral surface glycoproteins by the PC inhibitors such as dec-R-V-K-R-CMK completely abrogated the virus-induced cellular cytopathicity. The surface glycoproteins of other viruses, particularly the hemorrhagic fever viruses (Arenaviridae family), such as Lassa (Basak *et al.*, 2002; Lenz *et al.*, 2001), Crimean Congo hemorrhagic fever (Vincent *et al.*, 2003), and lymphocytic choriomeningitis (Beyer *et al.*, 2003), were shown to be cleaved by the convertase SKI-1 that cleaves at hydrophobic residues. Similarly, blocking of SKI-1 activity by a specific inhibitor has also shown to affect the processing and the stability of the glycoproteins of these viruses (Pullikotil *et al.*, 2004).

For all highly pathogenic avian influenza (HPAI) viruses of subtypes H5 and H7 known to date, the cleavage of HA occurs at the C-terminal R residue in the consensus multibasic motifs, such as R-X-K/R-R with R at position P4 and K-K/R-K/T-R with K at P4, and leads to systemic infection. Early studies demonstrated that the ubiquitously expressed furin and proprotein convertases (PCs 5 and 6) are activating proteases for HPAI viruses (Basak *et al.*, 2001; Stieneke-Grober *et al.*, 1992). Recently, ubiquitous type II transmembrane serine proteases, MSPL and its splice variant TMPRSS13, have been proposed as novel candidates for proteases processing HA proteins of HPAI (Okumura *et al.*, 2010).

Is it interesting to note that viral receptors can also be modified by proprotein convertase. Indeed, PCSK9 impedes hepatitis C virus infection *in vitro*, modulates liver CD81 expression, and enhances the degradation of the low-density lipoprotein receptor (LDLR) (Labonte *et al.*, 2009), bestowing on the proprotein convertase an additional role in controlling the fusogenicity of the envelope glycoprotein.

2. Fusion peptide

The exhibition and insertion of a hydrophobic fragment of 10–30 residues in the membrane, named "fusion peptide" or "fusion loop" is a crucial step of the fusion process (Epand, 2003). The fusion peptides in an N-terminal position (such as for the retrovirus or the influenza virus) is liberated for most viruses after envelope glycoprotein cleavage, and it can insert into the external layer of the membrane in an oblique manner, whereas the fusion loop (for the class II and III viruses) remains probably more superficial (see Table 4.3). The fusion peptide of the class I and II proteins is initially buried in the envelope glycoprotein trimer or dimer, respectively. For the class III, the fusion loop is present outside of the structure, most likely because the fusion peptide of class III fusion proteins is weakly hydrophobic and probably requires a cooperation between several loops to be functional and efficient. The simple picture of a viral fusion protein acting on cell and viral membranes by means of only two restricted segments, that is to say, the fusion peptide and the transmembrane domain, is too simplistic. Instead, a more complex concerted action of different membranotropic segments of the fusion proteins is necessary. More conformational changes are required to achieve a complete fusion of the two lipid bilayers. As described previously, the class I–III fusion proteins roughly share common refolding processes and formation of intermediates. Several regions of the fusion protein complex indirectly aid the fusion process, as for example, the "stem" regions (see below).

Table 4.3. Fusion Peptide Characteristics from the Fusion Protein from Different Classes

Fusion peptide	Class I	Class II	Class III
Initial situation	Buried in trimer interface	Buried in the dimer interface	Buried in the interface between different trimers
Localization	N-term (HIV, HA2, etc.) or internal (RSV, Ebola, etc.)	Internal loop embedded between 2 beta-strands	2 internal loop (segmented fusion peptide; nonobvious on primary sequence)
Structure flexibility	Alpha helix ↔ random coil/turn	Stable random ↔ coil and turn	Alpha helix ↔ random coil/turn
Interaction with membrane	Insert into one bilayer leaflet	Stay at the membrane surface (insert into hydrocarbon chains of the outer leaflet)	Stay at the membrane surface (insert into hydrocarbon chains of the outer leaflet)
Maturation to prefusion state through	Proteolytic processing of fusion protein (except Ebola, Sars)	Proteolytic processing of companion protein	No proteolytic process
Activated in the glycoprotein complex	2 proteins (PIV 5) 1 cleaved protein (HA) 1 uncleaved (Ebo,SARS)	2 identical or different proteins (SFV, TBEV)	1 uncleaved protein (G) 3 proteins (gB with gH/gL)

Contrary to the relatively simple and canonical organization of the fusion peptides for the influenza virus or the flavivirus E protein, the recently resolved structures of the gB glycoprotein of the simplex herpes type 1 virus (Heldwein *et al.*, 2006) and G protein (Roche *et al.*, 2006, 2007) of the vesicular stomatitis virus (VSV) indicated a bipartite fusion peptide composed of two hydrophobic loops, each loop being relatively nonpolar or very weakly hydrophobic (which rarely leads to the identification of a fusion peptide by fusion peptide prediction based on hydrophobic domain identification). These differences in the organization of the fusion peptides suggest differences of action of the fusion proteins, notably in the number required. According to experiments using neutralization assays, it seems that one HIV envelope glycoprotein is capable of inducing membrane fusion. However, it has been shown that the induced fusion by HA requires a collaboration of several complexes of envelope glycoprotein (8–9 trimers). In the same way, different networks of class II fusion proteins have been proposed, such as the hexagonal organization observed on the surface of liposomes or an association of five trimers in a structure similar to a volcano based on structure predictions (Stiasny and Heinz, 2006). Concerning the fusion protein of class III, a hexagonal structure has been proposed for rabies virus envelope glycoprotein and recently for VSV-G according to a modeling using its structure. The requirement for multiple fusion peptides may be compared to the number of receptors required. The assembly of a complex containing several receptors may be a prerequisite for the membrane fusion steps that require multiple EnvGP molecules to cooperatively participate in this process. For example, in the case of HIV-1, the presence of more than one CD4 in contact with the virus enhances the infectivity dramatically and reduces the concentration of coreceptors needed for infection (Platt et al., 1998). Further investigation of this system has implied that a critical complex containing approximately four to six coreceptors is a requirement for infection, although it is not known whether this complex performs a transient role and then disperses or is maintained throughout the membrane fusion process (Kuhmann et al., 2000). Despite some uncertainties, several lines of evidence have suggested that three to six hemagglutin trimers may cooperatively participate in the influenza A virus-mediated membrane fusion reaction (Blumenthal et al., 1996; Boulay et al., 1988) and that multiple envelope glycoprotein trimers are required for rabies virus-mediated membrane fusion (Roche and Gaudin, 2002). see Table 4.3.

3. The role of cytoplasmic tail in fusion and influence of its length

The cytoplasmic tails of envelope viruses harbor different motifs that are responsible for its trafficking and are variable in length. It is surprising to see that the cytoplasmic tails of fusion proteins are not exchangeable. For example, when the HCV E1 and E2 cytoplasmic tails or the F cytoplasmic tails of HRSV (Human Respiratory Syncytial virus) are substituted for that of VSV-G envelope glycoprotein (or CD4), the fusogenicity of these envelopes in cell–cell fusion assays and virus–cell assay (infection) is destroyed (Buonocore et al., 2002; Oomens et al., 2006). Similarly, when the HA glycoprotein is anchored by a GPI, the entry process is stopped at the hemifusion step (Kemble et al., 1994; Markosyan et al., 2000). The cellular localizations (e.g., cholesterol-rich microdomains) are sometimes modified, and biochemical modifications (glycosylation, oligomerization) can affect the properties of the VSV-G EnvGP (Kemble et al., 1994). Nevertheless, in their initial context, these cytoplasmic tails allow fusion.

For some viruses, regulation of fusion is mediated by the cleavage of the cytoplasmic tail. For γ-retrovirus, in addition to the SU–TM cleavage, a significant fraction of virion-associated TM is further processed by the viral protease removing the C-terminal 16 amino acids of the cytoplasmic domain, the R peptide. γ-Retrovirus virions assemble and bud from infected cells as immature particles that must undergo an additional proteolytic maturation to become infectious (Green et al., 1981; Lavillette et al., 1998, 2002; Rein et al., 1994). This maturation concerns the viral protease-dependent cleavage of the so-called peptide R at the C-terminus of the cytoplasmic tail. The R peptide inhibits fusion, and different hypothesis have been proposed. First, the R peptide contains an Y-X-X-internalization motif, and the removal of this motif following the cleavage of the R peptide might result in a higher amount of envelope at the surface membrane and thus more fusion (Song *et al.*, 2003). Second, following the R peptide cleavage, the remaining cyt tail forms a membrane-embedded amphiphilic alpha-helix domain that destabilizes the membrane (Rozenberg-Adler *et al.*, 2008; Zhao *et al.*, 1998). Third, it has been proposed that as the R peptide contains a palmitoylation, its removal induces the close trimerization of the cyt tail and drastic conformational changes in the ectodomain of Env (Aguilar *et al.*, 2003) which might influence Env fusogeneity by destabilizing SU–TM complexes. These conformational changes are necessary for the isomerization of the SU–TM disulfide MLV Env (Loving *et al.*, 2008). This R peptide cleavage is the last step leading to a fusion competent infectious MLV retrovirus, but this final modification does not exist in lentiviruses which harbor a long cytoplasmic tail. However, truncations of the long cytoplasmic domains of lentiviral Env proteins occur under certain culture conditions (Chakrabarti *et al.*, 1989) and increase Env fusogenicity in a similar way to mutated truncated versions of SIV, HIV-1, and HIV-2 Envs (Mulligan *et al.*, 1992; Spies *et al.*, 1994; Wilk *et al.*, 1992). This regulation is a hallmark of adaptation of endogenous retroviruses, as this cleavage is fulfilled by a cellular protease that activates the endogenous EnvGP HERV-W in relevant tissues involved in placenta development. It is interesting to note that for most viruses (and not only the γ-retrovirus), the truncation of the cytoplasmic tail increases the fusogenicity of the EnvGP, as is seen in paramyxoviruses (Moll *et al.*, 2002). In many cases, this truncation seems to increase the cell surface expression (since cellular trafficking signals in the cytoplasmic tails are modified), which may explain the increase of fusogenicity. For certain other Env, the truncation leads to conformational changes in the ectodomain which lowers the activation threshold for fusion, resulting in enhanced Env fusion activity and kinetics (Aguilar *et al.*, 2003; Cote *et al.*, 2008; Spies *et al.*, 1994; Zhao *et al.*, 1998). Finally, the cleavage of the cytoplasmic tail sometimes allows, at least partially, a cell–cell fusion at neutral pH (pH-independent), although the entry of the virus remains pH-dependent. The notion of pH-dependence, seems in this case, to be due to a specific conformation or a particular density of envelope glycoprotein (Cote *et al.*, 2008).

4. The transmembrane proximal region

Numerous studies on several viruses have highlighted the critical role of the pretransmembrane sequence (PTM), also called the membrane proximale region or the juxtamembrane domain (JMD), which is rich in aromatic amino acids. The class I fusion glycoproteins of coronavirus, lentivirus (HIV, SIV, FIV), ebola virus (Munoz-Barroso *et al.*, 1999; Saez-Cirion *et al.*, 2003), and many other viruses contain this short JMD region in the ectodomain between the end of HR-2 and the beginning of the transmembrane (TM) domain (Salzwedel et al., 1999). The class II (SFV, dengue, TEBV) and class III (VSV and the herpes virus) fusion proteins also possess these regions, although they are less rich in tryptophan than the class I JMD (Jeetendra et al., 2002, 2003; Roche et al., 2008). Although the JMDs of fusion proteins of enveloped viruses are rich in aromatic amino acids, the number, spacing, and sequence of the aromatic amino acids are quite variable; however, the function remains the same. These JMDs contribute to the conformational changes that occur during membrane fusion, interact with membranes, induce membrane destabilization, and/or facilitate membrane fusion (Munoz-Barroso et al., 1999). Therefore, the JMD of viral fusion proteins is a potential target for viral inhibitors. Entry inhibitors that target the JMD of class I fusion proteins include monoclonal antibodies that bind the JMD of gp41 to the FIV and to the HIV (Lorizate et al., 2006; Purtscher et al., 1994). Some peptides that mimic the JMD have been designed for EnvGP of FIV (Giannecchini et al., 2004), HIV (Moreno et al., 2006), Ebola virus (Saez-Cirion et al., 2003), and the SARS virus (Howard et al., 2008) and inhibit viral entry. This strategy has been also broadly used against many other paramyxoviruses such as the Sendai virus (Joshi et al., 1998; Rapaport et al., 1995), the Newcastle disease virus (Young et al., 1999), the human parainfluenza type 3 (HPIV-3) (Yao and Compans, 1996), the respiratory syncytial virus (RSV), and the measles virus (MV) (Lambert et al., 1996). In the same way, peptides that mimic the JMD from class II and III fusion proteins have also been developed against infection by Dengue virus (Hrobowski et al., 2005) and CMV (English et al., 2006; Lopper and Compton, 2004). In the case of HCV, some juxtamembrane domains have been proposed in E1 and E2 (Drummer and Poumbourios, 2004; Drummer et al., 2007). High-throughput screening (HTS) of peptides derived from E1 and E2 sequences has identified inhibitory peptides close to the transmembrane domain of E2, though they are not among the most inhibitory (Cheng et al., 2008). However, these strategies suffer from certain limitations. The derived peptides of these JMD often have the capacity to oligomerize, which inactivates them, and some stratagems need to be implemented to make the peptides more bioreactive against viruses. Moreover, these peptides that prevent the correct conformational change of the envelope glycoprotein, must act in a certain window of time and in a particular compartment compatible with their active structure. Indeed, the acid pH of some compartment is not compatible with the bioactive structure of the peptide. For the SARS-CoV virus, the peptide is able to inhibit the entry of the virus in the presence of protease at the cellular surface, but the peptide has little effect on the entry of the virus by endocytosis or on the activation by cathepsin L in the acid conditions of the endosome (Ujike et al., 2008). In the case of Influenza, it has not been possible to develop such inhibitory peptides targeting the juxtamembrane domain. Reasons suggested for this include the entry through acid compartments which modify protein structures, the entry by different pathways of endocytose, and the rapidity of the conformation changes of HA. Some modifications can be made to these peptides to increase their efficiency for viruses that fuse at the plasma membrane. For three paramyxoviruses, HPIV-3, a major cause of lower respiratory tract diseases in infants, and the emerging zoonotic viruses Hendra virus (HeV) and NiV, which cause lethal central nervous system (CNS) diseases, the addition of cholesterol to a paramyxovirus HRC-derived peptide (derived from the heptad repeat immediately preceding the transmembrane domain) increased antiviral potency by 2 log units (Porotto et al., 2010). This enhanced activity is the result of the targeting of the peptide to the plasma membrane. The cholesterol-tagged peptides on the cell surface create a protective antiviral shield, target the F protein directly at its site of action, and expand the potential utility of inhibitory peptides for paramyxoviruses.

III. HOW VIRUSES SUBVERT DIFFERENT CELL PROTEINS FOR ENTRY?

A. Definition of receptors, adsorption molecules, and cofactors

The definition of a receptor is very complicated and has limitations. It is sometimes difficult to distinguish between "simple" receptors that mediate adsorption or binding and that may not even initiate conformational changes, and "critical" receptors for fusion which, upon binding, will generate the conformational changes that will allow the exposure of the fusion peptide and lead to membrane fusion. Some cellular molecules are also involved in the localization or trafficking of viral receptors and these are important cofactors of entry. Another ambiguity is the role of enzymatic activities. Some are necessary to process the fusion protein inside producer cells, or inside the endosomes of target cells, and they are necessary to activate the potential of the EnvGP for membrane fusion. While it is inaccurate to consider them as "receptors", they are certainly critical cofactors.

1. Virus adsorption

All viruses likely bind at least weakly to multiple cell surface components such as heparan sulfate proteoglycans, DC-SIGN, integrins, or glycolipids (Bounou et al., 2002; Cantin et al., 1997; Fortin et al., 1997; Jinno-Oue et al., 2001; Mondor et al., 1998; Saphire et al., 1999, 2001). Although such binding substances probably do not induce conformational changes in EnvGP that are necessary for membrane fusion, they can enhance viral adsorption and substantially increase efficiencies of infections, thus contributing to pathogenesis (Alvarez *et al.*, 2002; Bounou *et al.*, 2002; Geijtenbeek *et al.*, 2000; Jinno-Oue *et al.*, 2001; Saphire *et al.*, 2001). Because such binding proteins contribute to infections, it can be difficult to unambiguously distinguish them from receptors that directly mediate the membrane fusion process, especially for viruses that bind to their authentic receptors only weakly (e.g., for retroviruses, in the cases of FeLV-T or polytropic MuLVs; Anderson *et al.*, 2000; Marin *et al.*, 1999; Temin, 1988). We emphasize this because pathogenic variants of different animal viruses have often been associated with abilities to bind to apparently novel cell surface components, and it has sometimes been inferred that the viruses have switched their receptor specificities. In these instances, it has generally not been established that the cell surface binding components are receptors that directly mediate infections. In the case of Ebola, for example, the receptor(s) that mediates its entry has yet to be definitively identified. C-type lectins such as DC-SIGN and DC-SIGNR are thought to serve as adherence factors for Ebola or Marburg virus (Marzi *et al.*, 2006; Matsuno *et al.*, 2010). Other plasma membrane-associated proteins have been implicated in EBOV uptake, including folate receptor alpha and the tyrosine kinase receptor Axl (Chan *et al.*, 2001; Shimojima *et al.*, 2006, 2007; Sinn *et al.*, 2003), but the physical interaction of EBOV GP and these proteins has not been demonstrated, and cells that do not express these proteins are permissive for EBOV GP-mediated virion uptake. Some previous studies have implicated the actin cytoskeleton in EBOV entry, where agents such as cytochalasin D and swinholide A that impair microfilament function, inhibited GP-mediated entry (Yonezawa *et al.*, 2005). Similarly, VSV was shown to bind ubiquitously to cells via phosphatidylserine (PS) (Schlegel *et al.*, 1983). However, a more recent study reports that PS is not a receptor for VSV, as no correlation was found between cell surface PS levels and VSV infection, and annexin V, which specifically binds PS, did not inhibit infection of VSV (Coil and Miller, 2004). Therefore, the cell surface receptors for VSV have not been identified, but it is generally thought that binding via the G-protein is rather unspecific and involves negative charges on the plasma membrane (Carneiro *et al.*, 2006; Coil and Miller, 2005).

Adsorption of viruses onto cultured cells from the medium is usually a very slow and inefficient process, principally because of the slow rates of their diffusion into contact with the cell surfaces (Allison and Valentine, 1960; Andreadis *et al.*, 2000). In general, the rate of contact cannot be significantly enhanced by mixing or stirring, because the boundary layer of the relatively stationary fluid that surrounds walls or other large objects (e.g., cells) in flowing liquids is substantial compared with the rate of virus diffusion, hence stirring does not increase the concentration of virus surrounding this boundary zone (Allison and Valentine, 1960). In the case of retroviruses, it has become especially clear that adsorption is a severely limiting step in infection of cultured cells. In classic studies in which virus samples were incubated with cells for several hours before washing with fresh medium and subsequently detecting the foci of infection, it was estimated that only 1/1000 or fewer of the virions in the medium were infectious. In contrast to previous interpretations, studies suggest that this low infectivity-to-virion ratio is principally caused by the inefficiency of adsorption (Andreadis et al., 2000). Accordingly, serial incubation of a virus-containing medium for 2-h periods with sequential cell cultures results in the same titers in each of the cultures after correction for spontaneous viral decay (Kabat et al., 1994). Furthermore, centrifuging the virus down onto the cultured cells (i.e., spinoculation) often increases retroviral titers by 1–2 log orders of magnitude (Bahnson et al., 1995).

Studies aiming to count retrovirions adsorbed onto cell surfaces by confocal immunofluoresence microscopy or by quantitative PCR methods (Marechal et al., 2001; Pizzato et al., 1999, 2001) have demonstrated that receptors for viral entry are irrelevant for initial adsorption of retrovirions onto surfaces of most cells (Pizzato et al., 1999, 2001). On the contrary, the initial steps of virus attachment seem to more critically depend on accessory cellular binding substances, such as heparin sulfates, integrins, or lectins, including DC-SIGN (Bounou et al., 2002; Guibinga et al., 2002; Jinno-Oue et al., 2001; Mondor et al., 1998; Saphire et al., 2001). By forming multivalent weak reversible bonds with such abundant cell surface components, a virus would become efficiently bound in a manner that would allow it to "graze" until it makes appropriate contact with a true receptor (Haywood, 1994; Park et al., 2000).

2. S–S shuffling

It is well known that for several viruses (Rubella togavirus, BVDV pestivirus, Newcastle disease paramyxovirus, HIV lentivirus, mouse hepatitis coronavirus (MHV)), rearrangements of the thiol content and of the disulfide bridges, induced by thioredoxine or protein disulfide isomerase (PDI), are essential to induce some big conformational changes necessary for the membrane fusion (Fenouillet et al., 2007).

Interestingly, the MLV and HTLV retroviruses developed an "internal" oxydoreduction activity by adapting a catalytic motif involved in disulfure bridge isomerization. The two SU and TM subunits can be linked in either a covalent or noncovalent manner. For HIV-1, the existence of the soluble gp120 protein indicates a noncovalent link between SU and TM (Kowalski et al., 1987). However, for most other retroviruses, a covalent link was described at one stage. In all cases, with the exception of MMTV and JSRV, a disulfide bond between the SU and the TM is formed between the highly conserved CX6CC motif of the TM and the CXXC of the SU (Pinter et al., 1997; Schulz et al., 1992; Sitbon et al., 1991). This CXXC motif is extremely rare in cellular proteins and is similar to a motif found in the catalytic site of enzymes involved in thiol isomerization, like PDI or thioredoxin (Pinter et al., 1997; Sanders, 2000). This motif in the SU has been shown to be part of an autocatalytic isomerization function of SU to destroy the initial bond between SU and TM generated during Env synthesis and create an intra-SU bond inside the CXXC motif (Li et al., 2008; Wallin et al., 2004). This disulfide bond isomerization is crucial for the fusogenicity of γ-retrovirus (MLV) (Fenouillet et al., 2007).

3. Fusion activation by proteases

Some viruses use the protease activity of particular cellular enzymes localized in the endosomes or at the cellular surface for activating their EnvGP. The Ebola filovirus (for which the GP1–GP2 envelope glycoprotein is cleaved by the furine in the producer cells) (Chandran et al., 2005; Schornberg et al., 2006), the HeV (Pager and Dutch, 2005), NiV (Diederich et al., 2009), or the SARS-CoV and MHV-2 coronaviruses (for which the S spikes is not cleaved in the producer cells) (Huang et al., 2006; Qiu et al., 2006; Simmons et al., 2005) require the cysteins lysosomal proteases, the L or B cathepsine (CatL or CatB) for their entry process. After virus uptake following angiotensin-converting enzyme 2 receptor binding, cathepsin L-mediated proteolysis induces conformational changes in the SARS-CoV S glycoprotein to trigger the endosomal membrane fusion process (Simmons et al., 2005). Likewise, cleavage of the Ebola glycoprotein by CatL cleavage removes a glycosylated glycan cap and mucin-like domain (MUC domain) and exposes the conserved core residues implicated in receptor binding (Chandran et al., 2005; Hood et al., 2010; Lee et al., 2008; Schornberg et al., 2006). Entry of this virus is pH-dependent and associated with the cleavage of GP by proteases, including CatL and/or CatB, in the endosome or cell membrane, which is required for entry into the host cell. However, the precise role of the cleavage of Ebola envelope glycoprotein, which is already cleaved intracellularly during its exit, is uncertain. The cleavage of the GP to a stable form of 18 kDa of GP1 may increase binding, suggesting that the cleavage facilitates the interaction with a cellular receptor (Dube et al., 2009; Kaletsky et al., 2007). Another possibility is that the CatB cleavage is required to facilitate the triggering of viral membrane fusion by destabilizing the prefusion conformation of Ebola envelope glycoprotein (Wong et al., 2010).

The cathepsins comprise a family of lysosomal protease enzymes whose primary function (i.e., protein degradation) plays a critical role in normal cellular homeostasis. Cathepsin L is one of the 11 members of human lysosomal cysteine proteases (i.e., B, C, F, H, K, L, O, S, V, W, and X) that fall in the C1 family (papain family) of the CA clan (Rossi et al., 2004). These enzymes were traditionally linked to nonspecific proteolytic activity within lysosomes. More recently, cathepsin L has been implicated in regulatory events relating to cancer, diabetes, immunological responses, degradation of the articular cartilage matrix, and other pathological processes (Vasiljeva *et al.*, 2007), including osteoporosis, rheumatoid arthritis, and tumor metastasis (Palermo and Joyce, 2008).

Interestingly, several host cell proteases appear to be able to prime fusion activation in the case of SARS-CoV, including cathepsin L, trypsin, factor Xa, thermolysin, plasmin, TMPRSS11a, and elastase. The proteolytic cleavage events in SARS-CoV S that lead to membrane fusion occur both at the S1/S2 boundary and adjacent to a fusion peptide in the S2 domain (Belouzard *et al.*, 2009). Elastase-mediated activation of SARS-CoV was originally reported by Taguchi and coworkers, and it has been proposed that elastase may have important implications for viral pathogenesis (Matsuyama *et al.*, 2005). Elastase is known to be secreted by neutrophils as part of an inflammatory response to a viral infection and is also produced by opportunistic bacteria (e.g., *Pseudomonas aeruginosa*) that can colonize virally infected respiratory tissue. As such, it has been considered that elastase-mediated activation of SARS-CoV might be an important factor in the severe pneumonia seen in SARS-CoV-infected patients (Belouzard *et al.*, 2010).

In conclusion, in order to better classify the receptors, the receptors must be differentiated according to their precise function: those that permit a nonspecific adsorption, such as the glycosaminoglycans (used by Dengue, TBE, HSV-1), the type C lectin receptors, such as L-SIGN, DC-SIGN, the hMGL, the LSECtin, and the asialoglycoprotein receptor (used by HCV, Ebola, Marburg); those that permit a specific adsorption receptor (binding receptor) allowing the sorting (toward particular endosomes and intracytoplasmic compartment) and the initial conformational changes (such as the CD4 to HIV, the integrins or the laminin receptor for the Dengue, Sindbis, or Lassa viruses); and finally, those that permit the exposition of the domains implicated in the destabilization of the membrane (like the fusion peptide) and are the latest receptors to act (such as the HIV coreceptors). Therefore, studying the kinetics of the conformation changes of the EnvGP and the kinetics of action and utilization of the receptors is essential to accurately categorize the receptors (nonspecific adsorption receptor, receptor of binding, receptor of fusion).

All the cellular proteins that allow the exposition, localization, and the trafficking of these receptors and/or endosomes should be considered as cofactors. With all the siRNA screens that are coming out, the list of such cofactors is dramatically increasing.

B. Use of multiple receptors—Receptor switch

The use of different receptors often correlates with the need of a virus to overcome barriers existing in the cell type or tissue that they infect. One well-studied example is the binding of Coxsackievirus B to decay-accelerating factor

(DAF) in the apical surface of epithelial cells, and subsequently to the Coxsack-ievirus and adenovirus receptor (CAR), which is localized in the tight junction region. DAF helps to bring the virus to the tight junctions, and CAR induces a conformational change and promotes endocytosis (Coyne and Bergelson, 2006).

Almost all animal viruses use receptors exclusively containing a single TM sequence (see Table 4.1). In striking contrast, cell surface receptors for γ-retroviruses have multiple transmembrane (TM) sequences, compatible with their identification in known instances as transporters for important solutes. Similarly, hepatitis C virus, in addition to LDL receptor, exclusively uses multi-transmembrane receptors:the Scavenger receptor class B type 1 (SRB1 with two transmembrane domains), the tetraspanin receptors CD81 and Claudins (4 TMs), and another tight junction protein Occludin-1 (4 transmembrane domains; Ploss and Rice, 2009).

Another surprise is that some viruses, including many γ-retroviruses, use not just one receptor but pairs of closely related receptors as alternatives (Tailor et al., 2003). This appears to have enhanced viral survival by severely limiting the likelihood of host escape mutations. All of the receptors used by γ-retroviruses contain hypervariable regions that are often heavily glycosylated and that control the viral host range properties, consistent with the idea that these sequences are battlegrounds of virus–host coevolution. However, in con-trast to previous assumptions, it is probable that γ-retroviruses have adapted to recognize conserved sites that are important for the receptor's natural function and that the hypervariable sequences have been elaborated by the hosts as defense bulwarks surrounding the conserved viral attachment sites.

The fact that all virus groups have been severely limited throughout evolution in the types of receptors they can employ, may initially appear incon-sistent with evidence that some viruses can switch their receptor specificities with apparent ease. This has been most dramatically suggested by shifts of influenza A viruses between animal reservoirs, which involve single amino acid changes in the viral hemagglutinin, enabling recognition of different sialic acid structures (e.g., N-acetyl or N-glycolyl neuraminic acids in α2,6 or α2,3 linkages to galactose) that predominate in the different host species (Baranowski et al., 2001; Gambaryan et al., 2005; Skehel and Wiley, 2000). Similarly, slight changes in specificity for receptors accompanied the emergence, in 1978, of the canine parvovirus (Parker et al., 2001). However, these are small shifts in receptor specificities rather than global jumps to dissimilar receptors. Similar slight shifts are involved in the change from the CCR5 to CXCR4 coreceptor usage during AIDS progression (Scarlatti et al., 1997). Small shifts in usages of highly similar receptors have also been reported during cell culture selections of subgroup B, D, and E avian leukosis viruses that all use polymorphic variants of the same TVB receptor (Taplitz and Coffin, 1997) and during cell culture selections of HIV-1 variants (Platt et al., 1998). Therefore, despite the rarity of receptor repertoire expansions throughout millions of years of retrovirus evolution, limited switches can occur within single infected animals. For example, there is an evolution of coreceptor usage in HIV-1/AIDS, some *in vivo* adaptations of ecotropic MuLVs and formation of polytropic MuLVs, and some evolution of altered receptor usages in domestic cats infected with FeLV-A (Tailor *et al.*, 2003).

Several viruses have been reported to use multiple alternative receptors or even alternative pathways for infection of cells. For example, MV isolates appear to be capable of using CD46 or SLAM, which both contain single TM domains (Baranowski *et al.*, 2001; Oldstone *et al.*, 1999; Tatsuo *et al.*, 2000). Complex viruses such as herpesviruses that contain several distinct EnvGP are also typically able to bind to several cell surface components (Baranowski *et al.*, 2001; Borza and Hutt-Fletcher, 2002). The foot-and-mouth disease picornavirus (FMDV) may also use multiple receptors, including heparan sulfates and integrins, and may, in addition, be able to invade cells via immunoglobulin Fc receptors when the virus is coated with antibodies (Baranowski *et al.*, 2001; Mason *et al.*, 1994). This alternative entry route is also used by the dengue flavivirus, which may explain the extremely strong pathogenesis that occurs when it reinfects previously exposed individuals (Baranowski *et al.*, 2001). In the case of FMDV, it has not been established whether heparin sulfate is a true receptor that directly mediates infection or merely a binding factor that influences infection indirectly by enhancing virus adsorption. HIV-1 infections are also strongly stimulated by accessory cell surface binding components including heparan sulfates, glycolipids, and DC-SIGN (Bounou *et al.*, 2002; Geijtenbeek *et al.*, 2000; Pohlmann *et al.*, 2001). Similarly, a paralysis-inducing neurotropic variant of Friend MuLV binds more strongly than the parental virus to heparan sulfate, and thereby becomes more infectious for brain capillary endothelial cells while still remaining dependent on the CAT1 receptor (Jinno-Oue *et al.*, 2001). These examples illustrate how changes in affinities for accessory binding substances can dramatically alter cellular tropisms and pathogenesis of viruses, and why it has often been difficult to distinguish such accessory binding factors from true receptors or coreceptors that are essential for infections. On the basis of these considerations, we believe that the available evidence strongly supports our proposal that all virus groups have been severely constrained in the types of receptors they can employ for infection of cells. However, some viruses have evolved several pathways for infection, and viruses such as HIV-1 have evolved distinct sites in a single SU glycoprotein for recognition of dissimilar receptors and coreceptors.

C. Separation of the binding and fusion functions

The complexity of EnvGP is variable. An example of a very simple fusion protein is the FAST proteins of Orthoreovirus that do not belong, however, to the family of the enveloped virus (Barry *et al.*, 2010; Shmulevitz and Duncan, 2000). The

Orthoreovirus genus includes some no-enveloped viruses that cause the cell–cell fusion of infected cells. This fusion activity is due to the small no-structural membrane viral proteins named FAST proteins (fusion-associated small transmembrane protein), in size ranging between ~10 and 15 kDa. The FAST protein ectodomains are very small, with extreme cases of only 20 residues, and contain hydrophobic short regions with/without acid properties, and a myristoylation site. Though their small size challenges their ability to form a hairpin structure, these FAST proteins expressed alone are sufficient to induce the membrane fusion (Barry et al., 2010; Shmulevitz and Duncan, 2000). Surprisingly, they are able to traffic as far as the plasma membrane without inducing intracellular disorder. Indeed, by opposition, the expression of retroviral TM or orthomyxovirus HA2 alone does not lead to cell surface membrane expression, and these subunits are blocked intracellularly. The fusion induced by the FAST protein is very broad, which questions their use of a particular protein receptor. Thus, some fusion proteins are simple and possess solely the fusion function, and others comprise several domains in addition to the domain involved in fusion. The function of some of these domains has remained more or less independent from the fusion domain, and sometimes, they can be naturally separated in different proteins (i.e., H and F from paramyxoviruses) or experimentally on different fragments, as explain below. Such is the case of the γ-retroviruses for which the function of binding to the receptor can be separated from the function of fusion. This line of inquiry was initiated by the studies of Bae et al. (1997) relating to a conserved PHQ motif that occurs near the amino-terminal ends of SU glycoproteins in all γ-retroviruses. Mutation of this PHQ motif blocked membrane fusion but had no effect on receptor attachment. Subsequently, we discovered that noninfectious γ-retrovirions lacking this histidine could be transactivated by addition of a soluble SU or an amino-terminal fragment of SU, called the receptor-binding domain (RBD), to the cultured cells (Lavillette et al., 2000). Interestingly, studies by Overbaugh and coworkers have demonstrated that similar transactivation processes can occur in natural infections by γ-retroviruses. Specifically, they found that infections by the immunosuppressive FeLV-T virus, which has a Pro in place of His in its PHQ motif, require transactivation either by a soluble FeLV-B-related SU glycoprotein termed FELIX that is endogenously expressed in cat T cells or by an FeLV-B SU glycoprotein (Anderson et al., 2000). This activating process parallels that of herpes simplex virus for the transmission of a fusogenic signal among the EnvGP of the herpes simplex virus on receptor binding by glycoprotein gD. The soluble gD ectodomain has been shown to allow entry of engineered HSV-1 virus particles that lack gD (i.e., gD-null mutants; Cocchi et al., 2004). The evidence reviewed provides very strong support for the hypothesis that attachment of viruses to their receptors initiates a pathway that obligatorily contains intermediate steps. These intermediate steps very likely include viral association with multiple receptors, cooperative conformational changes within Env glycoprotein, and cross-talk between Env on the viral surfaces. In the case of the retroviruses, this evidence suggests that virus binding to receptors does not directly induce irreversible structural changes in SU–TM complexes as was previously believed. Rather, it implies that the binding to receptors induces SU–SU interactions that are pre-requisites for later steps in a highly coordinated membrane fusion pathway. We anticipate that similar intermediate steps are likely to be involved in infections by other groups of retroviruses and perhaps in infections by other membrane-enveloped viruses.

Other viruses can be activated by soluble receptors that, by interacting with the envelope glycoprotein, induce some of the conformational changes necessary to trigger fusion. This example indicates that binding and fusion steps can be separated and can take place at different locations. In these cases, MHV, avian retrovirus ASLV, and the herpes simplex 1 (HSV-1) can infect some cells that do not harbor binding receptors provided that the soluble form is present in the infection media (the soluble CEACAM1 receptor for MHVR, the soluble Tva receptor for ASLV, and the soluble nectin-1 for HSV-1; Kwon et al., 2006; Matsuyama and Taguchi, 2002; Mothes et al., 2000).

IV. BASIC AND TRANSLATIONAL RESEARCH EXPLOITING ENTRY PROPERTIES OF VIRUSES

A. Tropism properties and use of pseudoparticles in gene therapy

Vectors derived from retroviruses such as lentiviruses and oncoretroviruses are probably among the most suitable tools to achieve long-term gene transfer, since they allow stable integration of a transgene and its propagation in daughter cells. Lentiviral vectors should be the preferred gene-delivery vehicles over vectors derived from oncoretroviruses (MLV) since, in contrast to the latter, they can transduce nonproliferating target cells. Moreover, lentiviral vectors that have the capacity to deliver transgenes into specific tissues are expected to be of great value for various gene transfer approaches in vivo (Frecha et al., 2008b). To achieve such in vivo gene transfer, innovative approaches have been developed to upgrade lentiviral vectors for tissue or cell targeting and which have potential for in vivo gene delivery. One strategy is to develop vectors that harbor EnvGP with selective tropisms. Vectors derived from retroviruses offer particularly flexible properties in gene transfer applications, given the numerous possible associations of various viral surface glycoproteins (determining cell tropism) with viral cores. Selective tropisms were achieved by taking advantage of the natural tropisms of EnvGP from other membrane-enveloped viruses. For instance, the use of surface glycoproteins derived from viruses that cause lung infection and infect via the airway epithelia, such as Ebola virus or Influenza virus, may prove useful for gene therapy of the human airway (Kobinger et al., 2001). Exclusive transduction of retinal pigmented epithelium could be achieved following sub-retinal inoculations of some vector pseudotypes in rat eyes (Duisit et al., 2002). High transgene expression was detected in dermal fibroblasts transduced with VSV-G-, EboZ-, or MuLV-pseudotyped HIV vector and effectively targeted quiescent epidermal stem cells which underwent terminal differentiation result-ing in transgene expression in their progenies (Hachiya et al., 2007). Important-ly, several viral EnvGP target lentiviral vector to the CNS such as rabies (Wong et al., 2004), mokola (Watson et al., 2002; Wong et al., 2004), lymphocytic choriomeningitis virus envelope (LCMV) (Miletic et al., 2004; Stein et al., 2005), or Ross River (Kang et al., 2002) viral EnvGP that even permit transduc-tion of specific cell types within the CNS. Likewise, screening of a large panel of pseudotyped vectors established the superiority of the gibbon ape leukemia virus (GALV) (Sandrin et al., 2002; Stitz et al., 2000) and the cat endogenous retroviral-modified glycoproteins (RD114) (Sandrin et al., 2002) for transduction of progenitor and differentiated hematopoietic cells. Recently, a new LV carry-ing the MV EnvGP on its surface was able to overcome vector restrictions in both quiescent T and B cells (Frecha et al., 2008a, 2009). Importantly, naive as well as memory T and B cells were efficiently transduced, while no apparent activation, cell-cycle entry, or phenotypic switching were detected, opening the door to a multitude of gene therapy and immunotherapy applications. Vectors derived from HIV pseudotyped with Sendai virus fusion protein F (Kowolik and Yee, 2002) or E1E2 from hepatitis C virus (Bartosch et al., 2003), and such vectors are able to transduce human hepatoma cells and primary human hepa-tocytes efficiently, although they are unable to enter nonliver cells. The GP64 glycoprotein from baculovirus *Autographa californica* multinuclear polyhedrosis virus pseudotyped FIV efficiently and also showed excellent hepatocyte tropism (Kang et al., 2005).

B. Identification of viral cell entry receptors using pseudoparticles

The screening of cDNA libraries has emerged as a powerful tool to identify and clone viral entry receptors. It is an alternative to the use of human/Chinese hamster radiation hybrid panels of cells. In order to clone an unknown receptor by complementation screens, cDNA from a cell permissive to infection by a certain virus is introduced into a nonpermissive cell. As some recessive cell lines are poorly transfectable, the use of pseudoparticles provides a good tool to transduce the cDNA library for this cloning strategy. A retroviral cDNA library approach, involving transfer and expression of cDNAs from highly infectable cells to nonpermissive cells, has been used to clone and identify the MuLV polytropic X-receptor (Battini et al., 1999; Tailor et al., 1999a; Yang et al., 1999), the RD114 ASCT2 receptor (Rasko *et al.*, 1999; Tailor *et al.*, 1999b), FeLV-C FLVCR1 receptor from human and domestic cat cDNA libraries (Quigley *et al.*, 2000; Tailor *et al.*, 1999c), and two closely related human proteins, PHuR-A1 and PHuR-A2, that function as receptors for PERV-A (Ericsson *et al.*, 2003). Briefly, when a cell line that is not susceptible to a particular pseudotype retrovirus vector harboring an envelope for which the receptor is not known, a cDNA library from a cell line highly susceptible to transduction, was constructed by cloning the cDNA into a retroviral expression vector. Afterward, the cDNA retroviral expression library was transduced into nonsusceptible cells by infection at a relatively low multiplicity of infection so that the majority of infectants would contain single-copy provirus inserts. The library-containing cells were then screened for susceptibility to pseudotype vector transduction through selection of drug-resistant cells after exposure to the vector carrying a resistance gene. Of drug-resistant clones obtained from the primary screen and using PCR primers specific for vector sequences, cDNA products from clones with conferred susceptibility were identified after nested PCR was performed on DNA extracted from reinfectable clones.

However, for many viruses, initial attempts using a retroviral cDNA library were unsuccessful due to an inherent background of nonspecific infection with pseudoparticles. In fact, no cell line was completely nonpermissive to even "no envelope" pseudoparticles bearing no viral envelope proteins, indicating the existence of nonspecific uptake mechanisms. In the screens, this resulted in a high background of drug-resistant colonies, independent of glycoprotein-mediated cell entry. Thus, unless the entry factor cDNA was highly represented in the library, a single round of transduction/challenge would not suffice. To deal with the high background observed in screens, methods that would allow multiple rounds of selection and enrichment have been developed. Recently, the use of a cyclic packaging rescue (CPR) system using non-self-inactivating vectors has been shown to increase the efficiency of receptor cloning with powerful iterative screening methods (Evans *et al.*, 2007; Ploss *et al.*, 2009). Most retroviral vectors commonly used for gene-delivery applications are self-inactivating vectors that contain deletions in the long terminal repeat (LTR) elements. No packaging competent retroviral RNA transcripts are generated from such integrated proviruses; instead, transgene expression is driven by an internal nonretroviral promoter. In contrast, if the cDNA library is constructed in a provirus that retains the complete LTR elements, the retroviral promoter is active in transduced cells and a full-length viral RNA is expressed. Expression of the packaging components, gag-pol and vesicular stomatitis virus VSV-G envelope (that will efficiently infect recessive cell lines), in these cells allows packaging of the RNA into pseudoparticles capable of transducing naive cells. This approach, termed CPR (Bhattacharya *et al.*, 2002; Koh *et al.*, 2002), allows retrieval of the library after selection has been performed, followed by transduction of a naive cell population, concluded by a new round of selection. This process can be repeated sequentially for an unlimited number of selection/enrichment steps.

For an additional level of selection, different challenge virus genomes can be used, each encoding a different drug-selectable marker (e.g., puromycin (PuroR) or zeocin (ZeoR) resistance), in self-inactivating retroviruses. Thus, after challenge and selection of a library-transduced population with one pseudoparticle-packaged selection cassette (e.g., PuroR), the population can be pooled and rechallenged with the second-selectable pseudovirus (e.g., ZeoR). Then, during CPR, only the full-length retroviral transcripts from the non-self-inactivating provirus that encodes the library but not the self-inactivating challenge virus genomes are repackaged and transferred to the naive cell population. This enables researchers to perform multiple rounds of selection, thereby overcoming the background of nonspecific pseudoparticle uptake. Moreover, using this scheme, underrepresented cDNA will be enriched. Once the final round of selection of pseudotype-susceptible cells has been achieved, genomic DNA can be prepared from selected clones and used as a template in a PCR across the provirus cDNA-cloning site.

The nonpermissive target cell line for the cDNA screen adheres to several stringent criteria. As stated above, the primary requirement is that (1) to minimize nonspecific background, cell lines with minimal uptake of pseudoparticle of interest and "no envelope" pseudoparticles were preferred. (2) To ensure that nonpermissiveness was a phenotype due to the lack of a pseudoparticle-specific entry factor(s) rather than poor infection by pseudotypes in general, chosen cell lines should be highly permissive to an unrelated pseudoparticle (VSVGpp, rhabdovirus) infection. In addition, this also ensures that the target cell line would be easily transduced with the cDNA library. (3) For selection, candidate cell lines also had to be susceptible to the desired drug selections. (4) To perform multiple rounds of screening involving CPR, the ideal cell line had to demonstrate this method well and be highly transfectable. (5) Finally, to facilitate the screen, the chosen cell line needed to be relatively fast growing and clone efficiently.

C. Determining the endocytosis pathway of entry and the different cell proteins involved in entry by RNA interference screens

To gain insights into virus entry, it is necessary to examine several inhibitors of pathway-mediated endocytosis in terms of their role in blocking infection mediated by pseudotypes with different EnvGP. An advantage of pseudoparticles is that the entry process of pathogens of BSL4 can be studied in BSL2 conditions. For example, to gain insights into Ebola virus entry, inhibitors against different endocytosis pathways have been examined for their ability to block infection mediated by HIV pseudotyped with the Ebola EnvGP (Bhattacharyya *et al.*, 2010). The use of control pseudoparticles (e.g., pseudotyped with Vesicular Stomatitis Virus EnvGP (VSV G)) can be used as controls to assess cell viability and specificity of inhibition. Inhibition of clathrin function traditionally relied on three principal approaches: drugs that inhibit acidification of endosomes (such as BafA1, a specific, nonreversible endosomal proton pump inhibitor), as well as commonly used lysosomotropic agents (such as ammonium chloride (NH_4Cl) and chloroquine); potassium depletion; and finally, treatment of cells with brefeldin A (BFA) or chlorpromazine (Sieczkarski and Whittaker, 2002). As such, these drugs have multiple effects on cell function, and their use to inhibit virus infection should be treated with some caution. Thanks to pseudotypes, some other studies have implicated the actin cytoskeleton in Ebola virus entry, where agents such as cytochalasin D and swinholide A that impair microfilament function were shown to inhibit EnvGP-mediated entry (Yonezawa *et al.*, 2005). Ebola enters cells through a low-pH-dependent, endocytosis-mediated process. A large body of evidence indicates that Ebola viruses enter cells by clathrin-mediated endocytosis (Sanchez, 2007), but lipid raft-associated, caveolin-mediated endocytosis has also been proposed as an alternative mechanism of Ebola virus uptake (Empig and Goldsmith, 2002). Low-pH events lead to cathepsin-dependent cleavage of Ebola virus EnvGP that is required for productive uptake of the virus (Chandran *et al.*, 2005; Kaletsky *et al.*, 2007; Schornberg *et al.*, 2006). Other low-pH-dependent events have been postulated to be required as well (Schornberg *et al.*, 2006). Furthermore, Ebola virus likely uses a Rho-mediated pathway, as is seen in VSV virus, suggesting that this may be a route of entry utilized by many different viruses (Quinn *et al.*, 2009).

More recently, proteins interfering with endocytosis, such as the use of dominant-negative Eps15, or RNA interference (RNAi) have been developed, and such approaches target the different pathways with higher specificity (Mercer and Helenius, 2009; Mercer *et al.*, 2010b). The RNAi approach allows researchers to perform screens to identify previously unrecognized host factors that are required for viral replication. RNAi screens rely on either short interfering RNAs (siRNAs) or short hairpin RNAs (shRNAs) to knock down the function of a particular gene in a cell. Researchers can then infect the cells with specific viruses and monitor levels of viral replication. If viral replication is reduced, then the knocked-down gene might be necessary for the virus to replicate itself or function within the host cell.

Some small-scale screens have been developed using wild-type viruses (Kolokoltsov *et al.*, 2007) or pseudoparticles (Trotard *et al.*, 2009). Pseudoparticles can be exploited to focus a siRNA screen specifically on the entry step of a virus. Indeed, only the entry steps are governed by the EnvGP, whereas all the uncoating, integration, and expression steps of the transgene depend on retroviral proteins. Having said that, this suffers an inherent drawback in that some steps are dependent on retroviral proteins and therefore limits the scope of the screen. One limitation to the siRNA screen is specificity and cell toxicity, which can be overcome by the use of pseudotypes. Indeed, a siRNA screen can be done with different pseudotypes in parallel. If, contrary to the pseudotypes of interest, some pseudotypes are not affected by siRNA, this indicates that the siRNA specifically inhibits the virus of interest and, moreover, that the siRNA is not toxic. To better characterize the entry pathway of the hepatitis C virus, a small interfering RNA library dedicated to membrane trafficking and remodeling was screened in the context of the Huh-7.5.1 liver cell line cells infected by HCV pseudoparticles (HCVpp) (Trotard et al., 2009). Results showed that the downregulation of different factors implicated in clathrin-mediated endocytosis inhibit HCVpp cell infection. In addition, knockdown of the phosphatidylinositol 4-kinase type III-alpha (PI4KIIIalpha) prevented infection by HCVpp, and the presence of PI4KIIIbeta in the host cells influenced their susceptibility to HCVpp infection. This library screening using pseudoparticles identified two kinases, PI4KIIIalpha and beta, as being involved in the HCV life cycle. These results have been confirmed in published works of the identification of cellular factors required for the HCV life cycle using siRNA libraries screening either over 4500 drugable genes (Ng et al., 2007; Vaillancourt et al., 2009), 140 cellular membrane-trafficking genes (Berger et al., 2009; Coller et al., 2009), kinome (Supekova et al., 2008), or the entire genome (Li et al., 2009; Tai et al., 2009). These works are much more time- and cost-consuming, and even though pseudoparticles may appear to be an overly simple model to study and identify host factors necessary for viral infection, they are also powerful and flexible tools that have led to major discoveries, as these examples have shown.

Recently, however, in order to identify host factors necessary for viral infections, researchers are turning to genome-wide genetic screens. Indeed, the sequencing of the human genome and the emergence of technologies that allow the silencing of individual genes in cells one by one using siRNA, when combined with the use of genome-wide siRNA libraries and automated HTS methodology, allows molecular information to be gained about virtually every critical intracellular event occurring during virus infection. Within the last few years, there have been several publications describing such genome-wide siRNA screens that have been applied to virus infections in tissue-cultured cells. RNAi screens have now been preformed for several viruses, including HIV (Brass et al., 2008; Konig et al., 2008; Yeung et al., 2009; Zhou et al., 2008), West Nile Virus (Krishnan et al., 2008), Influenza (Karlas et al., 2010; Konig et al., 2008), and, in drosophila cells, Dengue (Sessions et al., 2009) and vaccinia virus (Moser et al., 2010). From the hundreds of cellular factors that have been implicated in the outcome of infection, some were already known from other studies, but the majority are entirely novel, and the validation and analysis of data is ongoing in many laboratories. However, a poor overlap between results from different screens published on the same viruses has been observed. The differences are probably linked not only to technical variations but also to the inherent risk of false positives through off-target effects and false negatives due to inefficient silencing. Cell toxicity is a common complication, as well as cell-type specificity. Following-up such screens with pseudoparticle studies is useful to validate the cell proteins that are involved in the entry step only and facilitate the analysis, compared to the use of a replication competent virus.

D. Screening and development of entry inhibitors using pseudoparticles

Resistance to individual antivirals is likely to develop for most specific therapeutics targeting particular viral proteins, thus making therapy consisting of a combination of drugs targeting different stages of the viral life cycle highly desirable. The entry process represents another series of potential targets for therapeutic intervention. It has not been extensively explored due to high-throughput experimental limitations for some viruses that require biosafety level 3 or 4 or for the viruses with no robust infection system.

Pseudoparticle infection systems, utilizing different reporter genes or proteins (i.e., firefly luciferase or GFP), can be developed for HTS of small molecule libraries, peptide libraries, or neutralizing antibodies for their entry inhibitor properties. In order to facilitate the screen, assay performance can be improved by modifying the properties of both the parental host cell line and the pseudovirus. For example, cells with improved pseudoparticle entry and cell spreading can be selected. Pseudoparticles can be improved by using EnvGP that increase infectivity either by selecting a specific variant or by expressing a human codon-optimized envelope glycoprotein sequence. Finally, the pseudoviruses can be engineered to express a human codon-optimized reporter to improve the sensitivity of the assay. Using these modifications, HTS can be developed using 96- or 384-well plates. Compounds found to inhibit pseudoparticle infection can then be counterscreened against pseudoparticles containing other variants to characterize the cross-reactivity of the molecules in a virus family. Moreover, molecules can then be counterscreened against pseudoparticles harboring EnvGP from other virus families to evaluate the specificity of the inhibitor or its large spectrum.

A wide variety of entry inhibitors, namely, peptides, chemical compounds, or antibodies, exist. They can interfere with the different steps of the entry process, such as the binding to receptors or the conformational changes of the fusion proteins, by acting on the envelope protein itself, on the acidification of the endosomes, if necessary, or on the endocytosis.

For HIV, most marketed drugs for treating AIDS are inhibitors of HIV-1 reverse transcriptase or protease enzymes, but new targets include the integrase enzyme, cell–surface interactions, membrane fusion, and also virus particle maturation and assembly (Kaushik-Basu et al., 2008). Entry inhibition entails preventing HIV-1 breaching the cell, either as a strategy to prevent infection altogether or to curtail infection of new cells in an HIV-positive individual. Several strategies have proved effective in HIV-1 entry inhibition either *in vitro* (using pseudoparticles or replication competent viruses) or *in vivo*: binding to viral surface proteins gp120 and gp41, binding to human cell surface receptor CD4, and binding to human cell surface coreceptors, CCR5 and CXCR4 (Leonard and Roy, 2006; Liu et al., 2007). In particular, the synthesized peptide T-20 is believed to act by binding to gp41 (Wild et al., 1993, 1994) and is currently in clinical use. Yet, several more recent studies have revealed that T-20 does not block the six-helix bundle prehairpin formation (Liu et al., 2005). Another peptide, C37, derived from the C-terminus of gp41 (and nearly identical to the widely reported C-peptide C34; Liu et al., 2005; Root et al., 2001), is also described as having strong anti-HIV entry activity due to the tight binding of the gp41 N-terminal helices. Maraviroc is a small antiretroviral compound known to be a CCR5 antagonist, which blocks R5-tropic HIV entry into CD4 cells (Hunt and Romanelli, 2009). Several studies have established that synergy can occasionally be observed when two fusion inhibitors are combined in a viral assay or *in vitro* fusion assay. For example, some synergy is observed in the combination of CCR5 antibodies with T-20 (Ji et al., 2007; Murga et al., 2006), the combination of small molecular antagonists of coreceptors with T-20 (Tremblay et al., 2000, 2002), the combination of small molecular antagonists of CCR5 with CCR5 antibodies (Ji et al., 2007; Murga et al., 2006), and the combination of small molecular antagonists of CCR5 with chemokines (Murga et al., 2006; Tremblay et al., 2002).

In the case of HCV, entry into hepatocytes is a multistep process, involving at least four cellular receptors, leading to virion endocytosis and fusion of the viral and cellular membranes. Unlike the HCV replication process, these steps have not been thoroughly exploited as targets for antiviral intervention. Recently, with the development of HCVpp and the JFH1 infectious molecular clone, it has become possible to test drugs against entry. *In vitro*, proof-of-concept studies for inhibiting the HCV entry process have been demonstrated using cyanovirin-N that targets the N-linked glycans of the viral envelope proteins and prevents E2–CD81 interaction (Helle et al., 2006), neutralizing antibodies directed against the HCV E1 and E2 proteins (Broering et al., 2009; Habersetzer et al., 1998; Keck et al., 2008; Law et al., 2008; Owsianka et al., 2005; Perotti et al., 2008; Schofield et al., 2005; Yu et al., 2004), antibodies against cellular receptors CD81 (Bartosch et al., 2003; Cormier et al., 2004; Lavillette et al., 2005; Molina et al., 2008), SR-BI (Catanese et al., 2007; Zeisel et al., 2007), Claudin-1

(Krieger *et al.*, 2010), and agents that block endosomal acidification (Bartosch *et al.*, 2003; Hsu *et al.*, 2003; Koutsoudakis *et al.*, 2006; Lavillette *et al.*, 2005). *In vivo* studies using humanized trimera mice model or human liver-u-PA-SCID mice have also demonstrated prophylactic efficacy of monoclonal anti-E2 (Eren *et al.*, 2006) and anti-CD81 antibodies (Meuleman *et al.*, 2008), respectively. Some broad-spectrum antiviral drugs have also been tested (Pecheur *et al.*, 2007), but more recent studies have used the HCVpp system in order to undertake a screening campaign that led to the discovery of a class of small molecule HCV-specific inhibitors consisting of several structurally related compounds defined by a common triazine core (Baldick *et al.*, 2010). Inhibition of entry was confirmed by using time-of-addition experiments to demonstrate that inhibitor activity is confined to the first 3 h of infection, with inhibition occurring postattachment and closely linked to the inhibition kinetics of the endosomal acidification inhibitor bafilomycin.

Pseudoparticles can also be used to identify the mode of action of molecules identified using a replication competent virus in a cell-based screening system involving multiple rounds of infection in a 96-well format. After analysis of a publicly available library of 446 clinically approved drugs, the impact of 33 identified compounds on viral entry was tested using HCVpp infection to recapitulate HCV particle adsorption, internalization, and viral envelope-mediated fusion (Gastaminza *et al.*, 2010). Many of the candidates were lysoso-motropic compounds that inhibited HCV entry with differential efficacy.

Both pseudotype (Larson *et al.*, 2008) and infectious virus screening (Bolken *et al.*, 2006) identified broadly active arenavirus entry inhibitors. Isolation and mapping of resistant viruses, as well as chimeras between sensitive and resistant strains, mapped the target of activity to the GP2 subunit of the G envelope protein complex, specifically to the interface between the C-terminal stem and TMD domains. Mechanistic studies showed that these arenavirus inhibitors prevented low-pH-induced fusion by blocking reorganization between the GP2 stem with N-terminal domains of the G protein complex (York *et al.*, 2008).

Another alternative for developing entry inhibitor compounds is to target endosomal protease necessary for entry of certain viruses. Inhibitors of cathepsin L block viral entry of severe acute respiratory syndrome coronavirus (SARS-CoV) and Ebola virus and impair conversion of HeV glycoprotein into the mature, active form (Chandran *et al.*, 2005; Pager and Dutch, 2005; Simmons *et al.*, 2005). With respect to the development of antiviral agents, inhibitors of human cathepsin L are not subject to resistance arising from rapid mutations of the viral genome (Shah *et al.*, 2010), making Cathepsin L an attractive target for drug development. HTS for cathepsin L inhibitors identified a novel thiocarba-zate compound exhibiting potent inhibition against cathepsin L (Beavers *et al.*, 2008; Myers *et al.*, 2008; Shah *et al.*, 2008). This compound prevented 293T cell infection by the Ebola and SARS-CoV pseudotypes, respectively. In addition, the thiocarbazate inhibited *in vitro* propagation of malaria parasite *Plasmodium falciparum* and inhibited *Leishmania* major (Shah *et al.*, 2008).

Finally, another natural entry inhibitor is provided by monoclonal neutralizing antibodies. Many monoclonal antibodies can be isolated from immortalized B-cells recovered from patients or from mice hybridomas following immunization. The antibodies can be selected by ELISA assay using soluble envelope proteins or pseudoparticles. The neutralizing potential of these antibodies can be easily screened using pseudoparticles with high-throughput infection assays, and the inhibitory effect can be verified with replicating viruses. In the case of HCV, using an antibody antigen-binding fragment phagedisplay library generated from a donor chronically infected with HCV, of 115 clones showing specific binding to HCV E2 glycoprotein, 5 monoclonal antibodies presented neutralizing activities against cell-culture HCV (HCVcc), JFH-1 virus, and a panel of HCVpp displaying E1–E2 from diverse genotypes (Law *et al.*, 2008). Overall, neutralizing antibodies can inhibit the different steps of the entry process from binding to membrane fusion by targeting the domains involved in this step or by limiting the conformational changes of the envelope complex. Interestingly, one study has recently highlighted the feasibility of targeting short-lived envelope glycoprotein intermediates for inhibition of membrane fusion using monoclonal antibodies (York *et al.*, 2010). This action is very similar to the peptides against HIV membrane fusion on the market. Such strategies to effectively target fusion peptide function in the endosome may lead to the discovery of novel classes of antiviral agents, and screens using pseudoparticles will provide an easily wielded system to identify such infrequent antibodies.

V. CONCLUSIONS

All viruses have developed varied mechanisms to reach the same goal. They vary greatly in structure, but all seem to have a common mechanism of action, in which a ligand-triggered large-scale conformational change in the fusion protein is coupled to apposition and merger of the two bilayers. In spite of the different mechanisms to activate the fusion peptide, fusion proteins are distributed into three classes based on their structural homologies. Future experiments must aim to elucidate the molecular mechanisms and the dynamics of the conformational changes driving virus entry. This will require the development of new approaches to study the rapid conformational changes of a small number of membrane interacting protein molecules and domains. A more realistic goal is the determination of all the structures of proteins that mediate the entry of all human viruses, either at a prebinding or postfusion stage.

Another challenge will be the identification of the cognate cellular receptors. Identification of all the cellular receptors for human viruses would be an important contribution to our understanding of virus tropism and pathogenesis. Once known, the role of receptors in entry, as a specific or nonspecific binding factor, or as receptors needed for conformational changes, or for routing the virus to the right compartment, will have to be established. Moreover, for most viruses, the EnvGP appear to function in an autonomous manner and can permit fusion without a requirement for receptors at acidic pH. Thus, the role of varied receptors remains enigmatic for many viruses. It is difficult to predict the roles of a receptor in fusion, sorting and routing the virus toward a particular favorable compartment in the pursuit of the infectious cycle based upon its family and its structure. The activation process of EnvGP and the postbinding events are early steps that are crucial to understand, as they could provide targets for the development of therapeutics. The better understanding of the envelope–receptor interaction also raises hopes for the possibility of designing entry machines that can deliver genes and other molecules to any cell of choice.

Recently, the use of whole genome siRNA screen has become more and more widely used for different viruses in order to identified factors important for entry. It should drastically increase our knowledge of the factors necessary for entry of viruses. Similarly, many high-throughput interactome studies will identify cellular proteins interacting with the different viral components. Altogether, the comparison of all these high-throughput screens should help us to identify cellular proteins and pathways common to different viruses which may help, with rational structure/mechanism-based design of entry inhibitors, to develop inhibitors that cross-react with different pathogens. In a similar fashion, the design of vaccine immunogens that are capable of eliciting potent, broadly neutralizing antibodies of known epitopes, is expected to contribute toward the development of vaccines. In parallel, the development of panels of human monoclonal antibodies against every entry-related protein from all pathogenic human viruses could accelerate our understanding of entry mechanisms and help to fight viral diseases. If research continues at the present pace, most of these goals could be accomplished within the next few decades.

Acknowledgments

Our research is supported by INSERM, ENS Lyon, CNRS, and UCB Lyon-I and by grants from the European Union (LSHB-CT-2004-005246 "COMPUVAC"), the European Research Council (ERC Advanced Grant to FLC No. 233130 "HEPCENT"), the Agence Nationale de Recherches sur le SIDA et les Hepatites Virales (ANRS), and the FINOVI foundation. We thank Sarah Kabani for helpful comments on this chapter. We apologize to the many authors of important original contributions that were not directly cited. Lastly, we thank the past and present members of our laboratory for their many experimental and intellectual contributions.

References

Aguilar, H. C., Anderson, W. F., and Cannon, P. M. (2003). Cytoplasmic tail of Moloney murine leukemia virus envelope protein influences the conformation of the extracellular domain: implications for mechanism of action of the R Peptide. *J. Virol.* **77**, 1281–1291.

Allison, A. C., and Valentine, R. C. (1960). Virus particle adsorption. II. Adsorption of vaccinia and fowl plague viruses to cells in suspension. *Biochim. Biophys. Acta* **40**, 393–399.

Alvarez, C. P., Lasala, F., Carrillo, J., Muniz, O., Corbi, A. L., and Delgado, R. (2002). C-type lectins DC-SIGN and L-SIGN mediate cellular entry by Ebola virus in cis and in trans. *J. Virol.* **76**, 6841–6844.

Anderson, M. M., Lauring, A. S., Burns, C. C., and Overbaugh, J. (2000). Identification of a cellular cofactor required for infection by feline leukemia virus. *Science* **287**, 1828–1830.

Andreadis, S., Lavery, T., Davis, H. E., Le Doux, J. M., Yarmush, M. L., and Morgan, J. R. (2000). Toward a more accurate quantitation of the activity of recombinant retroviruses: alternatives to titer and multiplicity of infection. *J. Virol.* **74**, 1258–1266.

Backovic, M., and Jardetzky, T. S. (2009). Class III viral membrane fusion proteins. *Curr. Opin. Struct. Bol.* **19**, 189–196.

Backovic, M., Longnecker, R., and Jardetzky, T. S. (2009). Class III viral membrane fusion proteins. *Proc. Natl. Acad. Sci. USA* **106**, 2880–2885.

Bae, Y., Kingsman, S. M., and Kingsman, A. J. (1997). Functional dissection of the Moloney murine leukemia virus envelope protein gp70. *J. Virol.* **71**, 2092–2099.

Bahnson, A. B., Dunigan, J. T., Baysal, B. E., Mohney, T., Atchison, R. W., Nimgaonkar, M. T., Ball, E. D., and Barranger, J. A. (1995). Centrifugal enhancement of retroviral mediated gene transfer. *J. Virol. Methods* **54**, 131–143.

Baldick, C. J., Wichroski, M. J., Pendri, A., Walsh, A. W., Fang, J., Mazzucco, C. E., Pokornowski, K. A., Rose, R. E., Eggers, B. J., Hsu, M., Zhai, W., Zhai, G., et al. (2010). A novel small molecule inhibitor of hepatitis C virus entry. *PLoS Pathog.* **6**, 100–110.

Balliet, J. W., and Bates, P. (1998). Efficient infection mediated by viral receptors incorporated into retroviral particles. *J. Virol.* **72**, 671–676.

Baranowski, E., Ruiz-Jarabo, C. M., and Domingo, E. (2001). Evolution of cell recognition by viruses. *Science* **292**, 1102–1105.

Barry, C., Key, T., Haddad, R., and Duncan, R. (2010). Features of a spatially constrained cystine loop in the p10 FAST protein ectodomain define a new class of viral fusion peptides. *J. Biol. Chem.* **285**, 16424–16433.

Bartosch, B., Dubuisson, J., and Cosset, F. L. (2003). Infectious hepatitis C virus pseudo-particles containing functional E1-E2 envelope protein complexes. *J. Exp. Med.* **197**, 633–642.

Basak, A., Zhong, M., Munzer, J. S., Chretien, M., and Seidah, N. G. (2001). Implication of the proprotein convertases furin, PC5 and PC7 in the cleavage of surface glycoproteins of Hong Kong, Ebola and respiratory syncytial viruses: a comparative analysis with fluorogenic peptides. *Biochem. J.* **353**, 537–545.

Basak, A., Chretien, M., and Seidah, N. G. (2002). A rapid fluorometric assay for the proteolytic activity of SKI-1/S1P based on the surface glycoprotein of the hemorrhagic fever Lassa virus. *FEBS Lett.* **514**, 333–339.

Battini, J. L., Rasko, J. E., and Miller, A. D. (1999). A human cell-surface receptor for xenotropic and polytropic murine leukemia viruses: possible role in G protein-coupled signal transduction. *Proc. Natl. Acad. Sci. USA* **96**, 1385–1390.

Beavers, M. P., Myers, M. C., Shah, P. P., Purvis, J. E., Diamond, S. L., Cooperman, B. S., Huryn, D. M., and Smith, A. B., 3rd. (2008). Molecular docking of cathepsin L inhibitors in the binding site of papain. *J. Chem. Inf. Model.* **48**, 1464–1472.

Belouzard, S., Chu, V. C., and Whittaker, G. R. (2009). Activation of the SARS coronavirus spike protein via sequential proteolytic cleavage at two distinct sites. *Proc. Natl. Acad. Sci. USA* **106,** 5871–5876.

Belouzard, S., Madu, I., and Whittaker, G. R. (2010). Elastase-mediated activation of the severe acute respiratory syndrome coronavirus spike protein at discrete sites within the S2 domain. *J. Biol. Chem.* **285,** 22758–22763.

Berger, K. L., Cooper, J. D., Heaton, N. S., Yoon, R., Oakland, T. E., Jordan, T. X., Mateu, G., Grakoui, A., and Randall, G. (2009). Roles for endocytic trafficking and phosphatidylinositol 4-kinase III alpha in hepatitis C virus replication. *Proc. Natl. Acad. Sci. USA* **106,** 7577–7582.

Bergeron, E., Vincent, M. J., Wickham, L., Hamelin, J., Basak, A., Nichol, S. T., Chretien, M., and Seidah, N. G. (2005). Implication of proprotein convertases in the processing and spread of severe acute respiratory syndrome coronavirus. *Biochem. Biophys. Res. Commun.* **326,** 554–563.

Bertrand, P., Cote, M., Zheng, Y. M., Albritton, L. M., and Liu, S. L. (2008). Jaagsiekte sheep retrovirus utilizes a pH-dependent endocytosis pathway for entry. *J. Virol.* **82,** 2555–2559.

Beyer, W. R., Popplau, D., Garten, W., von Laer, D., and Lenz, O. (2003). Endoproteolytic processing of the lymphocytic choriomeningitis virus glycoprotein by the subtilase SKI-1/S1P. *J. Virol.* **77,** 2866–2872.

Bhattacharya, D., Logue, E. C., Bakkour, S., DeGregori, J., and Sha, W. C. (2002). Identification of gene function by cyclical packaging rescue of retroviral cDNA libraries. *Proc. Natl. Acad. Sci. USA* **99,** 8838–8843.

Bhattacharyya, S., Warfield, K. L., Ruthel, G., Bavari, S., Aman, M. J., and Hope, T. J. (2010). Ebola virus uses clathrin-mediated endocytosis as an entry pathway. *Virology* **401,** 18–28.

Blumenthal, R., Sarkar, D. P., Durell, S., Howard, D. E., and Morris, S. J. (1996). Dilation of the influenza hemagglutinin fusion pore revealed by the kinetics of individual cell-cell fusion events. *J. Cell Biol.* **135,** 63–71.

Bolken, T. C., Laquerre, S., Zhang, Y., Bailey, T. R., Pevear, D. C., Kickner, S. S., Sperzel, L. E., Jones, K. F., Warren, T. K., Amanda Lund, S., Kirkwood-Watts, D. L., King, D. S., *et al.* (2006). Identification and characterization of potent small molecule inhibitor of hemorrhagic fever New World arenaviruses. *Antiviral Res.* **69,** 86–97.

Borza, C. M., and Hutt-Fletcher, L. M. (2002). Alternate replication in B cells and epithelial cells switches tropism of Epstein-Barr virus. *Nat. Med.* **8,** 594–599.

Boulay, F., Doms, R. W., Webster, R. G., and Helenius, A. (1988). Posttranslational oligomerization and cooperative acid activation of mixed influenza hemagglutinin trimers. *J. Cell Biol.* **106,** 629–639.

Bounou, S., Leclerc, J. E., and Tremblay, M. J. (2002). Presence of host ICAM-1 in laboratory and clinical strains of human immunodeficiency virus type 1 increases virus infectivity and CD4(+)-T-cell depletion in human lymphoid tissue, a major site of replication in vivo. *J. Virol.* **76,** 1004–1014.

Brass, A. L., Dykxhoorn, D. M., Benita, Y., Yan, N., Engelman, A., Xavier, R. J., Lieberman, J., and Elledge, S. J. (2008). Identification of host proteins required for HIV infection through a functional genomic screen. *Science* **319,** 921–926.

Broering, T. J., Garrity, K. A., Boatright, N. K., Sloan, S. E., Sandor, F., Thomas, W. D., Jr., Szabo, G., Finberg, R. W., Ambrosino, D. M., and Babcock, G. J. (2009). Identification and characterization of broadly neutralizing human monoclonal antibodies directed against the E2 envelope glycoprotein of hepatitis C virus. *J. Virol.* **83,** 12473–12482.

Buonocore, L., Blight, K. J., Rice, C. M., and Rose, J. K. (2002). Characterization of vesicular stomatitis virus recombinants that express and incorporate high levels of hepatitis C virus glycoproteins. *J. Virol.* **76,** 6865–6872.

Cantin, R., Fortin, J. F., Lamontagne, G., and Tremblay, M. (1997). The presence of host-derived HLA-DR1 on human immunodeficiency virus type 1 increases viral infectivity. *J. Virol.* **71**, 1922–1930.

Cantin, C., Holguera, J., Ferreira, L., Villar, E., and Munoz-Barroso, I. (2007). Newcastle disease virus may enter cells by caveolae-mediated endocytosis. *J. Gen. Virol.* **88**, 559–569.

Carneiro, F. A., Lapido-Loureiro, P. A., Cordo, S. M., Stauffer, F., Weissmuller, G., Bianconi, M. L., Juliano, M. A., Juliano, L., Bisch, P. M., and Da Poian, A. T. (2006). Probing the interaction between vesicular stomatitis virus and phosphatidylserine. *Eur. Biophys. J.* **35**, 145–154.

Catanese, M. T., Graziani, R., von Hahn, T., Moreau, M., Huby, T., Paonessa, G., Santini, C., Luzzago, A., Rice, C. M., Cortese, R., Vitelli, A., and Nicosia, A. (2007). High-avidity monoclonal antibodies against the human scavenger class B type I receptor efficiently block hepatitis C virus infection in the presence of high-density lipoprotein. *J. Virol.* **81**, 8063–8071.

Chakrabarti, L., Emerman, M., Tiollais, P., and Sonigo, P. (1989). The cytoplasmic domain of simian immunodeficiency virus transmembrane protein modulates infectivity. *J. Virol.* **63**, 4395–4403.

Chan, S. Y., Empig, C. J., Welte, F. J., Speck, R. F., Schmaljohn, A., Kreisberg, J. F., and Goldsmith, M. A. (2001). Folate receptor-alpha is a cofactor for cellular entry by Marburg and Ebola viruses. *Cell* **106**, 117–126.

Chandran, K., Sullivan, N. J., Felbor, U., Whelan, S. P., and Cunningham, J. M. (2005). Endosomal proteolysis of the Ebola virus glycoprotein is necessary for infection. *Science* **308**, 1643–1645.

Cheng, G., Montero, A., Gastaminza, P., Whitten-Bauer, C., Wieland, S. F., Isogawa, M., Fredericksen, B., Selvarajah, S., Gallay, P. A., Ghadiri, M. R., and Chisari, F. V. (2008). A virocidal amphipathic {alpha}-helical peptide that inhibits hepatitis C virus infection in vitro. *Proc. Natl. Acad. Sci. USA* **105**, 3088–3093.

Chernomordik, L. V., and Kozlov, M. M. (2003). Protein-lipid interplay in fusion and fission of biological membranes. *Annu. Rev. Biochem.* **72**, 175–207.

Chernomordik, L. V., and Kozlov, M. M. (2005). Membrane hemifusion: crossing a chasm in two leaps. *Cell* **123**, 375–382.

Chernomordik, L. V., and Kozlov, M. M. (2008). Mechanics of membrane fusion. *Nat. Struct. Mol. Biol.* **15**, 675–683.

Cocchi, F., Fusco, D., Menotti, L., Gianni, T., Eisenberg, R. J., Cohen, G. H., and Campadelli-Fiume, G. (2004). The soluble ectodomain of herpes simplex virus gD contains a membrane-proximal pro-fusion domain and suffices to mediate virus entry. *Proc. Natl. Acad. Sci. USA* **101**, 7445–7450.

Coil, D. A., and Miller, A. D. (2004). Phosphatidylserine is not the cell surface receptor for vesicular stomatitis virus. *J. Virol.* **78**, 10920–10926.

Coil, D. A., and Miller, A. D. (2005). Enhancement of enveloped virus entry by phosphatidylserine. *J. Virol.* **79**, 11496–11500.

Coller, K. E., Berger, K. L., Heaton, N. S., Cooper, J. D., Yoon, R., and Randall, G. (2009). RNA interference and single particle tracking analysis of hepatitis C virus endocytosis. *PLoS Pathog.* **5**, e1000702.

Cormier, E. G., Tsamis, F., Kajumo, F., Durso, R. J., Gardner, J. P., and Dragic, T. (2004). CD81 is an entry coreceptor for hepatitis C virus. *Proc. Natl. Acad. Sci. USA* **101**, 7270–7274.

Cote, M., Zheng, Y. M., Albritton, L. M., and Liu, S. L. (2008). Fusogenicity of Jaagsiekte sheep retrovirus envelope protein is dependent on low pH and is enhanced by cytoplasmic tail truncations. *J. Virol.* **82**, 2543–2554.

Coyne, C. B., and Bergelson, J. M. (2006). Virus-induced Abl and Fyn kinase signals permit coxsackievirus entry through epithelial tight junctions. *Cell* **124**, 119–131.

Decroly, E., Wouters, S., Di Bello, C., Lazure, C., Ruysschaert, J. M., and Seidah, N. G. (1996). Identification of the paired basic convertases implicated in HIV gp160 processing based on in vitro assays and expression in CD4(+) cell lines. *J. Biol. Chem.* **271**, 30442–30450.

DeTulleo, L., and Kirchhausen, T. (1998). The clathrin endocytic pathway in viral infection. *EMBO J.* **17,** 4585–4593.

Diederich, S., Dietzel, E., and Maisner, A. (2009). Nipah virus fusion protein: influence of cleavage site mutations on the cleavability by cathepsin L, trypsin and furin. *Virus Res.* **145,** 300–306.

Drummer, H. E., and Poumbourios, P. (2004). Hepatitis C virus glycoprotein E2 contains a membrane-proximal heptad repeat sequence that is essential for E1E2 glycoprotein heterodimerization and viral entry. *J. Biol. Chem.* **279,** 30066–30072.

Drummer, H. E., Boo, I., and Poumbourios, P. (2007). Mutagenesis of a conserved fusion peptide-like motif and membrane-proximal heptad-repeat region of hepatitis C virus glycoprotein E1. *J. Virol.* **88,** 1144–1148.

Dube, D., Brecher, M. B., Delos, S. E., Rose, S. C., Park, E. W., Schornberg, K. L., Kuhn, J. H., and White, J. M. (2009). The primed ebolavirus glycoprotein (19-kilodalton GP1,2): sequence and residues critical for host cell binding. *J. Virol.* **83,** 2883–2891.

Duisit, G., Conrath, H., Saleun, S., Folliot, S., Provost, N., Cosset, F. L., Sandrin, V., Moullier, P., and Rolling, F. (2002). Five recombinant simian immunodeficiency virus pseudotypes lead to exclusive transduction of retinal pigmented epithelium in rat. *Mol. Ther.* **6,** 446–454.

Earp, L. J., Delos, S. E., Park, H. E., and White, J. M. (2005). The many mechanisms of viral membrane fusion proteins. *Curr. Top. Microbiol. Immunol.* **285,** 25–66.

Empig, C. J., and Goldsmith, M. A. (2002). Association of the caveola vesicular system with cellular entry by filoviruses. *J. Virol.* **76,** 5266–5270.

Endres, M. J., Jaffer, S., Haggarty, B., Turner, J. D., Doranz, B. J., O'Brien, P. J., Kolson, D. L., and Hoxie, J. A. (1997). Targeting of HIV- and SIV-infected cells by CD4-chemokine receptor pseudotypes. *Science* **278,** 1462–1464.

English, E. P., Chumanov, R. S., Gellman, S. H., and Compton, T. (2006). Rational development of beta-peptide inhibitors of human cytomegalovirus entry. *J. Biol. Chem.* **281,** 2661–2667.

Epand, R. M. (2003). Fusion peptides and the mechanism of viral fusion. *Biochim. Biophys. Acta* **1614,** 116–121.

Eren, R., Landstein, D., Terkieltaub, D., Nussbaum, O., Zauberman, A., Ben-Porath, J., Gopher, J., Buchnick, R., Kovjazin, R., Rosenthal-Galili, Z., Aviel, S., Ilan, E., *et al.* (2006). Preclinical evaluation of two neutralizing human monoclonal antibodies against hepatitis C virus (HCV): a potential treatment to prevent HCV reinfection in liver transplant patients. *J. Virol.* **80,** 2654–2664.

Ericsson, T. A., Takeuchi, Y., Templin, C., Quinn, G., Farhadian, S. F., Wood, J. C., Oldmixon, B. A., Suling, K. M., Ishii, J. K., Kitagawa, Y., Miyazawa, T., Salomon, D. R., *et al.* (2003). Identification of receptors for pig endogenous retrovirus. *Proc. Natl. Acad. Sci. USA* **100,** 6759–6764.

Evans, M. J., von Hahn, T., Tscherne, D. M., Syder, A. J., Panis, M., Wolk, B., Hatziioannou, T., McKeating, J. A., Bieniasz, P. D., and Rice, C. M. (2007). Claudin-1 is a hepatitis C virus co-receptor required for a late step in entry. *Nature* **446,** 801–805.

Fenouillet, E., Barbouche, R., and Jones, I. M. (2007). Cell entry by enveloped viruses: redox considerations for HIV and SARS-coronavirus. *Antioxid. Redox Signal.* **9,** 1009–1034.

Fortin, J. F., Cantin, R., Lamontagne, G., and Tremblay, M. (1997). Host-derived ICAM-1 glycoproteins incorporated on human immunodeficiency virus type 1 are biologically active and enhance viral infectivity. *J. Virol.* **71,** 3588–3596.

Frampton, A. R., Jr., Stolz, D. B., Uchida, H., Goins, W. F., Cohen, J. B., and Glorioso, J. C. (2007). Equine herpesvirus 1 enters cells by two different pathways, and infection requires the activation of the cellular kinase ROCK1. *J. Virol.* **81,** 10879–10889.

Frecha, C., Costa, C., Levy, C., Negre, D., Russell, S. J., Maisner, A., Salles, G., Peng, K. W., Cosset, F. L., and Verhoeyen, E. (2009). Efficient and stable transduction of resting B lymphocytes and primary chronic lymphocyte leukemia cells using measles virus gp displaying lentiviral vectors. *Blood* **114,** 3173–3180.

Frecha, C., Costa, C., Negre, D., Gauthier, E., Russell, S. J., Cosset, F. L., and Verhoeyen, E. (2008a). Stable transduction of quiescent T cells without induction of cycle progression by a novel lentiviral vector pseudotyped with measles virus glycoproteins. *Blood* **112,** 4843–4852.

Frecha, C., Szecsi, J., Cosset, F. L., and Verhoeyen, E. (2008b). Strategies for targeting lentiviral vectors. *Curr. Gene Ther.* **8,** 449–460.

Gambaryan, A., Yamnikova, S., Lvov, D., Tuzikov, A., Chinarev, A., Pazynina, G., Webster, R., Matrosovich, M., and Bovin, N. (2005). Receptor specificity of influenza viruses from birds and mammals: new data on involvement of the inner fragments of the carbohydrate chain. *Virology* **334,** 276–283.

Garoff, H., Sjoberg, M., and Cheng, R. H. (2004). Budding of alphaviruses. *Virus Res.* **106,** 103–116.

Gastaminza, P., Whitten-Bauer, C., and Chisari, F. V. (2010). Unbiased probing of the entire hepatitis C virus life cycle identifies clinical compounds that target multiple aspects of the infection. *Proc. Natl. Acad. Sci. USA* **107,** 291–296.

Geijtenbeek, T. B., Kwon, D. S., Torensma, R., van Vliet, S. J., van Duijnhoven, G. C., Middel, J., Cornelissen, I. L., Nottet, H. S., KewalRamani, V. N., Littman, D. R., Figdor, C. G., and van Kooyk, Y. (2000). DC-SIGN, a dendritic cell-specific HIV-1-binding protein that enhances trans-infection of T cells. *Cell* **100,** 587–597.

Giannecchini, S., Bonci, F., Pistello, M., Matteucci, D., Sichi, O., Rovero, P., and Bendinelli, M. (2004). The membrane-proximal tryptophan-rich region in the transmembrane glycoprotein ectodomain of feline immunodeficiency virus is important for cell entry. *Virology* **320,** 156–166.

Gianni, T., Forghieri, C., and Campadelli-Fiume, G. (2006). The herpesvirus glycoproteins B and H. L are sequentially recruited to the receptor-bound gD to effect membrane fusion at virus entry. *Proc. Natl. Acad. Sci. USA* **103,** 14572–14577.

Green, N., Shinnick, T. M., Witte, O., Ponticelli, A., Sutcliffe, J. G., and Lerner, R. A. (1981). Sequence-specific antibodies show that maturation of Moloney leukemia virus envelope poly-protein involves removal of a COOH-terminal peptide. *Proc. Natl. Acad. Sci. USA* **78,** 6023–6027.

Guibinga, G. H., Miyanohara, A., Esko, J. D., and Friedmann, T. (2002). Cell surface heparan sulfate is a receptor for attachment of envelope protein-free retrovirus-like particles and VSV-G pseu-dotyped MLV-derived retrovirus vectors to target cells. *Mol. Ther.* **5,** 538–546.

Habersetzer, F., Fournillier, A., Dubuisson, J., Rosa, D., Abrignani, S., Wychowski, C., Nakano, I., Trepo, C., Desgranges, C., and Inchauspe, G. (1998). Characterization of human monoclonal antibodies specific to the hepatitis C virus glycoprotein E2 with in vitro binding neutralization properties. *Virology* **249,** 32–41.

Hachiya, A., Sriwiriyanont, P., Patel, A., Saito, N., Ohuchi, A., Kitahara, T., Takema, Y., Tsuboi, R., Boissy, R. E., Visscher, M. O., Wilson, J. M., and Kobinger, G. P. (2007). Gene transfer in human skin with different pseudotyped HIV-based vectors. *Gene Ther.* **14,** 648–656.

Harrison, S. C. (2005). Mechanism of membrane fusion by viral envelope proteins. *Adv. Virus Res.* **64,** 231–261.

Harrison, S. C. (2008). Viral membrane fusion. *Nat. Struct. Mol. Biol.* **15,** 690–698.

Haywood, A. M. (1994). Virus receptors: binding, adhesion strengthening, and changes in viral structure. *J. Virol.* **68,** 1–5.

Heldwein, E. E., Lou, H., Bender, F. C., Cohen, G. H., Eisenberg, R. J., and Harrison, S. C. (2006). Crystal structure of glycoprotein B from herpes simplex virus 1. *Science* **313,** 217–220.

Helenius, A., Kartenbeck, J., Simons, K., and Fries, E. (1980). On the entry of Semliki forest virus into BHK-21 cells. *J. Cell Biol.* **84,** 404–420.

Helle, F., Wychowski, C., Vu-Dac, N., Gustafson, K. R., Voisset, C., and Dubuisson, J. (2006). Cyanovirin-N inhibits hepatitis C virus entry by binding to envelope protein glycans. *J. Biol. Chem.* **281,** 25177–25183.

Hood, C. L., Abraham, J., Boyington, J. C., Leung, K., Kwong, P. D., and Nabel, G. J. (2010). Biochemical and structural characterization of cathepsin L-processed Ebola virus glycoprotein: implications for viral entry and immunogenicity. *J. Virol.* **84**, 2972–2982.

Howard, M. W., Travanty, E. A., Jeffers, S. A., Smith, M. K., Wennier, S. T., Thackray, L. B., and Holmes, K. V. (2008). Aromatic amino acids in the juxtamembrane domain of severe acute respiratory syndrome coronavirus spike glycoprotein are important for receptor-dependent virus entry and cell-cell fusion. *J. Virol.* **82**, 2883–2894.

Hrobowski, Y. M., Garry, R. F., and Michael, S. F. (2005). Peptide inhibitors of dengue virus and West Nile virus infectivity. *Virol. J.* **2**, 49.

Hsu, M., Zhang, J., Flint, M., Logvinoff, C., Cheng-Mayer, C., Rice, C. M., and McKeating, J. A. (2003). Hepatitis C virus glycoproteins mediate pH-dependent cell entry of pseudotyped retroviral particles. *Proc. Natl. Acad. Sci. USA* **100**, 7271–7276.

Huang, I. C., Bosch, B. J., Li, F., Li, W., Lee, K. H., Ghiran, S., Vasilieva, N., Dermody, T. S., Harrison, S. C., Dormitzer, P. R., Farzan, M., Rottier, P. J., *et al.* (2006). SARS coronavirus, but not human coronavirus NL63, utilizes cathepsin L to infect ACE2-expressing cells. *J. Biol. Chem.* **281**, 3198–3203.

Hunt, J. S., and Romanelli, F. (2009). Maraviroc, a CCR5 coreceptor antagonist that blocks entry of human immunodeficiency virus type 1. *Pharmacotherapy* **29**, 295–304.

Hunter, E. (1997). Viral entry and receptors. *In* "Retroviruses" (J. Coffin, S. Hughes, and H. Varmus, eds.), pp. 71–120. Cold Spring Harbor Laboratory Press, Cold Spring Harbor, NY.

Jeetendra, E., Robison, C. S., Albritton, L. M., and Whitt, M. A. (2002). The membrane-proximal domain of vesicular stomatitis virus G protein functions as a membrane fusion potentiator and can induce hemifusion. *J. Virol.* **76**, 12300–12311.

Jeetendra, E., Ghosh, K., Odell, D., Li, J., Ghosh, H. P., and Whitt, M. A. (2003). The membrane-proximal region of vesicular stomatitis virus glycoprotein G ectodomain is critical for fusion and virus infectivity. *J. Virol.* **77**, 12807–12818.

Ji, C., Zhang, J., Dioszegi, M., Chiu, S., Rao, E., Derosier, A., Cammack, N., Brandt, M., and Sankuratri, S. (2007). CCR5 small-molecule antagonists and monoclonal antibodies exert potent synergistic antiviral effects by cobinding to the receptor. *Mol. Pharmacol.* **72**, 18–28.

Jin, M., Park, J., Lee, S., Park, B., Shin, J., Song, K. J., Ahn, T. I., Hwang, S. Y., Ahn, B. Y., and Ahn, K. (2002). Hantaan virus enters cells by clathrin-dependent receptor-mediated endocytosis. *Virology* **294**, 60–69.

Jinno-Oue, A., Oue, M., and Ruscetti, S. K. (2001). A unique heparin-binding domain in the envelope protein of the neuropathogenic PVC-211 murine leukemia virus may contribute to its brain capillary endothelial cell tropism. *J. Virol.* **75**, 12439–12445.

Joshi, S. B., Dutch, R. E., and Lamb, R. A. (1998). A core trimer of the paramyxovirus fusion protein: parallels to influenza virus hemagglutinin and HIV-1 gp41. *Virology* **248**, 20–34.

Kabat, D., Kozak, S. L., Wehrly, K., and Chesebro, B. (1994). Differences in CD4 dependence for infectivity of laboratory-adapted and primary patient isolates of human immunodeficiency virus type 1. *J. Virol.* **68**, 2570–2577.

Kadlec, J., Loureiro, S., Abrescia, N. G., Stuart, D. I., and Jones, I. M. (2008). The postfusion structure of baculovirus gp64 supports a unified view of viral fusion machines. *Nat. Struct. Mol. Biol.* **15**, 1024–1030.

Kaletsky, R. L., Simmons, G., and Bates, P. (2007). Proteolysis of the Ebola virus glycoproteins enhances virus binding and infectivity. *J. Virol.* **81**, 13378–13384.

Kang, Y., Stein, C. S., Heth, J. A., Sinn, P. L., Penisten, A. K., Staber, P. D., Ratliff, K. L., Shen, H., Barker, C. K., Martins, I., Sharkey, C. M., Sanders, D. A., *et al.* (2002). In vivo gene transfer using a nonprimate lentiviral vector pseudotyped with Ross River Virus glycoproteins. *J. Virol.* **76**, 9378–9388.

Kang, Y., Xie, L., Tran, D. T., Stein, C. S., Hickey, M., Davidson, B. L., and McCray, P. B., Jr. (2005). Persistent expression of factor VIII in vivo following nonprimate lentiviral gene transfer. *Blood* **106,** 1552–1558.

Karlas, A., Machuy, N., Shin, Y., Pleissner, K. P., Artarini, A., Heuer, D., Becker, D., Khalil, H., Ogilvie, L. A., Hess, S., Maurer, A. P., Muller, E., *et al.* (2010). Genome-wide RNAi screen identifies human host factors crucial for influenza virus replication. *Nature* **463,** 818–822.

Kaushik-Basu, N., Basu, A., and Harris, D. (2008). Peptide inhibition of HIV-1: current status and future potential. *BioDrugs* **22,** 161–175.

Keck, Z. Y., Machida, K., Lai, M. M., Ball, J. K., Patel, A. H., and Foung, S. K. (2008). Therapeutic control of hepatitis C virus: the role of neutralizing monoclonal antibodies. *Curr. Top. Microbiol. Immunol.* **317,** 1–38.

Kee, S. H., Cho, E. J., Song, J. W., Park, K. S., Baek, L. J., and Song, K. J. (2004). Effects of endocytosis inhibitory drugs on rubella virus entry into VeroE6 cells. *Microbiol. Immunol.* **48,** 823–829.

Kemble, G. W., Danieli, T., and White, J. M. (1994). Lipid-anchored influenza hemagglutinin promotes hemifusion, not complete fusion. *Cell* **76,** 383–391.

Kielian, M., and Rey, F. A. (2006). Virus membrane-fusion proteins: more than one way to make a hairpin. *Nat. Rev. Microbiol.* **4,** 67–76.

Kobinger, G. P., Weiner, D. J., Yu, Q. C., and Wilson, J. M. (2001). Filovirus-pseudotyped lentiviral vector can efficiently and stably transduce airway epithelia in vivo. *Nat. Biotechnol.* **19,** 225–230.

Koh, E. Y., Chen, T., and Daley, G. Q. (2002). Novel retroviral vectors to facilitate expression screens in mammalian cells. *Nucleic Acids Res.* **30,** e142.

Kolokoltsov, A. A., Deniger, D., Fleming, E. H., Roberts, N. J., Jr., Karpilow, J. M., and Davey, R. A. (2007). Small interfering RNA profiling reveals key role of clathrin-mediated endocytosis and early endosome formation for infection by respiratory syncytial virus. *J. Virol.* **81,** 7786–7800.

Konig, R., Zhou, Y., Elleder, D., Diamond, T. L., Bonamy, G. M., Irelan, J. T., Chiang, C. Y., Tu, B. P., De Jesus, P. D., Lilley, C. E., Seidel, S., Opaluch, A. M., *et al.* (2008). Global analysis of host-pathogen interactions that regulate early-stage HIV-1 replication. *Cell* **135,** 49–60.

Koutsoudakis, G., Kaul, A., Steinmann, E., Kallis, S., Lohmann, V., Pietschmann, T., and Bartenschlager, R. (2006). Characterization of the early steps of hepatitis C virus infection by using luciferase reporter viruses. *J. Virol.* **80,** 5308–5320.

Kowalski, M., Potz, J., Basiripour, L., Dorfman, T., Goh, W. C., Terwilliger, E., Dayton, A., Rosen, C., Haseltine, W., and Sodroski, J. (1987). Functional regions of the envelope glycoprotein of human immunodeficiency virus type 1. *Science* **237,** 1351–1355.

Kowolik, C. M., and Yee, J. K. (2002). Preferential transduction of human hepatocytes with lentiviral vectors pseudotyped by Sendai virus F protein. *Mol. Ther.* **5,** 762–769.

Krieger, S. E., Zeisel, M. B., Davis, C., Thumann, C., Harris, H. J., Schnober, E. K., Mee, C., Soulier, E., Royer, C., Lambotin, M., Grunert, F., Dao Thi, V. L., *et al.* (2010). Inhibition of hepatitis C virus infection by anti-claudin-1 antibodies is mediated by neutralization of E2-CD81-claudin-1 associations. *Hepatology* **51,** 1144–1157.

Krishnan, M. N., Ng, A., Sukumaran, B., Gilfoy, F. D., Uchil, P. D., Sultana, H., Brass, A. L., Adametz, R., Tsui, M., Qian, F., Montgomery, R. R., Lev, S., *et al.* (2008). RNA interference screen for human genes associated with West Nile virus infection. *Nature* **455,** 242–245.

Kuhmann, S. E., Platt, E. J., Kozak, S. L., and Kabat, D. (2000). Cooperation of multiple CCR5 coreceptors is required for infections by human immunodeficiency virus type 1. *J. Virol.* **74,** 7005–7015.

Kuhn, R. J., Zhang, W., Rossmann, M. G., Pletnev, S. V., Corver, J., Lenches, E., Jones, C. T., Mukhopadhyay, S., Chipman, P. R., Strauss, E. G., Baker, T. S., and Strauss, J. H. (2002). Structure of dengue virus: implications for flavivirus organization, maturation, and fusion. *Cell* **108,** 717–725.

Kwon, H., Bai, Q., Baek, H. J., Felmet, K., Burton, E. A., Goins, W. F., Cohen, J. B., and Glorioso, J. C. (2006). Soluble V domain of Nectin-1/HveC enables entry of herpes simplex virus type 1 (HSV-1) into HSV-resistant cells by binding to viral glycoprotein D. *J. Virol.* **80,** 138–148.

Labonte, P., Begley, S., Guevin, C., Asselin, M. C., Nassoury, N., Mayer, G., Prat, A., and Seidah, N. G. (2009). PCSK9 impedes hepatitis C virus infection in vitro and modulates liver CD81 expression. *Hepatology* **50,** 17–24.

Labonte, P., and Seidah, N. G. (2008). Emerging viruses: risk of pandemic. *Expert Rev Anti Infect Ther* **6,** 581–583.

Lahlil, R., Calvo, F., and Khatib, A. M. (2009). The potential anti-tumorigenic and anti-metastatic side of the proprotein convertases inhibitors. *Recent Pat. Anticancer Drug Discov.* **4,** 83–91.

Lambert, D. M., Barney, S., Lambert, A. L., Guthrie, K., Medinas, R., Davis, D. E., Bucy, T., Erickson, J., Merutka, G., and Petteway, S. R., Jr. (1996). Peptides from conserved regions of paramyxovirus fusion (F) proteins are potent inhibitors of viral fusion. *Proc. Natl. Acad. Sci. USA* **93,** 2186–2191.

Larson, R. A., Dai, D., Hosack, V. T., Tan, Y., Bolken, T. C., Hruby, D. E., and Amberg, S. M. (2008). Identification of a broad-spectrum arenavirus entry inhibitor. *J. Virol.* **82,** 10768–10775.

Lavillette, D., Maurice, M., Roche, C., Russell, S. J., Sitbon, M., and Cosset, F. L. (1998). A proline-rich motif downstream of the receptor binding domain modulates conformation and fusogenicity of murine retroviral envelopes. *J. Virol.* **72,** 9955–9965.

Lavillette, D., Ruggieri, A., Russell, S. J., and Cosset, F. L. (2000). Activation of a cell entry pathway common to type C mammalian retroviruses by soluble envelope fragments. *J. Virol.* **74,** 295–304.

Lavillette, D., Marin, M., Ruggieri, A., Mallet, F., Cosset, F. L., and Kabat, D. (2002). The envelope glycoprotein of human endogenous retrovirus type W uses a divergent family of amino acid transporters/cell surface receptors. *J. Virol.* **76,** 6442–6452.

Lavillette, D., Tarr, A. W., Voisset, C., Donot, P., Bartosch, B., Bain, C., Patel, A. H., Dubuisson, J., Ball, J. K., and Cosset, F. L. (2005). Characterization of host-range and cell entry properties of the major genotypes and subtypes of hepatitis C virus. *Hepatology* **41,** 265–274.

Law, M., Maruyama, T., Lewis, J., Giang, E., Tarr, A. W., Stamataki, Z., Gastaminza, P., Chisari, F. V., Jones, I. M., Fox, R. I., Ball, J. K., McKeating, J. A., *et al.* (2008). Broadly neutralizing antibodies protect against hepatitis C virus quasispecies challenge. *Nat. Med.* **14,** 25–27.

Lee, J. E., Fusco, M. L., Hessell, A. J., Oswald, W. B., Burton, D. R., and Saphire, E. O. (2008). Structure of the Ebola virus glycoprotein bound to an antibody from a human survivor. *Nature* **454,** 177–182.

Lenz, O., ter Meulen, J., Klenk, H. D., Seidah, N. G., and Garten, W. (2001). The Lassa virus glycoprotein precursor GP-C is proteolytically processed by subtilase SKI-1/S1P. *Proc. Natl. Acad. Sci. USA* **98,** 12701–12705.

Leonard, J. T., and Roy, K. (2006). The HIV entry inhibitors revisited. *Curr. Med. Chem.* **13,** 911–934.

Lescar, J., Roussel, A., Wien, M. W., Navaza, J., Fuller, S. D., Wengler, G., and Rey, F. A. (2001). The Fusion glycoprotein shell of Semliki Forest virus: an icosahedral assembly primed for fusogenic activation at endosomal pH. *Cell* **105,** 137–148.

Li, M., Yang, C., Tong, S., Weidmann, A., and Compans, R. W. (2002). Palmitoylation of the murine leukemia virus envelope protein is critical for lipid raft association and surface expression. *J. Virol.* **76,** 11845–11852.

Li, K., Zhang, S., Kronqvist, M., Wallin, M., Ekstrom, M., Derse, D., and Garoff, H. (2008). Intersubunit disulfide isomerization controls membrane fusion of human T-cell leukemia virus Env. *J. Virol.* **82,** 7135–7143.

Li, Q., Brass, A. L., Ng, A., Hu, Z., Xavier, R. J., Liang, T. J., and Elledge, S. J. (2009). A genome-wide genetic screen for host factors required for hepatitis C virus propagation. *Proc. Natl. Acad. Sci. USA* **106**, 16410–16415.

Liu, S., Lu, H., Niu, J., Xu, Y., Wu, S., and Jiang, S. (2005). Different from the HIV fusion inhibitor C34, the anti-HIV drug Fuzeon (T-20) inhibits HIV-1 entry by targeting multiple sites in gp41 and gp120. *J. Biol. Chem.* **280**, 11259–11273.

Liu, S., Wu, S., and Jiang, S. (2007). HIV entry inhibitors targeting gp41: from polypeptides to small-molecule compounds. *Curr. Pharm. Des.* **13**, 143–162.

Lopper, M., and Compton, T. (2004). Coiled-coil domains in glycoproteins B and H are involved in human cytomegalovirus membrane fusion. *J. Virol.* **78**, 8333–8341.

Lorizate, M., Cruz, A., Huarte, N., Kunert, R., Perez-Gil, J., and Nieva, J. L. (2006). Recognition and blocking of HIV-1 gp41 pre-transmembrane sequence by monoclonal 4E10 antibody in a Raft-like membrane environment. *J. Biol. Chem.* **281**, 39598–39606.

Loving, R., Li, K., Wallin, M., Sjoberg, M., and Garoff, H. (2008). R-Peptide cleavage potentiates fusion-controlling isomerization of the intersubunit disulfide in Moloney murine leukemia virus Env. *J. Virol.* **82**, 2594–2597.

Madshus, I. H., Sandvig, K., Olsnes, S., and van Deurs, B. (1987). Effect of reduced endocytosis induced by hypotonic shock and potassium depletion on the infection of Hep 2 cells by picornaviruses. *J. Cell. Physiol.* **131**, 14–22.

Marechal, V., Prevost, M. C., Petit, C., Perret, E., Heard, J. M., and Schwartz, O. (2001). Human immunodeficiency virus type 1 entry into macrophages mediated by macropinocytosis. *J. Virol.* **75**, 11166–11177.

Marin, M., Tailor, C. S., Nouri, A., Kozak, S. L., and Kabat, D. (1999). Polymorphisms of the cell surface receptor control mouse susceptibilities to xenotropic and polytropic leukemia viruses. *J. Virol.* **73**, 9362–9368.

Markosyan, R. M., Cohen, F. S., and Melikyan, G. B. (2000). The lipid-anchored ectodomain of influenza virus hemagglutinin (GPI-HA) is capable of inducing nonenlarging fusion pores. *Mol. Biol. Cell* **11**, 1143–1152.

Marsh, M., and Helenius, A. (2006). Virus entry: open sesame. *Cell* **124**, 729–740.

Marzi, A., Akhavan, A., Simmons, G., Gramberg, T., Hofmann, H., Bates, P., Lingappa, V. R., and Pohlmann, S. (2006). The signal peptide of the ebolavirus glycoprotein influences interaction with the cellular lectins DC-SIGN and DC-SIGNR. *J. Virol.* **80**, 6305–6317.

Mason, P. W., Rieder, E., and Baxt, B. (1994). RGD sequence of foot-and-mouth disease virus is essential for infecting cells via the natural receptor but can be bypassed by an antibody-dependent enhancement pathway. *Proc. Natl. Acad. Sci. USA* **91**, 1932–1936.

Matlin, K. S., Reggio, H., Helenius, A., and Simons, K. (1981). Infectious entry pathway of influenza virus in a canine kidney cell line. *J. Cell Biol.* **91**, 601–613.

Matsuno, K., Kishida, N., Usami, K., Igarashi, M., Yoshida, R., Nakayama, E., Shimojima, M., Feldmann, H., Irimura, T., Kawaoka, Y., and Takada, A. (2010). Different potential of C-type lectin-mediated entry between Marburg virus strains. *J. Virol.* **84**, 5140–5147.

Matsuyama, S., and Taguchi, F. (2002). Receptor-induced conformational changes of murine coronavirus spike protein. *J. Virol.* **76**, 11819–11826.

Matsuyama, S., Ujike, M., Morikawa, S., Tashiro, M., and Taguchi, F. (2005). Protease-mediated enhancement of severe acute respiratory syndrome coronavirus infection. *Proc. Natl. Acad. Sci. USA* **102**, 12543–12547.

Mayor, S., and Pagano, R. E. (2007). Pathways of clathrin-independent endocytosis. *Nat. Rev. Mol. Cell Biol.* **8**, 603–612.

Mebatsion, T., Finke, S., Weiland, F., and Conzelmann, K. K. (1997). A CXCR4/CD4 pseudotype rhabdovirus that selectively infects HIV-1 envelope protein-expressing cells. *Cell* **90**, 841–847.

Melikyan, G. B. (2008). Common principles and intermediates of viral protein-mediated fusion: the HIV-1 paradigm. *Retrovirology* **5,** 111.

Melikyan, G. B., Barnard, R. J., Abrahamyan, L. G., Mothes, W., and Young, J. A. (2005). Imaging individual retroviral fusion events: from hemifusion to pore formation and growth. *Proc. Natl. Acad. Sci. USA* **102,** 8728–8733.

Mercer, J., and Helenius, A. (2008). Vaccinia virus uses macropinocytosis and apoptotic mimicry to enter host cells. *Science* **320,** 531–535.

Mercer, J., and Helenius, A. (2009). Virus entry by macropinocytosis. *Nat. Cell Biol.* **11,** 510–520.

Mercer, J., Knebel, S., Schmidt, F. I., Crouse, J., Burkard, C., and Helenius, A. (2010a). Vaccinia virus strains use distinct forms of macropinocytosis for host-cell entry. *Proc. Natl. Acad. Sci. USA* **107,** 9346–9351.

Mercer, J., Schelhaas, M., and Helenius, A. (2010b). Virus entry by endocytosis. *Annu. Rev. Biochem.* **79,** 803–833.

Meuleman, P., Hesselgesser, J., Paulson, M., Vanwolleghem, T., Desombere, I., Reiser, H., and Leroux-Roels, G. (2008). Anti-CD81 antibodies can prevent a hepatitis C virus infection in vivo. *Hepatology* **48,** 1761–1768.

Miletic, H., Fischer, Y. H., Neumann, H., Hans, V., Stenzel, W., Giroglou, T., Hermann, M., Deckert, M., and Von Laer, D. (2004). Selective transduction of malignant glioma by lentiviral vectors pseudotyped with lymphocytic choriomeningitis virus glycoproteins. *Hum. Gene Ther.* **15,** 1091–1100.

Miller, N., and Hutt-Fletcher, L. M. (1992). Epstein-Barr virus enters B cells and epithelial cells by different routes. *J. Virol.* **66,** 3409–3414.

Molina, S., Castet, V., Pichard-Garcia, L., Wychowski, C., Meurs, E., Pascussi, J. M., Sureau, C., Fabre, J. M., Sacunha, A., Larrey, D., Dubuisson, J., Coste, J., *et al.* (2008). Serum-derived hepatitis C virus infection of primary human hepatocytes is tetraspanin CD81 dependent. *J. Virol.* **82,** 569–574.

Moll, M., Klenk, H. D., and Maisner, A. (2002). Importance of the cytoplasmic tails of the measles virus glycoproteins for fusogenic activity and the generation of recombinant measles viruses. *J. Virol.* **76,** 7174–7186.

Mondor, I., Ugolini, S., and Sattentau, Q. J. (1998). Human immunodeficiency virus type 1 attachment to HeLa CD4 cells is CD4 independent and gp120 dependent and requires cell surface heparans. *J. Virol.* **72,** 3623–3634.

Moreno, M. R., Giudici, M., and Villalain, J. (2006). The membranotropic regions of the endo and ecto domains of HIV gp41 envelope glycoprotein. *Biochim. Biophys. Acta* **1758,** 111–123.

Moser, T. S., Jones, R. G., Thompson, C. B., Coyne, C. B., and Cherry, S. (2010). A kinome RNAi screen identified AMPK as promoting poxvirus entry through the control of actin dynamics. *PLoS Pathog.* **6,** e1000954.

Mothes, W., Boerger, A. L., Narayan, S., Cunningham, J. M., and Young, J. A. (2000). Retroviral entry mediated by receptor priming and low pH triggering of an envelope glycoprotein. *Cell* **103,** 679–689.

Mulligan, M. J., Yamshchikov, G. V., Ritter, G. D., Jr., Gao, F., Jin, M. J., Nail, C. D., Spies, C. P., Hahn, B. H., and Compans, R. W. (1992). Cytoplasmic domain truncation enhances fusion activity by the exterior glycoprotein complex of human immunodeficiency virus type 2 in selected cell types. *J. Virol.* **66,** 3971–3975.

Munoz-Barroso, I., Durell, S., Sakaguchi, K., Appella, E., and Blumenthal, R. (1998). Dilation of the human immunodeficiency virus-1 envelope glycoprotein fusion pore revealed by the inhibitory action of a synthetic peptide from gp41. *J. Cell Biol.* **140,** 315–323.

Munoz-Barroso, I., Salzwedel, K., Hunter, E., and Blumenthal, R. (1999). Role of the membrane-proximal domain in the initial stages of human immunodeficiency virus type 1 envelope glyco-protein-mediated membrane fusion. *J. Virol.* **73,** 6089–6092.

Murga, J. D., Franti, M., Pevear, D. C., Maddon, P. J., and Olson, W. C. (2006). Potent antiviral synergy between monoclonal antibody and small-molecule CCR5 inhibitors of human immunodeficiency virus type 1. *Antimicrob. Agents Chemother.* **50,** 3289–3296.

Myers, M. C., Shah, P. P., Beavers, M. P., Napper, A. D., Diamond, S. L., Smith, A. B., 3rd, and Huryn, D. M. (2008). Design, synthesis, and evaluation of inhibitors of cathepsin L: Exploiting a unique thiocarbazate chemotype. *Bioorg. Med. Chem. Lett.* **18,** 3646–3651.

Ng, T. I., Mo, H., Pilot-Matias, T., He, Y., Koev, G., Krishnan, P., Mondal, R., Pithawalla, R., He, W., Dekhtyar, T., Packer, J., Schurdak, M., *et al.* (2007). Identification of host genes involved in hepatitis C virus replication by small interfering RNA technology. *Hepatology* **45,** 1413–1421.

Ochsenbauer-Jambor, C., Miller, D. C., Roberts, C. R., Rhee, S. S., and Hunter, E. (2001). Palmitoylation of the Rous sarcoma virus transmembrane glycoprotein is required for protein stability and virus infectivity. *J. Virol.* **75,** 11544–11554.

Okumura, Y., Takahashi, E., Yano, M., Ohuchi, M., Daidoji, T., Nakaya, T., Bottcher, E., Garten, W., Klenk, H. D., and Kido, H. (2010). Novel type II transmembrane serine proteases, MSPL and TMPRSS13, Proteolytically activate membrane fusion activity of the hemagglutinin of highly pathogenic avian influenza viruses and induce their multicycle replication. *J. Virol.* **84,** 5089–5096.

Oldstone, M. B., Lewicki, H., Thomas, D., Tishon, A., Dales, S., Patterson, J., Manchester, M., Homann, D., Naniche, D., and Holz, A. (1999). Measles virus infection in a transgenic model: virus-induced immunosuppression and central nervous system disease. *Cell* **98,** 629–640.

Oomens, A. G., Bevis, K. P., and Wertz, G. W. (2006). The cytoplasmic tail of the human respiratory syncytial virus F protein plays critical roles in cellular localization of the F protein and infectious progeny production. *J. Virol.* **80,** 10465–10477.

Owens, R. J., Burke, C., and Rose, J. K. (1994). Mutations in the membrane-spanning domain of the human immunodeficiency virus envelope glycoprotein that affect fusion activity. *J. Virol.* **68,** 570–574.

Owsianka, A., Tarr, A. W., Juttla, V. S., Lavillette, D., Bartosch, B., Cosset, F. L., Ball, J. K., and Patel, A. H. (2005). Monoclonal antibody AP33 defines a broadly neutralizing epitope on the hepatitis C virus E2 envelope glycoprotein. *J. Virol.* **79,** 11095–11104.

Ozden, S., Lucas-Hourani, M., Ceccaldi, P. E., Basak, A., Valentine, M., Benjannet, S., Hamelin, J., Jacob, Y., Mamchaoui, K., Mouly, V., Despres, P., Gessain, A., *et al.* (2008). Inhibition of Chikungunya virus infection in cultured human muscle cells by furin inhibitors: impairment of the maturation of the E2 surface glycoprotein. *J. Biol. Chem.* **283,** 21899–21908.

Pager, C. T., and Dutch, R. E. (2005). Cathepsin L is involved in proteolytic processing of the Hendra virus fusion protein. *J. Virol.* **79,** 12714–12720.

Palermo, C., and Joyce, J. A. (2008). Cysteine cathepsin proteases as pharmacological targets in cancer. *Trends Pharmacol. Sci.* **29,** 22–28.

Park, P. W., Reizes, O., and Bernfield, M. (2000). Cell surface heparan sulfate proteoglycans: selective regulators of ligand-receptor encounters. *J. Biol. Chem.* **275,** 29923–29926.

Parker, J. S., Murphy, W. J., Wang, D., O'Brien, S. J., and Parrish, C. R. (2001). Canine and feline parvoviruses can use human or feline transferrin receptors to bind, enter, and infect cells. *J. Virol.* **75,** 3896–3902.

Pecheur, E. I., Lavillette, D., Alcaras, F., Molle, J., Boriskin, Y. S., Roberts, M., Cosset, F. L., and Polyak, S. J. (2007). Biochemical mechanism of hepatitis C virus inhibition by the broad-spectrum antiviral arbidol. *Biochemistry* **46,** 6050–6059.

Pernet, O., Pohl, C., Ainouze, M., Kweder, H., and Buckland, R. (2009). Nipah virus entry can occur by macropinocytosis. *Virology* **395,** 298–311.

Perotti, M., Mancini, N., Diotti, R. A., Tarr, A. W., Ball, J. K., Owsianka, A., Adair, R., Patel, A. H., Clementi, M., and Burioni, R. (2008). Identification of a broadly cross-reacting and neutralizing human monoclonal antibody directed against the hepatitis C virus E2 protein. *J. Virol.* **82,** 1047–1052.

Pinter, A., Kopelman, R., Li, Z., Kayman, S. C., and Sanders, D. A. (1997). Localization of the labile disulfide bond between SU and TM of the murine leukemia virus envelope protein complex to a highly conserved CWLC motif in SU that resembles the active-site sequence of thiol-disulfide exchange enzymes. *J. Virol.* **71**, 8073–8077.

Pizzato, M., Marlow, S. A., Blair, E. D., and Takeuchi, Y. (1999). Initial binding of murine leukemia virus particles to cells does not require specific Env-receptor interaction. *J. Virol.* **73**, 8599–8611.

Pizzato, M., Blair, E. D., Fling, M., Kopf, J., Tomassetti, A., Weiss, R. A., and Takeuchi, Y. (2001). Evidence for nonspecific adsorption of targeted retrovirus vector particles to cells.. *Gene Ther.* **8**, 1088–1096.

Pizzato, M., Popova, E., and Gottlinger, H. G. (2008). Nef can enhance the infectivity of receptor-pseudotyped human immunodeficiency virus type 1 particles. *J. Virol.* **82**, 10811–10819.

Platt, E. J., Wehrly, K., Kuhmann, S. E., Chesebro, B., and Kabat, D. (1998). Effects of CCR5 and CD4 cell surface concentrations on infections by macrophagetropic isolates of human immunodeficiency virus type 1. *J. Virol.* **72**, 2855–2864.

Ploss, A., and Rice, C. M. (2009). Human occludin is a hepatitis C virus entry factor required for infection of mouse cells. *EMBO Rep.* **10**, 1220–1227.

Ploss, A., Evans, M. J., Gaysinskaya, V. A., Panis, M., You, H., de Jong, Y. P., and Rice, C. M. (2009). Towards a small animal model for hepatitis C. *Nature* **457**, 882–886.

Pohlmann, S., Soilleux, E. J., Baribaud, F., Leslie, G. J., Morris, L. S., Trowsdale, J., Lee, B., Coleman, N., and Doms, R. W. (2001). DC-SIGNR, a DC-SIGN homologue expressed in endothelial cells, binds to human and simian immunodeficiency viruses and activates infection in trans. *Proc. Natl. Acad. Sci. USA* **98**, 2670–2675.

Porotto, M., Yokoyama, C. C., Palermo, L. M., Mungall, B., Aljofan, M., Cortese, R., Pessi, A., and Moscona, A. (2010). Viral entry inhibitors targeted to the membrane site of action. *J. Virol.* **84**, 6760–6768.

Pullikotil, P., Vincent, M., Nichol, S. T., and Seidah, N. G. (2004). Development of protein-based inhibitors of the proprotein of convertase SKI-1/S1P: processing of SREBP-2, ATF6, and a viral glycoprotein. *J. Biol. Chem.* **279**, 17338–17347.

Purtscher, M., Trkola, A., Gruber, G., Buchacher, A., Predl, R., Steindl, F., Tauer, C., Berger, R., Barrett, N., Jungbauer, A., et al. (1994). A broadly neutralizing human monoclonal antibody against gp41 of human immunodeficiency virus type 1. *AIDS Res. Hum. Retroviruses* **10**, 1651–1658.

Qiu, Z., Hingley, S. T., Simmons, G., Yu, C., Das Sarma, J., Bates, P., and Weiss, S. R. (2006). Endosomal proteolysis by cathepsins is necessary for murine coronavirus mouse hepatitis virus type 2 spike-mediated entry. *J. Virol.* **80**, 5768–5776.

Quigley, J. G., Burns, C. C., Anderson, M. M., Lynch, E. D., Sabo, K. M., Overbaugh, J., and Abkowitz, J. L. (2000). Cloning of the cellular receptor for feline leukemia virus subgroup C (FeLV-C), a retrovirus that induces red cell aplasia. *Blood* **95**, 1093–1099.

Quinn, K., Brindley, M. A., Weller, M. L., Kaludov, N., Kondratowicz, A., Hunt, C. L., Sinn, P. L., McCray, P. B., Jr., Stein, C. S., Davidson, B. L., Flick, R., Mandell, R., et al. (2009). Rho GTPases modulate entry of Ebola virus and vesicular stomatitis virus pseudotyped vectors. *J. Virol.* **83**, 10176–10186.

Rapaport, D., Ovadia, M., and Shai, Y. (1995). A synthetic peptide corresponding to a conserved heptad repeat domain is a potent inhibitor of Sendai virus-cell fusion: an emerging similarity with functional domains of other viruses. *EMBO J.* **14**, 5524–5531.

Rasko, J. E., Battini, J. L., Gottschalk, R. J., Mazo, I., and Miller, A. D. (1999). The RD114/simian type D retrovirus receptor is a neutral amino acid transporter. *Proc. Natl. Acad. Sci. USA* **96**, 2129–2134.

Rein, A., Mirro, J., Haynes, J. G., Ernst, S. M., and Nagashima, K. (1994). Function of the cytoplasmic domain of a retroviral transmembrane protein: p15E-p2E cleavage activates the membrane fusion capability of the murine leukemia virus env protein. *J. Virol.* **68**, 1773–1781.

Rey, F. A. (2006). Molecular gymnastics at the herpesvirus surface. *EMBO Rep.* **7,** 1000–1005.

Rey, F. A., Heinz, F. X., Mandl, C., Kunz, C., and Harrison, S. C. (1995). The envelope glycoprotein from tick-borne encephalitis virus at 2 A resolution. *Nature* **375,** 291–298.

Roche, S., and Gaudin, Y. (2002). Characterization of the equilibrium between the native and fusion-inactive conformation of rabies virus glycoprotein indicates that the fusion complex is made of several trimers. *Virology* **297,** 128–135.

Roche, S., Bressanelli, S., Rey, F. A., and Gaudin, Y. (2006). Crystal structure of the low-pH form of the vesicular stomatitis virus glycoprotein G. *Science* **313,** 187–191.

Roche, S., Rey, F. A., Gaudin, Y., and Bressanelli, S. (2007). Structure of the prefusion form of the vesicular stomatitis virus glycoprotein G. *Science* **315,** 843–848.

Roche, S., Albertini, A. A., Lepault, J., Bressanelli, S., and Gaudin, Y. (2008). Structures of vesicular stomatitis virus glycoprotein: membrane fusion revisited. *Cell. Mol. Life Sci.* **65,** 1716–1728.

Root, M. J., Kay, M. S., and Kim, P. S. (2001). Protein design of an HIV-1 entry inhibitor. *Science* **291,** 884–888.

Rossi, A., Deveraux, Q., Turk, B., and Sali, A. (2004). Comprehensive search for cysteine cathepsins in the human genome. *Biol. Chem.* **385,** 363–372.

Rozenberg-Adler, Y., Conner, J., Aguilar-Carreno, H., Chakraborti, S., Dimitrov, D. S., and Anderson, W. F. (2008). Membrane-proximal cytoplasmic domain of Moloney murine leukemia virus envelope tail facilitates fusion. *Exp. Mol. Pathol.* **84,** 18–30.

Saez-Cirion, A., Gomara, M. J., Agirre, A., and Nieva, J. L. (2003). Pre-transmembrane sequence of Ebola glycoprotein. Interfacial hydrophobicity distribution and interaction with membranes. *FEBS Lett.* **533,** 47–53.

Salminen, A., Wahlberg, J. M., Lobigs, M., Liljestrom, P., and Garoff, H. (1992). Membrane fusion process of Semliki Forest virus. II: Cleavage-dependent reorganization of the spike protein complex controls virus entry. *J. Cell Biol.* **116,** 349–357.

Salzwedel, K., West, J. T., and Hunter, E. (1999). A conserved tryptophan-rich motif in the membrane-proximal region of the human immunodeficiency virus type 1 gp41 ectodomain is important for Env-mediated fusion and virus infectivity. *J. Virol.* **73,** 2469–2480.

Sanchez, A. (2007). Analysis of filovirus entry into vero e6 cells, using inhibitors of endocytosis, endosomal acidification, structural integrity, and cathepsin (B and L) activity. *J. Infect. Dis.* **196** (Suppl. 2), S251–S258.

Sanders, D. A. (2000). Sulfhydryl involvement in fusion mechanisms. *Subcell. Biochem.* **34,** 483–514.

Sandrin, V., Boson, B., Salmon, P., Gay, W., Negre, D., Le Grand, R., Trono, D., and Cosset, F. L. (2002). Lentiviral vectors pseudotyped with a modified RD114 envelope glycoprotein show increased stability in sera and augmented transduction of primary lymphocytes and CD34+ cells derived from human and nonhuman primates. *Blood* **100,** 823–832.

Saphire, A. C., Bobardt, M. D., and Gallay, P. A. (1999). Host cyclophilin A mediates HIV-1 attachment to target cells via heparans. *EMBO J.* **18,** 6771–6785.

Saphire, A. C., Bobardt, M. D., Zhang, Z., David, G., and Gallay, P. A. (2001). Syndecans serve as attachment receptors for human immunodeficiency virus type 1 on macrophages. *J. Virol.* **75,** 9187–9200.

Scarlatti, G., Tresoldi, E., Bjorndal, A., Fredriksson, R., Colognesi, C., Deng, H. K., Malnati, M. S., Plebani, A., Siccardi, A. G., Littman, D. R., Fenyo, E. M., and Lusso, P. (1997). In vivo evolution of HIV-1 co-receptor usage and sensitivity to chemokine-mediated suppression. *Nat. Med.* **3,** 1259–1265.

Schlegel, R., Tralka, T. S., Willingham, M. C., and Pastan, I. (1983). Inhibition of VSV binding and infectivity by phosphatidylserine: is phosphatidylserine a VSV-binding site? *Cell* **32,** 639–646.

Schnell, M. J., Johnson, J. E., Buonocore, L., and Rose, J. K. (1997). Construction of a novel virus that targets HIV-1-infected cells and controls HIV-1 infection. *Cell* **90,** 849–857.

Schofield, D. J., Bartosch, B., Shimizu, Y. K., Allander, T., Alter, H. J., Emerson, S. U., Cosset, F. L., and Purcell, R. H. (2005). Human monoclonal antibodies that react with the E2 glycoprotein of hepatitis C virus and possess neutralizing activity. *Hepatology* **42**, 1055–1062.

Schornberg, K., Matsuyama, S., Kabsch, K., Delos, S., Bouton, A., and White, J. (2006). Role of endosomal cathepsins in entry mediated by the Ebola virus glycoprotein. *J. Virol.* **80**, 4174–4178.

Schulz, T. F., Jameson, B. A., Lopalco, L., Siccardi, A. G., Weiss, R. A., and Moore, J. P. (1992). Conserved structural features in the interaction between retroviral surface and transmembrane glycoproteins? *AIDS Res. Hum. Retroviruses* **8**, 1571–1580.

Sessions, O. M., Barrows, N. J., Souza-Neto, J. A., Robinson, T. J., Hershey, C. L., Rodgers, M. A., Ramirez, J. L., Dimopoulos, G., Yang, P. L., Pearson, J. L., and Garcia-Blanco, M. A. (2009). Discovery of insect and human dengue virus host factors. *Nature* **458**, 1047–1050.

Shah, P. P., Myers, M. C., Beavers, M. P., Purvis, J. E., Jing, H., Grieser, H. J., Sharlow, E. R., Napper, A. D., Huryn, D. M., Cooperman, B. S., Smith, A. B., 3rd, and Diamond, S. L. (2008). Kinetic characterization and molecular docking of a novel, potent, and selective slow-binding inhibitor of human cathepsin L. *Mol. Pharmacol.* **74**, 34–41.

Shah, P. P., Wang, T., Kaletsky, R. L., Myers, M. C., Purvis, J. E., Jing, H., Huryn, D. M., Greenbaum, D. C., Smith, A. B., 3rd, Bates, P., and Diamond, S. L. (2010). A small-molecule oxocarbazate inhibitor of human cathepsin L blocks severe acute respiratory syndrome and ebola pseudotype virus infection into human embryonic kidney 293T cells. *Mol. Pharmacol.* **78**, 319–324.

Shimojima, M., Takada, A., Ebihara, H., Neumann, G., Fujioka, K., Irimura, T., Jones, S., Feldmann, H., and Kawaoka, Y. (2006). Tyro3 family-mediated cell entry of Ebola and Marburg viruses. *J. Virol.* **80**, 10109–10116.

Shimojima, M., Ikeda, Y., and Kawaoka, Y. (2007). The mechanism of Axl-mediated Ebola virus infection. *J. Infect. Dis.* **196**(Suppl. 2), S259–S263.

Shmulevitz, M., and Duncan, R. (2000). A new class of fusion-associated small transmembrane (FAST) proteins encoded by the non-enveloped fusogenic reoviruses. *EMBO J.* **19**, 902–912.

Sieczkarski, S. B., and Whittaker, G. R. (2002). Dissecting virus entry via endocytosis. *J. Gen. Virol.* **83**, 1535–1545.

Simmons, G., Gosalia, D. N., Rennekamp, A. J., Reeves, J. D., Diamond, S. L., and Bates, P. (2005). Inhibitors of cathepsin L prevent severe acute respiratory syndrome coronavirus entry. *Proc. Natl. Acad. Sci. USA* **102**, 11876–11881.

Sinn, P. L., Hickey, M. A., Staber, P. D., Dylla, D. E., Jeffers, S. A., Davidson, B. L., Sanders, D. A., and McCray, P. B., Jr. (2003). Lentivirus vectors pseudotyped with filoviral envelope glycoproteins transduce airway epithelia from the apical surface independently of folate receptor alpha. *J. Virol.* **77**, 5902–5910.

Sitbon, M., d'Auriol, L., Ellerbrok, H., Andre, C., Nishio, J., Perryman, S., Pozo, F., Hayes, S. F., Wehrly, K., Tambourin, P., et al. (1991). Substitution of leucine for isoleucine in a sequence highly conserved among retroviral envelope surface glycoproteins attenuates the lytic effect of the Friend murine leukemia virus. *Proc. Natl. Acad. Sci. USA* **88**, 5932–5936.

Skehel, J. J., and Wiley, D. C. (2000). Receptor binding and membrane fusion in virus entry: the influenza hemagglutinin. *Annu. Rev. Biochem.* **69**, 531–569.

Song, C., Dubay, S. R., and Hunter, E. (2003). A tyrosine motif in the cytoplasmic domain of mason-pfizer monkey virus is essential for the incorporation of glycoprotein into virions. *J. Virol.* **77**, 5192–5200.

Spies, C. P., Ritter, G. D., Jr., Mulligan, M. J., and Compans, R. W. (1994). Truncation of the cytoplasmic domain of the simian immunodeficiency virus envelope glycoprotein alters the conformation of the external domain. *J. Virol.* **68**, 585–591.

Stadler, K., Allison, S. L., Schalich, J., and Heinz, F. X. (1997). Proteolytic activation of tick-borne encephalitis virus by furin. *J. Virol.* **71**, 8475–8481.

Stein, C. S., Martins, I., and Davidson, B. L. (2005). The lymphocytic choriomeningitis virus envelope glycoprotein targets lentiviral gene transfer vector to neural progenitors in the murine brain. *Mol. Ther.* **11**, 382–389.

Stiasny, K., and Heinz, F. X. (2006). Flavivirus membrane fusion. *J. Gen. Virol.* **87**, 2755–2766.

Stieneke-Grober, A., Vey, M., Angliker, H., Shaw, E., Thomas, G., Roberts, C., Klenk, H. D., and Garten, W. (1992). Influenza virus hemagglutinin with multibasic cleavage site is activated by furin, a subtilisin-like endoprotease. *EMBO J.* **11**, 2407–2414.

Stitz, J., Buchholz, C. J., Engelstadter, M., Uckert, W., Bloemer, U., Schmitt, I., and Cichutek, K. (2000). Lentiviral vectors pseudotyped with envelope glycoproteins derived from gibbon ape leukemia virus and murine leukemia virus 10A1. *Virology* **273**, 16–20.

Supekova, L., Supek, F., Lee, J., Chen, S., Gray, N., Pezacki, J. P., Schlapbach, A., and Schultz, P. G. (2008). Identification of human kinases involved in hepatitis C virus replication by small interference RNA library screening. *J. Biol. Chem.* **283**, 29–36.

Tai, A. W., Benita, Y., Peng, L. F., Kim, S. S., Sakamoto, N., Xavier, R. J., and Chung, R. T. (2009). A functional genomic screen identifies cellular cofactors of hepatitis C virus replication. *Cell Host Microbe* **5**, 298–307.

Tailor, C. S., Nouri, A., Lee, C. G., Kozak, C., and Kabat, D. (1999a). Cloning and characterization of a cell surface receptor for xenotropic and polytropic murine leukemia viruses. *Proc. Natl. Acad. Sci. USA* **96**, 927–932.

Tailor, C. S., Nouri, A., Zhao, Y., Takeuchi, Y., and Kabat, D. (1999b). A sodium-dependent neutral-amino-acid transporter mediates infections of feline and baboon endogenous retroviruses and simian type D retroviruses. *J. Virol.* **73**, 4470–4474.

Tailor, C. S., Willett, B. J., and Kabat, D. (1999c). A putative cell surface receptor for anemia-inducing feline leukemia virus subgroup C is a member of a transporter superfamily. *J. Virol.* **73**, 6500–6505.

Tailor, C. S., Lavillette, D., Marin, M., and Kabat, D. (2003). Cell surface receptors for gammare-troviruses. *Curr. Top. Microbiol. Immunol.* **281**, 29–106.

Takeda, M., Leser, G. P., Russell, C. J., and Lamb, R. A. (2003). Influenza virus hemagglutinin concentrates in lipid raft microdomains for efficient viral fusion. *Proc. Natl. Acad. Sci. USA* **100**, 14610–14617.

Taplitz, R. A., and Coffin, J. M. (1997). Selection of an avian retrovirus mutant with extended receptor usage. *J. Virol.* **71**, 7814–7819.

Tatsuo, H., Okuma, K., Tanaka, K., Ono, N., Minagawa, H., Takade, A., Matsuura, Y., and Yanagi, Y. (2000). Virus entry is a major determinant of cell tropism of Edmonston and wild-type strains of measles virus as revealed by vesicular stomatitis virus pseudotypes bearing their envelope proteins. *J. Virol.* **74**, 4139–4145.

Taylor, G. M., and Sanders, D. A. (1999). The role of the membrane-spanning domain sequence in glycoprotein-mediated membrane fusion. *Mol. Biol. Cell* **10**, 2803–2815.

Temin, H. M. (1988). Mechanisms of cell killing/cytopathic effects by nonhuman retroviruses. *Rev. Infect. Dis.* **10**, 399–405.

Tremblay, C. L., Kollmann, C., Giguel, F., Chou, T. C., and Hirsch, M. S. (2000). Strong in vitro synergy between the fusion inhibitor T-20 and the CXCR4 blocker AMD-3100. *J. Acquir. Immune Defic. Syndr.* **25**, 99–102.

Tremblay, C. L., Giguel, F., Kollmann, C., Guan, Y., Chou, T. C., Baroudy, B. M., and Hirsch, M. S. (2002). Anti-human immunodeficiency virus interactions of SCH-C (SCH 351125), a CCR5 antagonist, with other antiretroviral agents in vitro. *Antimicrob. Agents Chemother.* **46**, 1336–1339.

Trotard, M., Lepere-Douard, C., Regeard, M., Piquet-Pellorce, C., Lavillette, D., Cosset, F. L., Gripon, P., and Le Seyec, J. (2009). Kinases required in hepatitis C virus entry and replication highlighted by small interference RNA screening. *FASEB J.* **23**(11), 3780–3789.

Trujillo, J. R., Rogers, R., Molina, R. M., Dangond, F., McLane, M. F., Essex, M., and Brain, J. D. (2007). Noninfectious entry of HIV-1 into peripheral and brain macrophages mediated by the mannose receptor. *Proc. Natl. Acad. Sci. USA* **104**, 5097–5102.

Turner, A., Bruun, B., Minson, T., and Browne, H. (1998). Glycoproteins gB, gD, and gHgL of herpes simplex virus type 1 are necessary and sufficient to mediate membrane fusion in a Cos cell transfection system. *J. Virol.* **72**, 873–875.

Ugolini, S., Mondor, I., and Sattentau, Q. J. (1999). HIV-1 attachment: another look. *Trends Microbiol.* **7**, 144–149.

Ujike, M., Nishikawa, H., Otaka, A., Yamamoto, N., Matsuoka, M., Kodama, E., Fujii, N., and Taguchi, F. (2008). Heptad repeat-derived peptides block protease-mediated direct entry from the cell surface of severe acute respiratory syndrome coronavirus but not entry via the endosomal pathway. *J. Virol.* **82**, 588–592.

Umashankar, M., Sanchez San Martin, C., Liao, M., Reilly, B., Guo, A., Taylor, G., and Kielian, M. (2008). Differential Cholesterol Binding by Class Ii Fusion Proteins Determines Membrane Fusion Properties. *J. Virol.*

Vaillancourt, F. H., Pilote, L., Cartier, M., Lippens, J., Liuzzi, M., Bethell, R. C., Cordingley, M. G., and Kukolj, G. (2009). Identification of a lipid kinase as a host factor involved in hepatitis C virus RNA replication. *Virology* **387**, 5–10.

Vasiljeva, O., Reinheckel, T., Peters, C., Turk, D., Turk, V., and Turk, B. (2007). Emerging roles of cysteine cathepsins in disease and their potential as drug targets. *Curr. Pharm. Des.* **13**, 387–403.

Vincent, M. J., Sanchez, A. J., Erickson, B. R., Basak, A., Chretien, M., Seidah, N. G., and Nichol, S. T. (2003). Crimean-Congo hemorrhagic fever virus glycoprotein proteolytic processing by subtilase SKI-1. *J. Virol.* **77**, 8640–8649.

Wallin, M., Ekstrom, M., and Garoff, H. (2004). Isomerization of the intersubunit disulphide-bond in Env controls retrovirus fusion. *EMBO J.* **23**, 54–65.

Wang, H., Yang, P., Liu, K., Guo, F., Zhang, Y., Zhang, G., and Jiang, C. (2008). SARS coronavirus entry into host cells through a novel clathrin- and caveolae-independent endocytic pathway. *Cell Res.* **18**, 290–301.

Watson, D. J., Kobinger, G. P., Passini, M. A., Wilson, J. M., and Wolfe, J. H. (2002). Targeted transduction patterns in the mouse brain by lentivirus vectors pseudotyped with VSV, Ebola, Mokola, LCMV, or MuLV envelope proteins. *Mol. Ther.* **5**, 528–537.

Weissenhorn, W., Hinz, A., and Gaudin, Y. (2007). Virus membrane fusion. *FEBS Lett.* **581**, 2150–2155.

Wengler, G. (1989). Cell-associated West Nile flavivirus is covered with E+pre-M protein hetero-dimers which are destroyed and reorganized by proteolytic cleavage during virus release. *J. Virol.* **63**, 2521–2526.

Wild, C., Greenwell, T., and Matthews, T. (1993). A synthetic peptide from HIV-1 gp41 is a potent inhibitor of virus-mediated cell-cell fusion. *AIDS Res. Hum. Retroviruses* **9**, 1051–1053.

Wild, C. T., Shugars, D. C., Greenwell, T. K., McDanal, C. B., and Matthews, T. J. (1994). Peptides corresponding to a predictive alpha-helical domain of human immunodeficiency virus type 1 gp41 are potent inhibitors of virus infection. *Proc. Natl. Acad. Sci. USA* **91**, 9770–9774.

Wilk, T., Pfeiffer, T., and Bosch, V. (1992). Retained in vitro infectivity and cytopathogenicity of HIV-1 despite truncation of the C-terminal tail of the env gene product. *Virology* **189**, 167–177.

Wong, L. F., Azzouz, M., Walmsley, L. E., Askham, Z., Wilkes, F. J., Mitrophanous, K. A., Kingsman, S. M., and Mazarakis, N. D. (2004). Transduction patterns of pseudotyped lentiviral vectors in the nervous system. *Mol. Ther.* **9**, 101–111.

Wong, A. C., Sandesara, R. G., Mulherkar, N., Whelan, S. P., and Chandran, K. (2010). A forward genetic strategy reveals destabilizing mutations in the Ebolavirus glycoprotein that alter its protease dependence during cell entry. *J. Virol.* **84**, 163–175.

Yang, Y. L., Guo, L., Xu, S., Holland, C. A., Kitamura, T., Hunter, K., and Cunningham, J. M. (1999). Receptors for polytropic and xenotropic mouse leukaemia viruses encoded by a single gene at Rmc1. *Nat. Genet.* **21,** 216–219.

Yao, Q., and Compans, R. W. (1996). Peptides corresponding to the heptad repeat sequence of human parainfluenza virus fusion protein are potent inhibitors of virus infection. *Virology* **223,** 103–112.

Yeung, M. L., Houzet, L., Yedavalli, V. S., and Jeang, K. T. (2009). A genome-wide short hairpin RNA screening of jurkat T-cells for human proteins contributing to productive HIV-1 replication. *J. Biol. Chem.* **284,** 19463–19473.

Yonezawa, A., Cavrois, M., and Greene, W. C. (2005). Studies of ebola virus glycoprotein-mediated entry and fusion by using pseudotyped human immunodeficiency virus type 1 virions: involvement of cytoskeletal proteins and enhancement by tumor necrosis factor alpha. *J. Virol.* **79,** 918–926.

York, J., Dai, D., Amberg, S. M., and Nunberg, J. H. (2008). pH-induced activation of arenavirus membrane fusion is antagonized by small-molecule inhibitors. *J. Virol.* **82,** 10932–10939.

York, J., Berry, J. D., Stroher, U., Li, Q., Feldmann, H., Lu, M., Trahey, M., and Nunberg, J. H. (2010). An antibody directed against the fusion peptide of Junin virus envelope glycoprotein GPC inhibits pH-induced membrane fusion. *J. Virol.* **84,** 6119–6129.

Young, J. K., Li, D., Abramowitz, M. C., and Morrison, T. G. (1999). Interaction of peptides with sequences from the Newcastle disease virus fusion protein heptad repeat regions. *J. Virol.* **73,** 5945–5956.

Yu, M. Y., Bartosch, B., Zhang, P., Guo, Z. P., Renzi, P. M., Shen, L. M., Granier, C., Feinstone, S. M., Cosset, F. L., and Purcell, R. H. (2004). Neutralizing antibodies to hepatitis C virus (HCV) in immune globulins derived from anti-HCV-positive plasma. *Proc. Natl. Acad. Sci. USA* **101,** 7705–7710.

Zeisel, M. B., Koutsoudakis, G., Schnober, E. K., Haberstroh, A., Blum, H. E., Cosset, F. L., Wakita, T., Jaeck, D., Doffoel, M., Royer, C., Soulier, E., Schvoerer, E., *et al.* (2007). Scavenger receptor class B type I is a key host factor for hepatitis C virus infection required for an entry step closely linked to CD81. *Hepatology* **46,** 1722–1731.

Zeisel, M. B., Barth, H., Schuster, C., and Baumert, T. F. (2009). Hepatitis C virus entry: molecular mechanisms and targets for antiviral therapy. *Front. Biosci.* **14,** 3274–3285.

Zhang, X., Fugere, M., Day, R., and Kielian, M. (2003a). Furin processing and proteolytic activation of Semliki Forest virus. *J. Virol.* **77,** 2981–2989.

Zhang, Y., Corver, J., Chipman, P. R., Zhang, W., Pletnev, S. V., Sedlak, D., Baker, T. S., Strauss, J. H., Kuhn, R. J., and Rossmann, M. G. (2003b). Structures of immature flavivirus particles. *EMBO J.* **22,** 2604–2613.

Zhao, Y., Zhu, L., Benedict, C. A., Chen, D., Anderson, W. F., and Cannon, P. M. (1998). Functional domains in the retroviral transmembrane protein. *J. Virol.* **72,** 5392–5398.

Zhou, H., Xu, M., Huang, Q., Gates, A. T., Zhang, X. D., Castle, J. C., Stec, E., Ferrer, M., Strulovici, B., Hazuda, D. J., and Espeseth, A. S. (2008). Genome-scale RNAi screen for host factors required for HIV replication. *Cell Host Microbe* **4,** 495–504.

5

Molecular Signaling: How Do Axons Die?

Michael Coleman

The Babraham Institute, Babraham, Cambridge, United Kingdom

Advances in Genetics, Vol. 73
0065-2660/11 $35.00
DOI: 10.1016/B978-0-12-380860-8.00005-7

ABSTRACT

Axons depend critically on axonal transport both for supplying materials and for communicating with cell bodies. This chapter looks at each activity, asking what aspects are essential for axon survival. Axonal transport declines in neurodegenerative disorders, such as Alzheimer's disease, amyotrophic lateral sclerosis, and multiple sclerosis, and in normal ageing, but whether all cargoes are equally affected and what limits axon survival remains unclear. Cargoes can be differentially blocked in some disorders, either individually or in groups. Each missing protein cargo results in localized loss-of-function that can be partially modeled by disrupting the corresponding gene, sometimes with surprising results. The axonal response to losing specific proteins also depends on the rates of protein turnover and on whether the protein can be locally synthesized. Among cargoes with important axonal roles are components of the PI3 kinase, Mek/Erk, and Jnk signaling pathways, which help to communicate with cell bodies and to regulate axonal transport itself. Bidirectional trafficking of Bdnf, NT-3, and other neurotrophic factors contribute to intra- and intercellular signaling, affecting the axon's cellular environment and survival. Finally, several adhesion molecules and gangliosides are key determinants of axon survival, probably by mediating axon–glia interactions. Thus, failure of long-distance intracellular transport can deprive axons of one, few, or many cargoes. This can lead to axon degeneration either directly, through the absence of essential axonal proteins, or indirectly, through failures in communication with cell bodies and nonneuronal cells. © 2011, Elsevier Inc.

I. INTRODUCTION

Axons are more autonomous than we used to think. They can synthesize proteins locally (Giuditta et al., 2002), degrade them (Korhonen and Lindholm, 2004), replicate mitochondria (Amiri and Hollenbeck, 2008), carry out autophagy (Yue et al., 2008), form growth cones and respond to pathfinding cues (Campbell and Holt, 2001; Gumy et al., 2009), and even survive for several weeks without their cell bodies when a single gene is altered (Mack et al., 2001). When they do die, the mechanisms are distinct from those in cell bodies (Coleman, 2005). Ultimately, however, axons depend on two principle sources of support: neuronal cell bodies and glia. This chapter shows how molecular genetic studies are identifying which axonal molecules are most essential for survival.

Axon survival depends on continuous axonal transport, the bidirectional transport of materials from and toward cell bodies. However, some axonal transport cargoes are more essential than others, some have shorter half-lives than others, and in many cases, transport of different cargoes is mediated by different motor proteins, different scaffold proteins linking to these motors, and different regulatory mechanisms. Some proteins can also be synthesized locally within axons. The combined effect is that some proteins are completely dispensable for long-term axon survival, while a failure to deliver others kills axons in hours.

Molecular genetics gives us some insight into what is most essential. Spontaneous null mutations in man or model organisms and targeted gene disruptions help to identify proteins that axons cannot live without. Among them are many components of the axonal transport machinery itself, some mitochondrial proteins, adhesion molecules, and mediators of the ubiquitin proteasome system and autophagy. However, a surprising array of axonal proteins are not essential for survival. These include highly abundant axonal proteins such as neurofilament subunits, several disease-associated proteins, and key mediators of axon and synapse function. Some axons can even survive days without mitochondria.

Failure of glial support, protein aggregation, inflammatory demyelination, and other mechanisms also cause axon degeneration. These may act partly through impairment of axonal transport, and readers are referred to other recent reviews for more details on what triggers these mechanisms (Muchowski and Wacker, 2005; Nave and Trapp, 2008; Zhao et al., 2005). This chapter aims to move beyond observations that axonal transport is blocked in neurodegenerative disorders by asking which cargoes are delayed, which of the resulting deficiencies kill axons, and what determines how long this takes. Answers to these questions are essential to move toward treating axonopathies. They are also important for understanding the basic biology of this fascinating structure.

II. AXONAL TRANSPORT AND HOW TO BLOCK IT

A. Road closed! General defects in axonal transport

Axonal transport has to deliver proteins, vesicles, organelles, and other components over centimeter-to-meter long distances from their sites of origin in the cell body. This is a feat totally without comparison in any other cell type. Long-range transport is a microtubule-based mechanism mediated in the anterograde direction by a wide range of kinesin superfamily motor proteins and in the retrograde direction by the dynein motor complex. For the details of the emerging molecules and mechanisms, readers are referred to several excellent, recent reviews (Chevalier-Larsen and Holzbaur, 2006; De Vos et al., 2008; Hirokawa and Noda, 2008; Salinas et al., 2008).

What happens when axonal transport fails? After axon injury, the ultimate block of axonal transport, total and permanent isolation of a distal axon from its cell body results in Wallerian degeneration of the distal stump 1–2 days later. Wallerian degeneration involves a poorly understood latent phase, followed by characteristic granular disintegration of the axonal cytoskeleton, glial reaction, and loss of axon continuity. It can be delayed tenfold by the slow Wallerian degeneration protein (Wld^S) or by related proteins when they are expressed or overexpressed in transgenic animals (Coleman and Freeman, 2010). Wld^S also delays axon degeneration in some neurodegenerative disorders. Several of these involve disruption of axonal transport, strongly suggesting that the trigger for Wallerian degeneration is the failure to deliver an essential cargo (Coleman and Freeman, 2010). An alternative model, involving an injury signal generated by the lesion, cannot explain how Wallerian-like mechanisms can be triggered without physical injury.

Physical trauma or axon compression causes a similar, nonspecific block of axonal transport in several human disorders. High intraocular pressure disrupts the flow of materials at one end of the optic nerve (Howell et al., 2007; Martin et al., 2006), resulting in axon degeneration that Wld^S can delay (Beirowski et al., 2008; Howell et al., 2007). In some models of Alzheimer's disease pathogenesis, amyloid plaques physically compress nearby structures, including axons (Vickers et al., 2000). Spinal contusion injuries place chronic physical pressure on underlying axons, among other effects. Traumatic brain injury stretches axons, disrupting their cytoskeleton and axonal transport (Stone et al., 2004). Solid tumors, carpal tunnel syndrome, and other pressure palsies similarly restrict axonal transport due to physical pressure.

However, nonspecific impairment of axonal transport does not only result from mechanical disruption. Disruption of the microtubule "rails" along which long-range transport runs is the most obvious way to affect all cargoes. In progressive motor neuronopathy mice (pmn), for example, a loss-of-function mutation in the tubulin-specific chaperone e gene leads to a severe deficiency of microtubules in distal axons (Martin et al., 2002; Schaefer et al., 2007). Spastin, the protein mutated in the hereditary spastic paraplegia SPG4, also has critical roles in microtubule assembly and/or severing (Evans et al., 2005; Riano et al., 2009). Neurotoxic drugs such as Taxol and Vincristine probably alter axonal transport by directly targeting microtubules (Shemesh and Spira, 2009; Silva et al., 2006). Thus, little or nothing can be delivered without microtubules, their building blocks, or the chaperones that help put them together.

B. Traffic restrictions: Specific axonal transport defects

Disruption of axonal transport can be more specific in a number of ways. Ever since the discovery of fast anterograde transport (Weiss and Hiscoe, 1948), the complexity of axonal transport, as we understand it, has been increasing. First,

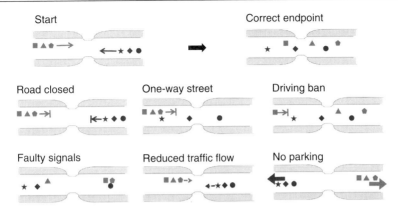

Figure 5.1. Specific defects in axonal transport. Movement of axonal transport cargoes anterogra-dely (blue) and retrogradely (red) over a fixed period of time. Top right: normal endpoint. Below: six ways in which axonal transport can fail to deliver all cargoes to their correct locations (see text for corresponding examples). (See Color Insert.)

nerve constituents were found to travel retrogradely as well as anterogradely (Lubinska, 1964). Then, slow anterograde transport was discovered and subdi-vided (Lasek, 1968). Today, we know a huge array of anterograde motor proteins of the kinesin superfamily carrying different cargoes at different speeds (Hirokawa and Noda, 2008), together with adapter proteins whose roles in regulating attachment of cargo to motor proteins are just beginning to emerge (Horiuchi et al., 2005; Wang and Schwarz, 2009). The range of specific faults that can affect axonal transport is correspondingly wide, resulting in unidirec-tional impairment, partial impairment, or failure to deliver specific cargoes or groups of cargoes (Fig. 5.1).

C. One-way street! Unidirectional transport impairment

The microtubule-associated protein tau has emerged as an important modulator of axonal transport, but it has differential effects on anterograde and retrograde transport. Tau overexpression slows anterograde transport of mitochondria more than retrograde in cell lines (Stamer et al., 2002), probably reflecting its ability to dissociate the anterograde motor kinesin from microtubules. In contrast, when the retrograde motor dynein encounters tau, it hesitates but does not dissociate (Dixit et al., 2008). In amyotrophic lateral sclerosis studies, primary motoneuro-nal cultures from the SOD1^{G93A} transgenic mouse model show reduced antero-grade transport of mitochondria and enhanced retrograde transport, which may deplete axons of mitochondria (De Vos et al., 2007). One intriguing possibility is that the paradoxical partial rescue of SOD1^{G93A} transgenic mice by a mutation affecting retrograde transport could reflect a rebalancing of anterograde and retrograde transport (Kieran et al., 2005). Molecular switches regulating the direction of axonal transport (Colin et al., 2008; De Vos et al., 2003) provide further scope for directional imbalances when they go wrong, but could also be targets for strategies to restore the balance.

D. Banned from driving: Cargo-specific defects

Cargoes travelling in the same direction can be affected to different extents. Viral overexpression of A53T alpha-synuclein in rat substantia nigra to model Parkinson's disease significantly decreases how much Kif1A, Kif1B, Kif2A, and Kif3A reach the striatum after 8 weeks, while Kif5 is barely changed (Chung et al., 2009). Conversely, mutation of Kif5a alone is sufficient to cause juvenile onset hereditary spastic paraplegia (SPG10) (Reid et al., 2002). Mutation of a Drosophila kinesin 3 family protein specifically blocks transport of synaptic vesicle precursors without altering mitochondria and other cargoes (Pack-Chung et al., 2007). Overexpression of K369I human tau in mice causes a large reduction in the amount of tyrosine hydroxylase, App, Gap43, and some other cargoes in the striatum, but no change in synaptophysin and synaptotagmin (Ittner et al., 2008). The motor proteins, Kif5B, Klc, and Kif1A, are also differentially affected. Finally, a mutant form of Huntingtin carrying an expanded polyglutamine repeat slows Bdnf transport but not that of mitochondria (Gauthier et al., 2004).

Specific axonal transport defects can also result from alterations to scaffold proteins linking motor proteins to their cargoes. Mutation or overexpression of one scaffold protein, the Drosophila Jip-1 homolog Aplip-1, perturbs anterograde transport of synaptobrevin-tagged vesicles but not of mitochondria, and causes axonal swellings and larval paralysis (Horiuchi et al., 2005). Another scaffold protein, milton, is critical for the anterograde axonal transport of mitochondria, but its loss leaves synaptic vesicles unaffected (Glater et al., 2006; Stowers et al., 2002).

Thus, disruption of microtubules affects all traffic by closing the road, while changes to the motors and regulators of axonal transport often stop only some "vehicles." An intriguing prospect is that different disorders may result from failure to transport different cargoes.

E. Faulty signals

It is not enough for axonal transport to deliver cargoes to axons. They must also be deposited in the correct location or risk ending up centimeters away from where they are needed. The signaling pathways controlling this are starting to become clear and provide further scope for axonal transport to go wrong. For example, the Rho GTPase *miro* confers calcium sensitivity on mitochondrial transport (Guo *et al.*, 2005; Wang and Schwarz, 2009). Under normal circumstances, this causes mitochondria to accumulate in regions of high calcium influx such as synapses where ATP synthesis and calcium sequestration are important (Macaskill *et al.*, 2009). Another consequence, however, is that mitochondrial movements stall with excessive calcium influx, as in excitotoxicity (Rintoul *et al.*, 2003). Continued movement of other cargoes in high calcium conditions shows the specificity of this effect (Brady *et al.*, 1984). NGF and Tnf also cause mitochondria to pause, probably acting through PI3 kinase and Jnk signaling pathways, respectively (Chada and Hollenbeck, 2004; Stagi *et al.*, 2006). NGF seems to preferentially affect anterograde transport and not stall other cargoes, whereas TNF disrupts mitochondrial transport in both directions along with synaptophysin transport. Finally, some proteins such as Gap43 are anchored to vesicle membranes by palmitoylation (El-Husseini *et al.*, 2001). This process is essential for their axonal targeting but likely has little effect on proteins that use other mechanisms.

F. Traffic congestion: Partial blockages and axonal swellings

Limited disruption of transport can take an entirely different form. Instead of a few specific cargoes being severely disrupted, the flow of many cargoes can be partially restricted. Organelle-filled swellings in kinesin or dynein-deficient *Drosophila* appear as "traffic jams," but mitochondria continue to move into and out just as a queue on the motorway has cars leaving at one end and joining at the other (Pilling *et al.*, 2006). Overexpression of tau in *Aplysia* causes similar axonal swellings, but live imaging shows that microtubule whorls within them continue to support transport (Shemesh *et al.*, 2008; Fig. 5.2). Only a fraction of cargoes become trapped. The same may be true in mammals. Axonal swellings near amyloid plaques in a mouse Alzheimer's disease model grow to the size of neuronal cell bodies and are packed with short, rounded mitochondria and markers of poor axonal transport (Fig. 5.2). Remarkably, however, more distal axon regions survive many months without degenerating, indicating that enough material flows through the swellings to support distal axons (Adalbert *et al.*, 2009).

G. Reduced traffic flow: Ageing and decline

Axonal transport falls dramatically with ageing, affecting fast and slow transport of many cargoes both in retrograde and anterograde directions (Castel *et al.*, 1994; Fernandez and Hodges-Savola, 1994; Frolkis *et al.*, 1997; Li *et al.*, 2004; McQuarrie *et al.*, 1989; Tashiro and Komiya, 1994; Viancour and Kreiter, 1993). In rats aged two years, the average velocity of axonal transport is less than half

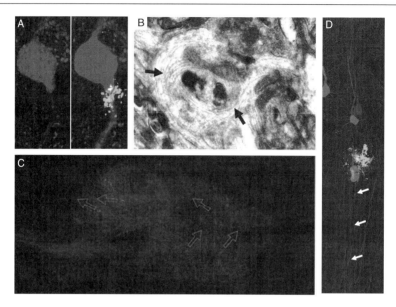

Figure 5.2. Partial blockage of axonal transport in amyloid or tau pathology. (A) Axons (green) swell next to amyloid plaques (red), and rounded mitochondria (blue) accumulate in these swellings. (B) Microtubule whorls (arrows) are evident in some swellings. (C) Similar microtubule diversions (green EB3 staining) form in *Aplysia* when tau is overexpressed (see also supplementary movie in Shemesh *et al.*, 2008). (D) Despite the extensive swelling, axons in the mouse amyloid model shown in A and B remain continuous and morphologically normal at more distal sites (white arrows) for several months, indicating continued flow of at least some transport. A, B, and D reproduced from Adalbert *et al.* (2009) with permission from Oxford Journals. C reproduced from Shemesh *et al.* (2008) with permission from John Wiley & Sons, Inc. (See Color Insert.)

that in young rats and may decline by as much as 71% (Frolkis *et al.*, 1997; Minoshima and Cross, 2008). Similar events occur in older primates (Kimura *et al.*, 2007). Until noninvasive methods can be applied in man, we can only guess how much more transport slows over the course of 80 years or more. As axonal transport impairment is clearly established as a cause of some neurodegenerative conditions (Martin *et al.*, 2002; Reid *et al.*, 2002), this age-related decline could predispose to a range of age-related disorders or even cause some.

Thus, young axons appear to transport far more material than they need for survival, because otherwise, an age-related decline would cause massive axon death. This could explain the survival of swollen axons in young amyloid mouse models (Adalbert *et al.*, 2009; Spires *et al.*, 2005; Fig. 5.2). If the defect were imposed on a lower basal level of transport, the consequences might be far worse

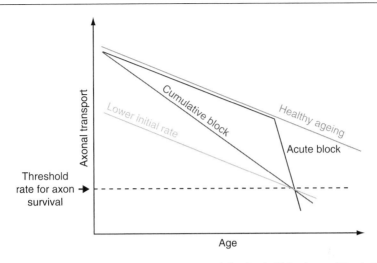

Figure 5.3. Age-related decline in axonal transport and the threshold for disease. The decline in axonal transport with age could predispose or even trigger neurodegenerative disease below a certain threshold (dashed line). The reasons for falling below this threshold may include acute events such as toxins, viruses, or demyelination that lead to a sudden decline in transport (red line), cumulative events such as metabolic disorders causing accelerated declines over a long period (green line), or a lower starting point, for example, due to polymorphisms in transport regulating proteins (orange line). (See Color Insert.)

(Fig. 5.3), just as amyloid pathology worsens when axonal transport is deliberately impaired (Stokin et al., 2005). Similarly, the decline in transport with age could underlie the age-associated axonal swelling found in YFP-H transgenic mice (Bridge et al., 2007).

H. No parking! Could increased axonal transport be too much of a "good" thing?

Finally, axonal disorders are so frequently associated with impaired transport that it is tempting to assume that enhancing transport can only be good for axonal health. However, when mitochondrial movement in axons is increased by deleting the mitochondrial docking protein syntaphilin, this reduces the density of mitochondria in axons and presynaptic terminals, thus impairing calcium buffering at nerve terminals (Kang et al., 2008). There are also subtle but significant motor defects.

Thus, axonal transport may have an optimum level such that too much could be as damaging as too little. Axonal components need to be distributed along the axon length not just delivered to the end, so failure to dissociate cargoes from the transport machinery would prevent cargoes reaching the correct destination just like failure to attach to transport machinery in the first place. A unidirectional increase also risks a gradual buildup of excess components at one end of the axon and a progressive deficiency at the other. Increased traffic may wear out the "road" or saturate motor complexes needed for other cargoes. Our understanding of how altered axonal transport leads to axon degeneration has come a very long way in recent years, but these and many other intriguing questions remain to be answered.

III. THE ESSENTIALS OF AXON SURVIVAL

Having established that axonal transport can be impaired in a cargo-specific manner, it makes sense to ask which cargoes are most essential for axon survival. Specific failure to deliver these would endanger axon survival more than when less essential cargoes are blocked. Alternatively, when transport declines generally, as in ageing or with mechanical pressures, these are the cargoes whose absence is most likely to precipitate axon degeneration. For individual proteins, molecular genetic studies can help to identify which are the most critical, as genetic disruption also stops their delivery to axons. Table 5.1 lists some proteins in this category.

Interestingly, these proteins can be clustered into a relatively small number of common themes that begin to tell us what functions are most critical for axon survival. Around half are connected with axonal transport or vesicle trafficking. Interestingly, not all motor proteins, scaffold proteins, or regulators of axonal transport are essential for axon survival (see Table 5.2). Probably, the motor proteins in Table 5.1 are those that carry essential cargoes. Microtubule integrity clearly needs to be closely regulated by chaperones and severing proteins as it probably influences all cargoes. Vesicle trafficking also clearly links to axonal transport, and the likely involvement of lipid rafts in axonal delivery of some membrane proteins illustrates how cholesterol and lipid metabolism could also fit this theme (el-Husseini Ael and Bredt, 2002). Both protein synthesis and degradation will influence the axonal levels of specific proteins, and the requirement for efficient protein degradation is also consistent with other reports that proteasome inhibition is toxic to axons (Kane et al., 2003; Laser et al., 2003). An important role for adhesion molecules also emerges, probably mediating essential axon–glia interactions (below).

Table 5.1. Proteins Needed for Axon Survival: Loss-of-Function or Dominant Negative Mutations Causing Axon Degeneration

Protein	Context	Phenotype	Reference
Axonal transport/vesicle trafficking			
Alsin	Human disorder, null mouse	ALS/HSP	Deng et al. (2007), Yamanaka et al. (2006)
Atlastin	Human disorder	Hereditary spastic paraplegia	Meijer et al. (2007)
Bpag1	Null mouse	Rapid axon degeneration	Guo et al. (1995)
Dynein	Mouse missense mutations	Motor/sensory axon loss	Hafezparast et al. (2003), Ilieva et al. (2008)
Huntingtin	Conditional null mouse, null fly	Progressive axon/synapse loss	Dragatsis et al. (2000)
Jnk1	Null mouse	Spinal cord axon degeneration	Chang et al. (2003)
Kif1b beta	Heterozygous null mouse	Peripheral neuropathy	Zhao et al. (2001)
Kif5a	Human disorder, null mouse	Hereditary spastic paraplegia	Reid et al. (2002), Xia et al. (2003)
Maspardin	Human disorder	Hereditary spastic paraplegia	Simpson et al. (2003)
p150^{Glued}	Human disorder, Tg mouse	Motor neuron disease	Laird et al. (2008), Puls et al. (2003)
Spartin	Human disorder	Hereditary spastic paraplegia	Patel et al. (2002)
Spastin	Human disorder, null mouse	CNS axonopathy	Tarrade et al. (2006)
Spastizin	Human disorder	Hereditary spastic paraplegia	Hanein et al. (2008)
Tbce	Mouse missense mutation	Early motor neuron disease	Martin et al. (2002)
Vps54	Mouse missense and gene trap	Motor neuron disease	Schmitt-John et al. (2005)
Cholesterol and other lipid metabolisms			
Cyp7b1	Human disorder	Hereditary spastic paraplegia	Tsaousidou et al. (2008)
iPla2β	Human disorder, null mouse	Neuroaxonal dystrophy	Shinzawa et al. (2008)
Npc1	Human disorder, null mouse	Axon swelling, lysosome defect	Pacheco et al. (2009)
Nte	Human toxicity, null mouse	Dying back axonopathy	Read et al. (2009)
Sap	Null mouse	Axonal spheroids	Oya et al. (1998)
Protein synthesis			
GARS	Human disorder	Peripheral neuropathy	Antonellis et al. (2006)
Smn	Human, null mouse, zebrafish MO	Motor neuron disease	Carrel et al. (2006), Cifuentes-Diaz et al. (2002)

(Continues)

Table 5.1. (*Continued*)

Protein	Context	Phenotype	Reference
Degradation pathways			
Atg7	Conditional null mouse	Axon swelling, loss of terminals	Komatsu et al. (2007)
Gigaxonin	Human disorder, null mouse	Giant axonal neuropathy	Allen et al. (2005), Bomont et al. (2000)
Trim2	Mouse gene trap	Axon swelling	Balastik et al. (2008)
Ube4b	Heterozygous null mouse	Gracile axonal dystrophy	Kaneko-Oshikawa et al. (2005)
Uch-l1	Mouse intragenic deletion	Gracile axonal dystrophy	Saigoh et al. (1999)
Adhesion			
α5 integrin	Conditional null mouse	CNS axon degeneration	McCarty et al. (2005)
Ankyrin$_B$	Null mouse	Optic nerve degeneration	Scotland et al. (1998)
Galgt1	Null mouse	CNS, PNS axon degeneration	Pan et al. (2005)
L1-Cam	Null mouse	Unmyelinated sensory axon loss	Haney et al. (1999)
Moca	Null mouse	CNS axonal dystrophy	Chen et al. (2009)
SLC33A1	Human disorder	Hereditary spastic paraplegia	Lin et al. (2008)
Mitochondria			
Gdap1	Human disorder	Charcot-Marie-Tooth Type 4A	Pedrola et al. (2008)
Hsp60	Human disorder	Hereditary spastic paraplegia	Bross et al. (2008)
Paraplegin	Human disorder, null mouse	Hereditary spastic paraplegia	Gelbard (2004)
Reep1	Human disorder	Hereditary spastic paraplegia	Beetz et al. (2008)
Other			
β-Spectrin	*C. elegans* Unc-70 mutants	Axons break	Hammarlund et al. (2007)
Grk5	Null mouse	CNS axonal swelling	Suo et al. (2007)

The table lists proteins with a likely axonal or axonal delivery role, whose genomic loss-of-function or dominant negative mutation results in axon degeneration. Defects that are clearly developmental have been excluded to focus on the maintenance of mature axons. Functional categories should be viewed only as a guide, as proteins may have multiple functions or their functions may not yet be fully characterized.

Table 5.2. Proteins Whose Functional Loss Does Not Cause Axon Degeneration

Protein	Context	Related deficiency in nulls/roles of protein	Reference
App	Null mouse	Reduced synaptic vesicle density	Wang et al. (2005)
Ataxin 3	Null mouse	Increased ubiquitination in the brain	Schmitt et al. (2007)
Bace	Null mouse	Hypomyelination	Willem et al. (2006)
Calpain 1	Null mouse	Calpain activated in Wallerian degeneration	Grammer et al. (2005)
Caspase 6	Null mouse	Necessary for some forms of axon degeneration	Zheng et al. (1999)
Cd38	Null mouse	Enhanced neuronal NAD$^+$ levels	Sasaki et al. (2009)
EphA4	Null mouse	Effects on axon regeneration	Goldshmit et al. (2004)
Erk5	Conditional null mouse	Role in retrograde survival signaling	Wang and Tournier (2006)
Imac/kinesin 3	Mutant *Drosophila* embryo	Synaptogenesis defects	Pack-Chung et al. (2007)
Jip1	Null mouse	Protects from excitotoxic stress	Whitmarsh et al. (2001)
Jnk2	Null mouse	Protects from apoptosis	Ries et al. (2008)
Jnk3	Null mouse	Protected from excitotoxicity	Yang et al. (1997)
K$_v$1.1	Null mouse	Increased excitability, epilepsy	Smart et al. (1998)
Milton	*Drosophila* second instar	Absence of axonal mitochondria	Stowers et al. (2002)
Miro	*Drosophila* third instar	Very few axonal mitochondria	Guo et al. (2005)
Nefh	Null mouse	Increased microtubule density	Zhu et al. (1998)
Nefl	Null mouse	Reduced caliber	Zhu et al. (1997)
Nefm	Null mouse	Reduced caliber	Elder et al. (1998)
Nogo66 receptor	Null mouse	Enhanced axon growth	Kim et al. (2004)
Nnos	Null mouse	Delayed axon regeneration	Keilhoff et al. (2002)
p110δ	Kinase dead knockin mouse	Poor axon regeneration	Eickholt et al. (2007)
Presenilin 1	Conditional null mouse	Inhibits amyloid pathology	Dewachter et al. (2002)
Presenilin 2	Null mouse		Donoviel et al. (1999)
Rab3	Quadruple null mouse	Evoked synaptic release deficiency	Schluter et al. (2004)
Scn2b	Null mouse	Reduced severity in EAE	O'Malley et al. (2009)

(Continues)

Table 5.2. (*Continued*)

Protein	Context	Related deficiency in nulls/roles of protein	Reference
Scn8a	Conditional null mouse	Impaired motor function	Levin *et al.* (2006)
Scn9a	Human null mutations	Insensitivity to pain	Cox *et al.* (2006)
Sir2	Adult fly comp. heterozygote		Avery *et al.* (2009)
Sod2	Mouse conditional null		Misawa *et al.* (2006)
Syntaphilin	Null mouse	Low mitochondrial density, short-term facilitation	Kang *et al.* (2008)
α-Synuclein	Spontaneous mouse null		Specht and Schoepfer (2004)
Synaptotagmin	Null neonatal mouse and null early adult *Drosophila*	Synaptic vesicle depletion, defective calcium-dependent release	Geppert *et al.* (1994), Loewen *et al.* (2001)
Tau	Null mouse	Exacerbates Npc1 and Map1b phenotypes	Dawson *et al.* (2001)

The table lists proteins with prominent roles in axonal or neuronal function that are dispensable for axon survival, at least in the experimental system indicated.

IV. TAKE IT OR LEAVE IT: WHAT AXONS DO NOT NEED TO SURVIVE

Conversely, there are many proteins whose functional absence has little effect on axon survival (Table 5.2). Null mouse or *Drosophila* studies, especially when limited to early developmental stages, cannot of course rule out degeneration over the longer term in larger human axons, but failure to deliver these proteins to axons does not precipitate axon degeneration within the confines of the experimental system listed. Some have other functional consequences, but axons survive.

There are a number of surprises in this list. Axons can survive in the absence of proteins essential for synaptic function, without some of their most abundant proteins, or without proteins closely associated with neurodegenerative diseases.

Redundancy is one obvious explanation why axons survive without some proteins. For example, mice lacking tau show no obvious axonal defects (Dawson *et al.*, 2001; Yuan *et al.*, 2008), but when Map1b is also removed, the axonal phenotype is far worse than for either individual deletion (Takei *et al.*, 2000). In this particular example, the question whether this is a degenerative phenotype as well as developmental still needs to be resolved, but it illustrates the point that the function of some axonal proteins can be filled by others.

Synaptic vesicle trafficking and function can be impaired without compromising axon survival. Rab3 proteins, a family of GTP-binding proteins with seemingly ubiquitous roles in vesicular transport, have essential roles in evoked release of synaptic vesicles. Structurally, however, there is no apparent change in brain or synapse structure even when all four isoforms are deleted simultaneously in mice (Schluter *et al.*, 2004). *Drosophila* embryos deficient in *Imac*, a kinesin 3 protein essential for synaptic vesicle transport, develop axons that contact their muscles, even if they lack extended branches and mature synapses (Pack-Chung *et al.*, 2007), and both *Drosophila* and mouse axons can develop without any signs of degeneration in the absence of synaptotagmin (Geppert *et al.*, 1994; Loewen *et al.*, 2001). Thus, while some aspects of vesicle trafficking are essential for axon survival (Table 5.1), profound defects in synaptic vesicle trafficking or function do not necessarily cause axon degeneration.

Axons are also remarkably tolerant of disruptions to neurofilaments, the major class of intermediate filaments in neurons, in contrast to the devastating effects of microtubule disruption (above). All three neurofilament subunits can be deleted without loss of axonal viability, although axonal caliber decreases without NF-M or NF-L (Elder *et al.*, 1998; Zhu *et al.*, 1997, 1998). Deletion of the peripheral nerve intermediate filament protein peripherin also affects only a small subset of the axons that express it (Lariviere *et al.*, 2002).

Several proteins not required for axon survival are mutated in neurodegenerative diseases. App, tau, α-synuclein, and ataxin 3 are mutated in some hereditary forms of Alzheimer's disease, frontotemporal dementia, Parkinson's disease, and spinal cerebellar ataxia, respectively. Furthermore, Bace and PS1 are responsible for cleaving the pathogenic Aβ1-42 species from its precursor App as a key step in Alzheimer's disease pathogenesis. Considering the major role of synapse loss played in many of these diseases, the ability of axons and synapses to survive without these proteins suggests that gain-of-toxic-function mechanisms (e.g., protein aggregation) are most likely to underlie their roles in disease. Alternatively, mice could be too short-lived, or their axons too short, to model the degenerative stages of the human disease in some cases.

Perhaps most remarkably of all, second instar larvae with the *Drosophila* mutation *milton* have axons that survive up to 5 days without mitochondria. Milton is a scaffold protein essential for anterograde axonal transport of mitochondria, but without it, axons develop and retain a normal ultrastructure and continue to transport other components, including synaptic vesicles (Glater *et al.*, 2006; Stowers *et al.*, 2002). Another mutant, *miro*, also has a severe depletion of axonal mitochondria but retains basal neurotransmitter release (Guo *et al.*, 2005). The larvae reach the third instar, extending axons as long as 1.5 mm (Tom Schwarz, personal communication). A possible explanation for how these axons generate enough ATP to survive comes from SCG primary culture studies, which show that glycolysis makes a significant contribution to axonal ATP synthesis (Tolkovsky and Suidan, 1987; Wakade *et al.*, 1985).

Together, these observations support the concept that axon survival depends much more on some cargoes than others, and Tables 5.1 and 5.2 begin to highlight which proteins are fundamentally important for axons to survive and which are dispensable.

V. PROTEIN TURNOVER FAST AND SLOW

Of course, the response of axons to long-term absence of proteins only partially models the deficiency caused by blocking axonal transport. For example, an acute block of axonal transport (e.g., by injury, ischemia, or toxins) impairs delivery of new cargo, but molecules or organelles already present in axons may be unaffected. Thus, even if their function is essential for axon survival (Table 5.1), the effect may not be felt until the axonal pool turns over. The effects of chronic transport impairment (e.g., in ageing, amyloid or other axonal swelling pathology (Fig. 5.2), or faulty axon–glia interactions) may also depend on half-life. This is because short-lived cargoes are less likely to reach distal axons than long-lived ones if the journey takes too long. Thus, a possible basis for "dying back" disorders (Cavanagh, 1979) is a delay in axonal transport of proteins that are both short-lived and essential for axons, depriving distal axons of the replenishment they constantly need.

Like any mixture of proteins, axonal proteins have a large range of half-lives. At one extreme, the light neurofilament protein has a half-life estimated at 3 weeks to 2.5 months, depending on the experimental system (Millecamps *et al.*, 2007; Yuan *et al.*, 2009). In contrast, App and Aplp2 have half-lives of around 4 h after being transported through the hamster optic nerve to the superior colliculus (Lyckman *et al.*, 1998). Some axonal proteins are even less stable. Ornithine decarboxylase has a half-life of only 5–30 min in many mammalian tissues, although its axonal half-life has not been reported (Iwami *et al.*, 1990).

It is hard to reconcile some of these half-lives with the time taken for proteins to reach axon termini. In the longest human axons, the most rapidly transported cargoes take around 2 days to reach the distal ends. Thus, a half-life of 4 h may suffice for App to reach the terminals of 1-cm hamster optic nerve axons, but little or none should reach neuromuscular junctions in our toes, and the viability of long blue whale axons starts to appear really questionable. For ornithine decarboxylase, a notoriously unstable protein, the problem is even greater, and yet it is abundant in motor axons and their terminals (Junttila *et al.*, 1993). Axonal protein synthesis could be one explanation for a subset of axonal proteins (see below). Another intriguing prospect is that cargoes are somehow "privileged" while they are being transported, so that the ubiquitin proteasome system does not "see" them. Once released from the transport machinery, their half-life could suddenly shorten.

In summary, diverse axonal proteins have half-lives ranging from months down to hours or even minutes. Mechanisms to understand how the more labile proteins reach the distal ends of longer axons require further investigation. However, most labile proteins are likely to be depleted quickly when transport fails, and distal regions are likely to suffer the greatest decline.

VI. AXONAL PROTEIN SYNTHESIS

After many years of controversy, it is now generally accepted that some axonal proteins can be translated locally using axonally targeted mRNAs (Giuditta *et al.*, 2002; Gumy *et al.*, 2009). This seems to concern only a subset of axonal proteins, and how much of these proteins is synthesized in axons remains unclear. However, there is a clear capacity to synthesize some proteins locally using a system that can respond to local events such as axonal damage (Gumy *et al.*, 2009; Perlson *et al.*, 2005) and extracellular stimuli (Willis *et al.*, 2007).

Axonal protein synthesis may partially answer how labile proteins are supplied to distal axons. For example, ornithine decarboxylase mRNA is present in axons, at least after a conditioning lesion (Willis *et al.*, 2007). Thus, the paradox that a large amount of this very labile protein exists in distal axons (above) could be explained if most of the protein derives from local synthesis.

Other mRNAs in both mammalian and *Aplysia* axons cluster around certain types of protein, in particular, those encoding protein translation machinery, cytoskeletal proteins, proteasome subunits, mitochondrial proteins, and heat shock proteins (Gumy *et al.*, 2009; Moccia *et al.*, 2003; Willis *et al.*, 2007), together with some membrane and secreted proteins (Merianda *et al.*, 2009). However, mRNAs encoding many of the essential proteins listed in Table 5.1 are conspicuously absent from the list of axonal mRNAs reported so far (Willis *et al.*, 2007). Thus, axonal protein synthesis does not in general appear to be designed to supply essential proteins that axonal transport fails to deliver. Instead, its main function may be to facilitate a rapid axonal response to local events such as trauma, neurotrophic signaling, or pathfinding cues (Campbell and Holt, 2001; Gumy *et al.*, 2009).

Nevertheless, there are indications that local protein synthesis does influence axon survival. First, spinal muscular atrophy, a lower motor neuron disease with early axon loss (Cifuentes-Diaz *et al.*, 2002), is caused by loss-of-function mutations in the survival motor neuron (smn) protein. Although there is some controversy, the essential function of smn does seem to be axonal, where it helps deliver mRNA for β-actin (Carrel *et al.*, 2006; Conforti *et al.*, 2007). Second, glycyl tRNA synthetase, a tRNA-charging enzyme whose loss-of-function causes Charcot-Marie-Tooth Disease Type 2D, is abundant in peripheral axons where it seems likely to function in local synthesis.

VII. ANTEROGRADE SURVIVAL SIGNALING

Together, the preceding sections tell us that (a) a subset of axonal proteins are critically important for survival, (b) some protein half-lives appear barely compatible with delivery to distal axons before these proteins degrade, and (c) a largely nonoverlapping subset of axonal proteins can be locally translated in axons. When delivery of many cargoes is impaired, those that limit axon survival are likely to be essential, short-lived, and unable to be locally synthesized (Fig. 5.4). Essential proteins with longer half-lives will remain abundant in axons until the dwindling supply of short-lived, essential proteins has already killed the axon. Locally synthesized proteins could be maintained for as long as the translation machinery remains active, which could be a long time if the protein synthesis machinery can be supplied by associated glia, as recently proposed (Court *et al.*, 2008). Thus, when the supply of specific cargoes fails, the life or death of an axon, and whether it dies quickly or slowly, will depend on whether the affected cargo is in this central category.

One example of such a protein has recently been reported. The NAD^+ biosynthesis enzyme nicotinamide mononucleotide adenylyltransferase 2 (Nmnat2) is a highly labile molecule whose presence in axons, at least in

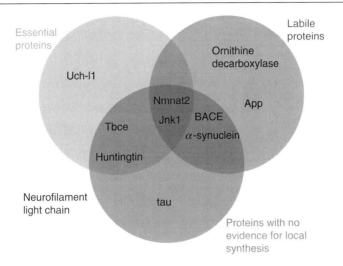

Figure 5.4. Venn diagram showing examples of how axonal proteins can be categorized into essential, labile, and not locally synthesized. For essential and nonessential proteins, see Tables 5.1 and 5.2. The list of axonal mRNAs detected by Willis *et al.* (2007) is used as a guide to locally synthesized proteins, but note that this list is growing (Gumy *et al.*, 2009, 2010), and validation at the protein level will be important. The full list in mature axons *in vivo* could also be different. Half-life data is based on the following references (Bennett *et al.*, 1999; Das *et al.*, 2007; Davis, 2000; Feuillette *et al.*, 2005; Gilley and Coleman, 2010; Iwami *et al.*, 1990; Kabuta *et al.*, 2008; Lyckman *et al.*, 1998; Martin *et al.*, 2002; Millecamps *et al.*, 2007; Tesco *et al.*, 2007; Yuan *et al.*, 2009). For the purposes of this figure, "short" half-life is defined as below 24 h, as this is theoretically too short to reach the ends of long human axons unless the protein is somehow stabilized. However, note that axonal half-lives could sometimes be different from those measured in cell cultures. (See Color Insert.)

primary culture, is essential for their survival (Gilley and Coleman, 2010). Its rapid turnover in axons is normally replaced by fast anterograde axonal transport of newly synthesized protein, but if transport or synthesis of Nmnat2 fails, this triggers Wallerian-like degeneration of the axon. An important future direction will be to identify more such proteins, particularly by knowing more about protein half-lives in axons. Manipulation of the expression, transport, or degradation of such proteins could open new therapeutic opportunities in axonal disorders. It would also be interesting to know whether neurons use similar mechanisms to regulate axonal survival in disease, after injury or during development.

Another approach to understanding anterograde survival signals is to study the axonal roles of proteins that mediate survival signaling in many cells. Although mechanisms of axon degeneration and cell death differ (Deckwerth and Johnson, 1994; Finn *et al.*, 2000), some steps could still be shared. Exposure of phosphatidyl serine on the external surface of degenerating axons directly mirrors one event during apoptosis (Ivins *et al.*, 1998; Sievers *et al.*, 2003), and the involvement of the apoptotic effector caspase 6 in axon death after trophic factor deprivation also suggests similarities (Nikolaev *et al.*, 2009). Thus, it is interesting that survival and stress signaling pathways involving PI3 kinases, Erk1/2, Erk5, and Jnk all have important axonal roles, including retrograde signaling of axonal damage, neurotrophin signaling, and regulating axonal transport (Cavalli *et al.*, 2005; Chada and Hollenbeck, 2004; Horiuchi *et al.*, 2005; Perlson *et al.*, 2005; Watson *et al.*, 2001). Mek/Erk signaling and Jnk signaling have also been reported to influence axon survival (Macinnis and Campenot, 2005; Miller *et al.*, 2009). An important future direction will be to understand the anterograde transport of these molecules and their regulators, and what influence this has on axon survival.

Neurotrophins are also now known to be transported anterogradely in axons, including Bdnf, NT-3, and Gdnf (von Bartheld *et al.*, 2001). This process influences the functions and survival of postsynaptic cells (Caleo *et al.*, 2003; Fawcett *et al.*, 1998), associated glia (Ng *et al.*, 2007), and the axon itself (Menna *et al.*, 2003), at least during development. Neurotrophin receptors involved in retrograde signaling clearly also have to be delivered by anterograde axonal transport. Thus, anterograde cargoes that may regulate axon survival include well-studied intracellular signaling proteins and intercellular ligands and their receptors whose delivery to axons influences cross-talk with supporting glia and other cells.

VIII. ADHESION MOLECULES

Adhesion molecules also have roles in survival signaling in many cell types and feature prominently in the list of proteins essential for axon survival (Table 5.1). Without efficient anterograde transport, these will also be unable to fulfill their important axonal roles such as contributing to axon–glial interactions. Mechanisms of glial support for axons, an absolute requirement for axon survival *in vivo*, have been extensively reviewed elsewhere (Nave and Trapp, 2008), so this chapter focuses on events on the axonal side.

Axonal L1 CAM is needed to maintain, but not to establish, Schwann cell ensheathment of unmyelinated sensory axons (Haney *et al.*, 1999), probably through interaction with Schwann cell $\beta1$-integrins (Itoh *et al.*, 2005). Without it, this axon subset is progressively lost. Interestingly, unmyelinated sympathetic fibers are unaffected, suggesting that different mechanisms mediate this interaction in different peripheral nerve axons. L1 also appears to be essential in the optic nerve prior to myelination, with axonal ankyrin$_B$ probably acting downstream. Despite normal development of L1 positive axons in ankyrin$_B$ null mice, L1 is lost unusually early in postnatal life. This is followed by catastrophic axon swelling and degeneration of the entire optic nerve within the space of 2 weeks (Scotland et al., 1998). Ankyrin$_B$ seems to have an essential role in stabilizing L1 and linking the membrane to the actin cytoskeleton to provide mechanical strength.

In myelinated axons, gangliosides appear to play a similar role. Gangliosides are glycosphingolipids in the axolemma that function as membrane-anchored ligands for myelin-associated glycoprotein (Mag), a glycoprotein located on adaxonal membranes of myelin-producing cells in CNS and PNS that is also required for axon survival (Nguyen et al., 2009; Yin et al., 1998). Mice lacking the complex ganglioside synthetic protein Galgt1 develop progressive axon degeneration in peripheral nerves and dorsal column of the spinal cord (Chiavegatto et al., 2000; Pan et al., 2005). A more severe phenotype develops in mice also lacking Siat9, which are unable to synthesize simple gangliosides too (Yamashita et al., 2005).

Thus, another function of anterograde axonal transport important for axon survival is to deliver these enzymes or their products into axons, in order to maintain axon–glia interactions.

IX. LIFE AFTER (CELL) DEATH: THE WLD^S PHENOTYPE

To return to a topic introduced at the beginning, it is interesting to reexamine the delayed Wallerian degeneration in Wld^S mice in the light of the discussions above. Recent advances in understanding the Wld^S mechanism are summarized elsewhere (Coleman and Freeman, 2010), but an important new point emerges here. Axon injury or cell death (Deckwerth and Johnson, 1994) interrupts the supply of all cargoes from the cell body. Some will be essential, some nonessential. Some will have shorter half-lives than others, and some will be locally synthesized in the distal axon stump. Among the essential missing factors, Nmnat2 seems to have a key role in triggering Wallerian degeneration when its level falls below a threshold needed for survival (Gilley and Coleman, 2010; see above). This may explain why injured axon stumps degenerate after a latent phase of around 36 h (Beirowski et al., 2005; Lubinska, 1977). Delaying this degenerative mechanism by substituting Nmnat2 with Wld^S allows axons survive for 2–3 weeks, indicating that axons do not need replenishment of most cargoes on this timescale. Thus, any protein whose loss triggers degeneration in 36 h should have a half-life that is a clear outlier from the population of essential axonal proteins. Moreover, to preserve injured or transport-impaired axons even longer, it would be useful to identify the next most labile, essential cargoes.

Further understanding of the mechanism by which Wld^S prolongs the survival of severed distal axon stumps should be an excellent way to shed light on the nature of anterograde axon survival signaling.

X. SUMMARY: AXONAL TRANSPORT AND AXON SURVIVAL SIGNALING

This chapter emphasizes that axonal proteins are not all equal in their essential nature, their transport mechanisms, their half-lives, and their capacity for local synthesis. Essential axonal proteins fall into a surprisingly narrow list of categories, while those that are dispensable for axon survival are not necessarily those we might have expected. There is an enormous range of protein half-lives, and some proteins will have difficulty making it to the ends of long axons unless there are mechanisms to stabilize them during transport or to synthesize them locally. Functionally, we can find clues to which proteins limit axon survival from molecular genetic studies. However, an equally promising area to explore is the axonal roles of signaling pathways that control survival in other cell types, as these same pathways have axonal functions but may be subject to different control mechanisms. Intracellular transport over such long distances is unique to axons and while this poses problems for delivering essential cargoes, it also provides a unique mechanism for regulating death programs. Long-range survival signaling and axonal transport may be two ways of looking at the same thing.

Acknowledgments

This chapter reflects memorable presentations and discussions with many colleagues in addition to the cited publications. These include but are not limited to Christine Beattie, Scott Brady, Anthony Brown, Felipe Court, James Fawcett, Marc Freeman, Paul Glynn, Larry Goldstein, Georg Haase, Peter Hollenbeck, Christine Holt, Erika Holzbaur, Keith Martin, Chris Miller, Thomas Misgeld, Hugh Perry, Evan Reid, Richard Ribchester, Tom Schwarz, Zu-Hang Sheng, Aviva Tolkovsy, Jeff Twiss, Xinnan Wang, and Dianna Willis. I am equally grateful to present former members of the Coleman laboratory for open and constructive discussions on many topics related to this chapter. I am also grateful to Hilda Tsang for provision of a literature summary of hereditary spastic paraplegia that contributed to Tables 5.1 and 5.2. The author is funded by the Biotechnology and Biological Sciences Research Council (BBSRC).

References

Adalbert, R., Nogradi, A., Babetto, E., Janeckova, L., Walker, S. A., Kerschensteiner, M., Misgeld, T., and Coleman, M. P. (2009). Severely dystrophic axons at amyloid plaques remain continuous and connected to viable cell bodies. *Brain* **132**, 402–416.

Allen, E., Ding, J., Wang, W., Pramanik, S., Chou, J., Yau, V., and Yang, Y. (2005). Gigaxonin-controlled degradation of MAP1B light chain is critical to neuronal survival. *Nature* **438**, 224–228.

Amiri, M., and Hollenbeck, P. J. (2008). Mitochondrial biogenesis in the axons of vertebrate peripheral neurons. *Dev. Neurobiol.* **68**, 1348–1361.

Antonellis, A., Lee-Lin, S. Q., Wasterlain, A., Leo, P., Quezado, M., Goldfarb, L. G., Myung, K., Burgess, S., Fischbeck, K. H., and Green, E. D. (2006). Functional analyses of glycyl-tRNA synthetase mutations suggest a key role for tRNA-charging enzymes in peripheral axons. *J. Neurosci.* **26**, 10397–10406.

Avery, M. A., Sheehan, A. E., Kerr, K. S., Wang, J., and Freeman, M. R. (2009). Wld S requires Nmnat1 enzymatic activity and N16–VCP interactions to suppress Wallerian degeneration. *J. Cell Biol.* **184**, 501–513.

Balastik, M., Ferraguti, F., Pires-da Silva, A., Lee, T. H., Alvarez-Bolado, G., Lu, K. P., and Gruss, P. (2008). Deficiency in ubiquitin ligase TRIM2 causes accumulation of neurofilament light chain and neurodegeneration. *Proc. Natl. Acad. Sci. USA* **105**, 12016–12021.

Beetz, C., Schule, R., Deconinck, T., Tran-Viet, K. N., Zhu, H., Kremer, B. P., Frints, S. G., van Zelst-Stams, W. A., Byrne, P., Otto, S., Nygren, A. O., Baets, J., et al. (2008). REEP1 mutation spectrum and genotype/phenotype correlation in hereditary spastic paraplegia type 31. *Brain* **131**, 1078–1086.

Beirowski, B., Adalbert, R., Wagner, D., Grumme, D., Addicks, K., Ribchester, R. R., and Coleman, M. P. (2005). The progressive nature of Wallerian degeneration in wild-type and slow Wallerian degeneration (WldS) nerves. *BMC Neurosci.* **6**, 6.

Beirowski, B., Babetto, E., Coleman, M. P., and Martin, K. R. (2008). The WldS gene delays axonal but not somatic degeneration in a rat glaucoma model. *Eur. J. Neurosci.* **28**, 1166–1179.

Bennett, M. C., Bishop, J. F., Leng, Y., Chock, P. B., Chase, T. N., and Mouradian, M. M. (1999). Degradation of alpha-synuclein by proteasome. *J. Biol. Chem.* **274**, 33855–33858.

Bomont, P., Cavalier, L., Blondeau, F., Ben Hamida, C., Belal, S., Tazir, M., Demir, E., Topaloglu, H., Korinthenberg, R., Tuysuz, B., Landrieu, P., Hentati, F., et al. (2000). The gene encoding gigaxonin, a new member of the cytoskeletal BTB/kelch repeat family, is mutated in giant axonal neuropathy. *Nat. Genet.* **26**, 370–374.

Brady, S. T., Lasek, R. J., Allen, R. D., Yin, H. L., and Stossel, T. P. (1984). Gelsolin inhibition of fast axonal transport indicates a requirement for actin microfilaments. *Nature* **310**, 56–58.

Bridge, K., Berg, N., Adalbert, R., Babetto, E., Dias, T., Spillantini, M. G., Ribchester, R. R., and Coleman, M. P. (2007). Late onset distal axonal swelling in YFP-H transgenic mice. *Neurobiol. Aging* (in press).

Bross, P., Naundrup, S., Hansen, J., Nielsen, M. N., Christensen, J. H., Kruhoffer, M., Palmfeldt, J., Corydon, T. J., Gregersen, N., Ang, D., Georgopoulos, C., and Nielsen, K. L. (2008). The Hsp60-(p.V98I) mutation associated with hereditary spastic paraplegia SPG13 compromises chaperonin function both in vitro and in vivo. *J. Biol. Chem.* **283**, 15694–15700.

Caleo, M., Medini, P., von Bartheld, C. S., and Maffei, L. (2003). Provision of brain-derived neurotrophic factor via anterograde transport from the eye preserves the physiological responses of axotomized geniculate neurons. *J. Neurosci.* **23**, 287–296.

Campbell, D. S., and Holt, C. E. (2001). Chemotropic responses of retinal growth cones mediated by rapid local protein synthesis and degradation. *Neuron* **32**, 1013–1026.

Carrel, T. L., McWhorter, M. L., Workman, E., Zhang, H., Wolstencroft, E. C., Lorson, C., Bassell, G. J., Burghes, A. H., and Beattie, C. E. (2006). Survival motor neuron function in motor axons is independent of functions required for small nuclear ribonucleoprotein biogenesis. *J. Neurosci.* **26**, 11014–11022.

Castel, M. N., Beaudet, A., and Laduron, P. M. (1994). Retrograde axonal transport of neurotensin in rat nigrostriatal dopaminergic neurons. Modulation during ageing and possible physiological role. *Biochem. Pharmacol.* **47**, 53–62.

Cavalli, V., Kujala, P., Klumperman, J., and Goldstein, L. S. (2005). Sunday Driver links axonal transport to damage signaling. *J. Cell Biol.* **168,** 775–787.

Cavanagh, J. B. (1979). The 'dying back' process. A common denominator in many naturally occurring and toxic neuropathies. *Arch. Pathol. Lab. Med.* **103,** 659–664.

Chada, S. R., and Hollenbeck, P. J. (2004). Nerve growth factor signaling regulates motility and docking of axonal mitochondria. *Curr. Biol.* **14,** 1272–1276.

Chang, L., Jones, Y., Ellisman, M. H., Goldstein, L. S., and Karin, M. (2003). JNK1 is required for maintenance of neuronal microtubules and controls phosphorylation of microtubule-associated proteins. *Dev. Cell* **4,** 521–533.

Chen, Q., Peto, C. A., Shelton, G. D., Mizisin, A., Sawchenko, P. E., and Schubert, D. (2009). Loss of modifier of cell adhesion reveals a pathway leading to axonal degeneration. *J. Neurosci.* **29,** 118–130.

Chevalier-Larsen, E., and Holzbaur, E. L. (2006). Axonal transport and neurodegenerative disease. *Biochim. Biophys. Acta* **1762,** 1094–1108.

Chiavegatto, S., Sun, J., Nelson, R. J., and Schnaar, R. L. (2000). A functional role for complex gangliosides: Motor deficits in GM2/GD2 synthase knockout mice. *Exp. Neurol.* **166,** 227–234.

Chung, C. Y., Koprich, J. B., Siddiqi, H., and Isacson, O. (2009). Dynamic changes in presynaptic and axonal transport proteins combined with striatal neuroinflammation precede dopaminergic neuronal loss in a rat model of AAV alpha-synucleinopathy. *J. Neurosci.* **29,** 3365–3373.

Cifuentes-Diaz, C., Nicole, S., Velasco, M. E., Borra-Cebrian, C., Panozzo, C., Frugier, T., Millet, G., Roblot, N., Joshi, V., and Melki, J. (2002). Neurofilament accumulation at the motor endplate and lack of axonal sprouting in a spinal muscular atrophy mouse model. *Hum. Mol. Genet.* **11,** 1439–1447.

Coleman, M. (2005). Axon degeneration mechanisms: Commonality amid diversity. *Nat. Rev. Neurosci.* **6,** 889–898.

Coleman, M. P., and Freeman, M.R. (2010). Wallerian degeneration, WldS and Nmnat. *Ann. Rev. Neurosci.* **33,** 245–267.

Colin, E., Zala, D., Liot, G., Rangone, H., Borrell-Pages, M., Li, X. J., Saudou, F., and Humbert, S. (2008). Huntingtin phosphorylation acts as a molecular switch for anterograde/retrograde transport in neurons. *EMBO J.* **27,** 2124–2134.

Conforti, L., Adalbert, R., and Coleman, M. P. (2007). Neuronal death: Where does the end begin? *Trends Neurosci.* **30,** 159–166.

Court, F. A., Hendriks, W. T., Macgillavry, H. D., Alvarez, J., and van Minnen, J. (2008). Schwann cell to axon transfer of ribosomes: Toward a novel understanding of the role of glia in the nervous system. *J. Neurosci.* **28,** 11024–11029.

Cox, J. J., Reimann, F., Nicholas, A. K., Thornton, G., Roberts, E., Springell, K., Karbani, G., Jafri, H., Mannan, J., Raashid, Y., Al-Gazali, L., Hamamy, H., *et al.* (2006). An SCN9A channelopathy causes congenital inability to experience pain. *Nature* **444,** 894–898.

Das, M., Jiang, F., Sluss, H. K., Zhang, C., Shokat, K. M., Flavell, R. A., and Davis, R. J. (2007). Suppression of p53-dependent senescence by the JNK signal transduction pathway. *Proc. Natl. Acad. Sci. USA* **104,** 15759–15764.

Davis, R. J. (2000). Signal transduction by the JNK group of MAP kinases. *Cell* **103,** 239–252.

Dawson, H. N., Ferreira, A., Eyster, M. V., Ghoshal, N., Binder, L. I., and Vitek, M. P. (2001). Inhibition of neuronal maturation in primary hippocampal neurons from tau deficient mice. *J. Cell Sci.* **114,** 1179–1187.

De Vos, K. J., Chapman, A. L., Tennant, M. E., Manser, C., Tudor, E. L., Lau, K. F., Brownlees, J., Ackerley, S., Shaw, P. J., McLoughlin, D. M., Shaw, C. E., Leigh, P. N., *et al.* (2007). Familial amyotrophic lateral sclerosis-linked SOD1 mutants perturb fast axonal transport to reduce axonal mitochondria content. *Hum. Mol. Genet.* **16,** 2720–2728.

De Vos, K. J., Grierson, A. J., Ackerley, S., and Miller, C. C. (2008). Role of axonal transport in neurodegenerative diseases. *Annu. Rev. Neurosci.* **31**, 151–173.

De Vos, K. J., Sable, J., Miller, K. E., and Sheetz, M. P. (2003). Expression of phosphatidylinositol (4, 5) bisphosphate-specific pleckstrin homology domains alters direction but not the level of axonal transport of mitochondria. *Mol. Biol. Cell* **14**, 3636–3649.

Deckwerth, T. L., and Johnson, E. M., Jr. (1994). Neurites can remain viable after destruction of the neuronal soma by programmed cell death (apoptosis). *Dev. Biol.* **165**, 63–72.

Deng, H. X., Zhai, H., Fu, R., Shi, Y., Gorrie, G. H., Yang, Y., Liu, E., Dal Canto, M. C., Mugnaini, E., and Siddique, T. (2007). Distal axonopathy in an alsin-deficient mouse model. *Hum. Mol. Genet.* **16**, 2911–2920.

Dewachter, I., Reverse, D., Caluwaerts, N., Ris, L., Kuiperi, C., Van den Haute, C., Spittaels, K., Umans, L., Serneels, L., Thiry, E., Moechars, D., Mercken, M., *et al.* (2002). Neuronal deficiency of presenilin 1 inhibits amyloid plaque formation and corrects hippocampal long-term potentiation but not a cognitive defect of amyloid precursor protein [V717I] transgenic mice. *J. Neurosci.* **22**, 3445–3453.

Dixit, R., Ross, J. L., Goldman, Y. E., and Holzbaur, E. L. (2008). Differential regulation of dynein and kinesin motor proteins by tau. *Science* **319**, 1086–1089.

Donoviel, D. B., Hadjantonakis, A. K., Ikeda, M., Zheng, H., Hyslop, P. S., and Bernstein, A. (1999). Mice lacking both presenilin genes exhibit early embryonic patterning defects. *Genes Dev.* **13**, 2801–2810.

Dragatsis, I., Levine, M. S., and Zeitlin, S. (2000). Inactivation of Hdh in the brain and testis results in progressive neurodegeneration and sterility in mice. *Nat. Genet.* **26**, 300–306.

Eickholt, B. J., Ahmed, A. I., Davies, M., Papakonstanti, E. A., Pearce, W., Starkey, M. L., Bilancio, A., Need, A. C., Smith, A. J., Hall, S. M., Hamers, F. P., Giese, K. P., *et al.* (2007). Control of axonal growth and regeneration of sensory neurons by the p110delta PI 3-kinase. *PLoS ONE* **2**, e869.

El-Husseini, A. E., Craven, S. E., Brock, S. C., and Bredt, D. S. (2001). Polarized targeting of peripheral membrane proteins in neurons. *J. Biol. Chem.* **276**, 44984–44992.

El-Husseini Ael, D., and Bredt, D. S. (2002). Protein palmitoylation: A regulator of neuronal development and function. *Nat. Rev. Neurosci.* **3**, 791–802.

Elder, G. A., Friedrich, V. L., Jr., Bosco, P., Kang, C., Gourov, A., Tu, P. H., Lee, V. M., and Lazzarini, R. A. (1998). Absence of the mid-sized neurofilament subunit decreases axonal calibers, levels of light neurofilament (NF-L), and neurofilament content. *J. Cell Biol.* **141**, 727–739.

Evans, K. J., Gomes, E. R., Reisenweber, S. M., Gundersen, G. G., and Lauring, B. P. (2005). Linking axonal degeneration to microtubule remodeling by Spastin-mediated microtubule severing. *J. Cell Biol.* **168**, 599–606.

Fawcett, J. P., Bamji, S. X., Causing, C. G., Aloyz, R., Ase, A. R., Reader, T. A., McLean, J. H., and Miller, F. D. (1998). Functional evidence that BDNF is an anterograde neuronal trophic factor in the CNS. *J. Neurosci.* **18**, 2808–2821.

Fernandez, H. L., and Hodges-Savola, C. A. (1994). Axoplasmic transport of calcitonin gene-related peptide in rat peripheral nerve as a function of age. *Neurochem. Res.* **19**, 1369–1377.

Feuillette, S., Blard, O., Lecourtois, M., Frebourg, T., Campion, D., and Dumanchin, C. (2005). Tau is not normally degraded by the proteasome. *J. Neurosci. Res.* **80**, 400–405.

Finn, J. T., Weil, M., Archer, F., Siman, R., Srinivasan, A., and Raff, M. C. (2000). Evidence that wallerian degeneration and localized axon degeneration induced by local neurotrophin deprivation do not involve caspases. *J. Neurosci.* **20**, 1333–1341.

Frolkis, V. V., Tanin, S. A., and Gorban, Y. N. (1997). Age-related changes in axonal transport. *Exp. Gerontol.* **32**, 441–450.

Gauthier, L. R., Charrin, B. C., Borrell-Pages, M., Dompierre, J. P., Rangone, H., Cordelieres, F. P., De Mey, J., MacDonald, M. E., Lessmann, V., Humbert, S., and Saudou, F. (2004). Huntingtin controls neurotrophic support and survival of neurons by enhancing BDNF vesicular transport along microtubules. *Cell* **118**, 127–138.

Gelbard, H. A. (2004). Synapses and Sisyphus: Life without paraplegin. *J. Clin. Investig.* **113**, 185–187.

Geppert, M., Goda, Y., Hammer, R. E., Li, C., Rosahl, T. W., Stevens, C. F., and Sudhof, T. C. (1994). Synaptotagmin I: A major Ca2+ sensor for transmitter release at a central synapse. *Cell* **79**, 717–727.

Gilley, J., and Coleman, M. P. (2010). Endogenous Nmnat2 is an essential survival factor for maintenance of healthy axons. *PLoS Biol.* **8**(1), e1000300.

Giuditta, A., Kaplan, B. B., van Minnen, J., Alvarez, J., and Koenig, E. (2002). Axonal and presynaptic protein synthesis: New insights into the biology of the neuron. *Trends Neurosci.* **25**, 400–404.

Glater, E. E., Megeath, L. J., Stowers, R. S., and Schwarz, T. L. (2006). Axonal transport of mitochondria requires milton to recruit kinesin heavy chain and is light chain independent. *J. Cell Biol.* **173**, 545–557.

Goldshmit, Y., Galea, M. P., Wise, G., Bartlett, P. F., and Turnley, A. M. (2004). Axonal regeneration and lack of astrocytic gliosis in EphA4-deficient mice. *J. Neurosci.* **24**, 10064–10073.

Grammer, M., Kuchay, S., Chishti, A., and Baudry, M. (2005). Lack of phenotype for LTP and fear conditioning learning in calpain 1 knock-out mice. *Neurobiol. Learn. Mem.* **84**, 222–227.

Gumy, L. F., Tan, C. L., and Fawcett, J. W. (2009). The role of local protein synthesis and degradation in axon regeneration. *Exp. Neurol.*

Guo, L., Degenstein, L., Dowling, J., Yu, Q. C., Wollmann, R., Perman, B., and Fuchs, E. (1995). Gene targeting of BPAG1: Abnormalities in mechanical strength and cell migration in stratified epithelia and neurologic degeneration. *Cell* **81**, 233–243.

Guo, X., Macleod, G. T., Wellington, A., Hu, F., Panchumarthi, S., Schoenfield, M., Marin, L., Charlton, M. P., Atwood, H. L., and Zinsmaier, K. E. (2005). The GTPase dMiro is required for axonal transport of mitochondria to Drosophila synapses. *Neuron* **47**, 379–393.

Hafezparast, M., Klocke, R., Ruhrberg, C., Marquardt, A., Ahmad-Annuar, A., Bowen, S., Lalli, G., Witherden, A. S., Hummerich, H., Nicholson, S., Morgan, P. J., Oozageer, R., *et al.* (2003). Mutations in dynein link motor neuron degeneration to defects in retrograde transport. *Science* **300**, 808–812.

Hammarlund, M., Jorgensen, E. M., and Bastiani, M. J. (2007). Axons break in animals lacking beta-spectrin. *J. Cell Biol.* **176**, 269–275.

Hanein, S., Martin, E., Boukhris, A., Byrne, P., Goizet, C., Hamri, A., Benomar, A., Lossos, A., Denora, P., Fernandez, J., Elleuch, N., Forlani, S., *et al.* (2008). Identification of the SPG15 gene, encoding spastizin, as a frequent cause of complicated autosomal-recessive spastic paraplegia, including Kjellin syndrome. *Am. J. Hum. Genet.* **82**, 992–1002.

Haney, C. A., Sahenk, Z., Li, C., Lemmon, V. P., Roder, J., and Trapp, B. D. (1999). Heterophilic binding of L1 on unmyelinated sensory axons mediates Schwann cell adhesion and is required for axonal survival. *J. Cell Biol.* **146**, 1173–1184.

Hirokawa, N., and Noda, Y. (2008). Intracellular transport and kinesin superfamily proteins, KIFs: Structure, function, and dynamics. *Physiol. Rev.* **88**, 1089–1118.

Horiuchi, D., Barkus, R. V., Pilling, A. D., Gassman, A., and Saxton, W. M. (2005). APLIP1, a kinesin binding JIP-1/JNK scaffold protein, influences the axonal transport of both vesicles and mitochondria in Drosophila. *Curr. Biol.* **15**, 2137–2141.

Howell, G. R., Libby, R. T., Jakobs, T. C., Smith, R. S., Phalan, F. C., Barter, J. W., Barbay, J. M., Marchant, J. K., Mahesh, N., Porciatti, V., Whitmore, A. V., Masland, R. H., *et al.* (2007). Axons of retinal ganglion cells are insulted in the optic nerve early in DBA/2J glaucoma. *J. Cell Biol.* **179**, 1523–1537.

Ilieva, H. S., Yamanaka, K., Malkmus, S., Kakinohana, O., Yaksh, T., Marsala, M., and Cleveland, D. W. (2008). Mutant dynein (Loa) triggers proprioceptive axon loss that extends survival only in the SOD1 ALS model with highest motor neuron death. *Proc. Natl. Acad. Sci. USA* **105,** 12599–12604.

Itoh, K., Fushiki, S., Kamiguchi, H., Arnold, B., Altevogt, P., and Lemmon, V. (2005). Disrupted Schwann cell-axon interactions in peripheral nerves of mice with altered L1-integrin interactions. *Mol. Cell. Neurosci.* **30,** 624–629.

Ittner, L. M., Fath, T., Ke, Y. D., Bi, M., van Eersel, J., Li, K. M., Gunning, P., and Gotz, J. (2008). Parkinsonism and impaired axonal transport in a mouse model of frontotemporal dementia. *Proc. Natl. Acad. Sci. USA* **105,** 15997–16002.

Ivins, K. J., Bui, E. T., and Cotman, C. W. (1998). Beta-amyloid induces local neurite degeneration in cultured hippocampal neurons: Evidence for neuritic apoptosis. *Neurobiol. Dis.* **5,** 365–378.

Iwami, K., Wang, J. Y., Jain, R., McCormack, S., and Johnson, L. R. (1990). Intestinal ornithine decarboxylase: Half-life and regulation by putrescine. *Am. J. Physiol.* **258,** G308–G315.

Junttila, T., Hietanen-Peltola, M., Rechardt, L., Persson, L., Hokfelt, T., and Pelto-Huikko, M. (1993). Ornithine decarboxylase-like immunoreactivity in rat spinal motoneurons and motoric nerves. *Brain Res.* **609,** 149–153.

Kabuta, T., Furuta, A., Aoki, S., Furuta, K., and Wada, K. (2008). Aberrant interaction between Parkinson disease-associated mutant UCH-L1 and the lysosomal receptor for chaperone-mediated autophagy. *J. Biol. Chem.* **283,** 23731–23738.

Kane, R. C., Bross, P. F., Farrell, A. T., and Pazdur, R. (2003). Velcade: U.S. FDA approval for the treatment of multiple myeloma progressing on prior therapy. *Oncologist* **8,** 508–513.

Kaneko-Oshikawa, C., Nakagawa, T., Yamada, M., Yoshikawa, H., Matsumoto, M., Yada, M., Hatakeyama, S., Nakayama, K., and Nakayama, K. I. (2005). Mammalian E4 is required for cardiac development and maintenance of the nervous system. *Mol. Cell. Biol.* **25,** 10953–10964.

Kang, J. S., Tian, J. H., Pan, P. Y., Zald, P., Li, C., Deng, C., and Sheng, Z. H. (2008). Docking of axonal mitochondria by syntaphilin controls their mobility and affects short-term facilitation. *Cell* **132,** 137–148.

Keilhoff, G., Fansa, H., and Wolf, G. (2002). Differences in peripheral nerve degeneration/regeneration between wild-type and neuronal nitric oxide synthase knockout mice. *J. Neurosci. Res.* **68,** 432–441.

Kieran, D., Hafezparast, M., Bohnert, S., Dick, J. R., Martin, J., Schiavo, G., Fisher, E. M., and Greensmith, L. (2005). A mutation in dynein rescues axonal transport defects and extends the life span of ALS mice. *J. Cell Biol.* **169,** 561–567.

Kim, J. E., Liu, B. P., Park, J. H., and Strittmatter, S. M. (2004). Nogo-66 receptor prevents raphespinal and rubrospinal axon regeneration and limits functional recovery from spinal cord injury. *Neuron* **44,** 439–451.

Kimura, N., Imamura, O., Ono, F., and Terao, K. (2007). Aging attenuates dynactin–dynein interaction: Down-regulation of dynein causes accumulation of endogenous tau and amyloid precursor protein in human neuroblastoma cells. *J. Neurosci. Res.* **85,** 2909–2916.

Komatsu, M., Wang, Q. J., Holstein, G. R., Friedrich, V. L., Jr., Iwata, J., Kominami, E., Chait, B. T., Tanaka, K., and Yue, Z. (2007). Essential role for autophagy protein Atg7 in the maintenance of axonal homeostasis and the prevention of axonal degeneration. *Proc. Natl. Acad. Sci. USA* **104,** 14489–14494.

Korhonen, L., and Lindholm, D. (2004). The ubiquitin proteasome system in synaptic and axonal degeneration: A new twist to an old cycle. *J. Cell Biol.* **165,** 27–30.

Laird, F. M., Farah, M. H., Ackerley, S., Hoke, A., Maragakis, N., Rothstein, J. D., Griffin, J., Price, D. L., Martin, L. J., and Wong, P. C. (2008). Motor neuron disease occurring in a mutant dynactin mouse model is characterized by defects in vesicular trafficking. *J. Neurosci.* **28,** 1997–2005.

Lariviere, R. C., Nguyen, M. D., Ribeiro-da-Silva, A., and Julien, J. P. (2002). Reduced number of unmyelinated sensory axons in peripherin null mice. *J. Neurochem.* **81,** 525–532.

Lasek, R. (1968). Axoplasmic transport in cat dorsal root ganglion cells: As studied with [3-H]-L-leucine. *Brain Res.* **7,** 360–377.

Laser, H., Mack, T. G., Wagner, D., and Coleman, M. P. (2003). Proteasome inhibition arrests neurite outgrowth and causes "dying-back" degeneration in primary culture. *J. Neurosci. Res.* **74,** 906–916.

Levin, S. I., Khaliq, Z. M., Aman, T. K., Grieco, T. M., Kearney, J. A., Raman, I. M., and Meisler, M. H. (2006). Impaired motor function in mice with cell-specific knockout of sodium channel Scn8a (NaV1.6) in cerebellar purkinje neurons and granule cells. *J. Neurophysiol.* **96,** 785–793.

Li, W., Hoffman, P. N., Stirling, W., Price, D. L., and Lee, M. K. (2004). Axonal transport of human alpha-synuclein slows with aging but is not affected by familial Parkinson's disease-linked mutations. *J. Neurochem.* **88,** 401–410.

Lin, P., Li, J., Liu, Q., Mao, F., Qiu, R., Hu, H., Song, Y., Yang, Y., Gao, G., Yan, C., Yang, W., Shao, C., *et al.* (2008). A missense mutation in SLC33A1, which encodes the acetyl-CoA transporter, causes autosomal-dominant spastic paraplegia (SPG42). *Am. J. Hum. Genet.* **83,** 752–759.

Loewen, C. A., Mackler, J. M., and Reist, N. E. (2001). Drosophila synaptotagmin I null mutants survive to early adulthood. *Genesis* **31,** 30–36.

Lubinska, L. (1964). Axoplasmic streaming in regenerating and in normal nerve fibres. *Prog. Brain Res.* **13,** 1–71.

Lubinska, L. (1977). Early course of Wallerian degeneration in myelinated fibres of the rat phrenic nerve. *Brain Res.* **130,** 47–63.

Lyckman, A. W., Confaloni, A. M., Thinakaran, G., Sisodia, S. S., and Moya, K. L. (1998). Post-translational processing and turnover kinetics of presynaptically targeted amyloid precursor superfamily proteins in the central nervous system. *J. Biol. Chem.* **273,** 11100–11106.

Macaskill, A. F., Rinholm, J. E., Twelvetrees, A. E., Arancibia-Carcamo, I. L., Muir, J., Fransson, A., Aspenstrom, P., Attwell, D., and Kittler, J. T. (2009). Miro1 is a calcium sensor for glutamate receptor-dependent localization of mitochondria at synapses. *Neuron* **61,** 541–555.

Macinnis, B. L., and Campenot, R. B. (2005). Regulation of Wallerian degeneration and nerve growth factor withdrawal-induced pruning of axons of sympathetic neurons by the proteasome and the MEK/Erk pathway. *Mol. Cell. Neurosci.* **28,** 430–439.

Mack, T. G., Reiner, M., Beirowski, B., Mi, W., Emanuelli, M., Wagner, D., Thomson, D., Gillingwater, T., Court, F., Conforti, L., Fernando, F. S., Tarlton, A., *et al.* (2001). Wallerian degeneration of injured axons and synapses is delayed by a Ube4b/Nmnat chimeric gene. *Nat. Neurosci.* **4,** 1199–1206.

Martin, K. R., Quigley, H. A., Valenta, D., Kielczewski, J., and Pease, M. E. (2006). Optic nerve dynein motor protein distribution changes with intraocular pressure elevation in a rat model of glaucoma. *Exp. Eye Res.* **83,** 255–262.

Martin, N., Jaubert, J., Gounon, P., Salido, E., Haase, G., Szatanik, M., and Guenet, J. L. (2002). A missense mutation in Tbce causes progressive motor neuronopathy in mice. *Nat. Genet.* **32,** 443–447.

McCarty, J. H., Lacy-Hulbert, A., Charest, A., Bronson, R. T., Crowley, D., Housman, D., Savill, J., Roes, J., and Hynes, R. O. (2005). Selective ablation of alphav integrins in the central nervous system leads to cerebral hemorrhage, seizures, axonal degeneration and premature death. *Development* **132,** 165–176.

McQuarrie, I. G., Brady, S. T., and Lasek, R. J. (1989). Retardation in the slow axonal transport of cytoskeletal elements during maturation and aging. *Neurobiol. Aging* **10,** 359–365.

Meijer, I. A., Dion, P., Laurent, S., Dupre, N., Brais, B., Levert, A., Puymirat, J., Rioux, M. F., Sylvain, M., Zhu, P. P., Soderblom, C., Stadler, J., et al. (2007). Characterization of a novel SPG3A deletion in a French-Canadian family. Ann. Neurol. **61**, 599–603.

Menna, E., Cenni, M. C., Naska, S., and Maffei, L. (2003). The anterogradely transported BDNF promotes retinal axon remodeling during eye specific segregation within the LGN. Mol. Cell. Neurosci. **24**, 972–983.

Merianda, T. T., Lin, A. C., Lam, J. S., Vuppalanchi, D., Willis, D. E., Karin, N., Holt, C. E., and Twiss, J. L. (2009). A functional equivalent of endoplasmic reticulum and Golgi in axons for secretion of locally synthesized proteins. Mol. Cell. Neurosci. **40**, 128–142.

Millecamps, S., Gowing, G., Corti, O., Mallet, J., and Julien, J. P. (2007). Conditional NF-L transgene expression in mice for in vivo analysis of turnover and transport rate of neurofilaments. J. Neurosci. **27**, 4947–4956.

Miller, B. R., Press, C., Daniels, R. W., Sasaki, Y., Milbrandt, J., and DiAntonio, A. (2009). A dual leucine kinase-dependent axon self-destruction program promotes Wallerian degeneration. Nat. Neurosci. **12**, 387–389.

Minoshima, S., and Cross, D. (2008). In vivo imaging of axonal transport using MRI: Aging and Alzheimer's disease. Eur. J. Nucl. Med. Mol. Imaging **35**(Suppl. 1), S89–S92.

Misawa, H., Nakata, K., Matsuura, J., Moriwaki, Y., Kawashima, K., Shimizu, T., Shirasawa, T., and Takahashi, R. (2006). Conditional knockout of Mn superoxide dismutase in postnatal motor neurons reveals resistance to mitochondrial generated superoxide radicals. Neurobiol. Dis. **23**, 169–177.

Moccia, R., Chen, D., Lyles, V., Kapuya, E. E. Y., Kalachikov, S., Spahn, C. M., Frank, J., Kandel, E. R., Barad, M., and Martin, K. C. (2003). An unbiased cDNA library prepared from isolated Aplysia sensory neuron processes is enriched for cytoskeletal and translational mRNAs. J. Neurosci. **23**, 9409–9417.

Muchowski, P. J., and Wacker, J. L. (2005). Modulation of neurodegeneration by molecular chaperones. Nat. Rev. Neurosci. **6**, 11–22.

Nave, K. A., and Trapp, B. D. (2008). Axon-glial signaling and the glial support of axon function. Annu. Rev. Neurosci. **31**, 535–561.

Ng, B. K., Chen, L., Mandemakers, W., Cosgaya, J. M., and Chan, J. R. (2007). Anterograde transport and secretion of brain-derived neurotrophic factor along sensory axons promote Schwann cell myelination. J. Neurosci. **27**, 7597–7603.

Nguyen, T., Mehta, N. R., Conant, K., Kim, K. J., Jones, M., Calabresi, P. A., Melli, G., Hoke, A., Schnaar, R. L., Ming, G. L., Song, H., Keswani, S. C., et al. (2009). Axonal protective effects of the myelin-associated glycoprotein. J. Neurosci. **29**, 630–637.

Nikolaev, A., McLaughlin, T., O'eary, D. D., and Tessier-Lavigne, M. (2009). APP binds DR6 to trigger axon pruning and neuron death via distinct caspases. Nature **457**, 981–989.

O'Malley, H. A., Shreiner, A. B., Chen, G. H., Huffnagle, G. B., and Isom, L. L. (2009). Loss of Na+ channel beta2 subunits is neuroprotective in a mouse model of multiple sclerosis. Mol. Cell. Neurosci. **40**, 143–155.

Oya, Y., Nakayasu, H., Fujita, N., and Suzuki, K. (1998). Pathological study of mice with total deficiency of sphingolipid activator proteins (SAP knockout mice). Acta Neuropathol. **96**, 29–40.

Pacheco, C. D., Elrick, M. J., and Lieberman, A. P. (2009). Tau deletion exacerbates the phenotype of Niemann-Pick type C mice and implicates autophagy in pathogenesis. Hum. Mol. Genet. **18**, 956–965.

Pack-Chung, E., Kurshan, P. T., Dickman, D. K., and Schwarz, T. L. (2007). A Drosophila kinesin required for synaptic bouton formation and synaptic vesicle transport. Nat. Neurosci. **10**, 980–989.

Pan, B., Fromholt, S. E., Hess, E. J., Crawford, T. O., Griffin, J. W., Sheikh, K. A., and Schnaar, R. L. (2005). Myelin-associated glycoprotein and complementary axonal ligands, gangliosides, mediate axon stability in the CNS and PNS: Neuropathology and behavioral deficits in single- and double-null mice. *Exp. Neurol.* **195,** 208–217.

Patel, H., Cross, H., Proukakis, C., Hershberger, R., Bork, P., Ciccarelli, F. D., Patton, M. A., McKusick, V. A., and Crosby, A. H. (2002). SPG20 is mutated in Troyer syndrome, an hereditary spastic paraplegia. *Nat. Genet.* **31,** 347–348.

Pedrola, L., Espert, A., Valdes-Sanchez, T., Sanchez-Piris, M., Sirkowski, E. E., Scherer, S. S., Farinas, I., and Palau, F. (2008). Cell expression of GDAP1 in the nervous system and pathogenesis of Charcot-Marie-Tooth type 4A disease. *J. Cell. Mol. Med.* **12,** 679–689.

Perlson, E., Hanz, S., Ben-Yaakov, K., Segal-Ruder, Y., Seger, R., and Fainzilber, M. (2005). Vimentin-dependent spatial translocation of an activated MAP kinase in injured nerve. *Neuron* **45,** 715–726.

Pilling, A. D., Horiuchi, D., Lively, C. M., and Saxton, W. M. (2006). Kinesin-1 and Dynein are the primary motors for fast transport of mitochondria in Drosophila motor axons. *Mol. Biol. Cell* **17,** 2057–2068.

Puls, I., Jonnakuty, C., LaMonte, B. H., Holzbaur, E. L., Tokito, M., Mann, E., Floeter, M. K., Bidus, K., Drayna, D., Oh, S. J., Brown, R. H., Jr., Ludlow, C. L., et al. (2003). Mutant dynactin in motor neuron disease. *Nat. Genet.* **33,** 455–456.

Read, D. J., Li, Y., Chao, M. V., Cavanagh, J. B., and Glynn, P. (2009). Neuropathy target esterase is required for adult vertebrate axon maintenance. *J. Neurosci.* **29,** 11594–11600.

Reid, E., Kloos, M., Ashley-Koch, A., Hughes, L., Bevan, S., Svenson, I. K., Graham, F. L., Gaskell, P. C., Dearlove, A., Pericak-Vance, M. A., Rubinsztein, D. C., and Marchuk, D. A. (2002). A kinesin heavy chain (KIF5A) mutation in hereditary spastic paraplegia (SPG10). *Am. J. Hum. Genet.* **71,** 1189–1194.

Riano, E., Martignoni, M., Mancuso, G., Cartelli, D., Crippa, F., Toldo, I., Siciliano, G., Di Bella, D., Taroni, F., Bassi, M. T., Cappelletti, G., and Rugarli, E. I. (2009). Pleiotropic effects of spastin on neurite growth depending on expression levels. *J. Neurochem.* **108,** 1277–1288.

Ries, V., Silva, R. M., Oo, T. F., Cheng, H. C., Rzhetskaya, M., Kholodilov, N., Flavell, R. A., Kuan, C. Y., Rakic, P., and Burke, R. E. (2008). JNK2 and JNK3 combined are essential for apoptosis in dopamine neurons of the substantia nigra, but are not required for axon degeneration. *J. Neurochem.* **107,** 1578–1588.

Rintoul, G. L., Filiano, A. J., Brocard, J. B., Kress, G. J., and Reynolds, I. J. (2003). Glutamate decreases mitochondrial size and movement in primary forebrain neurons. *J. Neurosci.* **23,** 7881–7888.

Saigoh, K., Wang, Y. L., Suh, J. G., Yamanishi, T., Sakai, Y., Kiyosawa, H., Harada, T., Ichihara, N., Wakana, S., Kikuchi, T., and Wada, K. (1999). Intragenic deletion in the gene encoding ubiquitin carboxy-terminal hydrolase in gad mice. *Nat. Genet.* **23,** 47–51.

Salinas, S., Bilsland, L. G., and Schiavo, G. (2008). Molecular landmarks along the axonal route: Axonal transport in health and disease. *Curr. Opin. Cell Biol.* **20,** 445–453.

Sasaki, Y., Vohra, B. P., Lund, F. E., and Milbrandt, J. (2009). Nicotinamide mononucleotide adenylyl transferase-mediated axonal protection requires enzymatic activity but not increased levels of neuronal nicotinamide adenine dinucleotide. *J. Neurosci.* **29,** 5525–5535.

Schaefer, M. K., Schmalbruch, H., Buhler, E., Lopez, C., Martin, N., Guenet, J. L., and Haase, G. (2007). Progressive motor neuronopathy: A critical role of the tubulin chaperone TBCE in axonal tubulin routing from the Golgi apparatus. *J. Neurosci.* **27,** 8779–8789.

Schluter, O. M., Schmitz, F., Jahn, R., Rosenmund, C., and Sudhof, T. C. (2004). A complete genetic analysis of neuronal Rab3 function. *J. Neurosci.* **24,** 6629–6637.

Schmitt, I., Linden, M., Khazneh, H., Evert, B. O., Breuer, P., Klockgether, T., and Wuellner, U. (2007). Inactivation of the mouse Atxn3 (ataxin-3) gene increases protein ubiquitination. *Biochem. Biophys. Res. Commun.* **362**, 734–739.

Schmitt-John, T., Drepper, C., Mussmann, A., Hahn, P., Kuhlmann, M., Thiel, C., Hafner, M., Lengeling, A., Heimann, P., Jones, J. M., Meisler, M. H., and Jockusch, H. (2005). Mutation of Vps54 causes motor neuron disease and defective spermiogenesis in the wobbler mouse. *Nat. Genet.* **37**, 1213–1215.

Scotland, P., Zhou, D., Benveniste, H., and Bennett, V. (1998). Nervous system defects of AnkyrinB (−/−) mice suggest functional overlap between the cell adhesion molecule L1 and 440-kD AnkyrinB in premyelinated axons. *J. Cell Biol.* **143**, 1305–1315.

Shemesh, O. A., Erez, H., Ginzburg, I., and Spira, M. E. (2008). Tau-induced traffic jams reflect organelles accumulation at points of microtubule polar mismatching. *Traffic* **9**, 458–471.

Shemesh, O. A., and Spira, M. E. (2009). Paclitaxel induces axonal microtubules polar reconfiguration and impaired organelle transport: Implications for the pathogenesis of paclitaxel-induced polyneuropathy. *Acta Neuropathol.*

Shinzawa, K., Sumi, H., Ikawa, M., Matsuoka, Y., Okabe, M., Sakoda, S., and Tsujimoto, Y. (2008). Neuroaxonal dystrophy caused by group VIA phospholipase A2 deficiency in mice: A model of human neurodegenerative disease. *J. Neurosci.* **28**, 2212–2220.

Sievers, C., Platt, N., Perry, V. H., Coleman, M. P., and Conforti, L. (2003). Neurites undergoing Wallerian degeneration show an apoptotic-like process with Annexin V positive staining and loss of mitochondrial membrane potential. *Neurosci. Res.* **46**, 161–169.

Silva, A., Wang, Q., Wang, M., Ravula, S. K., and Glass, J. D. (2006). Evidence for direct axonal toxicity in vincristine neuropathy. *J. Peripher. Nerv. Syst.* **11**, 211–216.

Simpson, M. A., Cross, H., Proukakis, C., Pryde, A., Hershberger, R., Chatonnet, A., Patton, M. A., and Crosby, A. H. (2003). Maspardin is mutated in mast syndrome, a complicated form of hereditary spastic paraplegia associated with dementia. *Am. J. Hum. Genet.* **73**, 1147–1156.

Smart, S. L., Lopantsev, V., Zhang, C. L., Robbins, C. A., Wang, H., Chiu, S. Y., Schwartzkroin, P. A., Messing, A., and Tempel, B. L. (1998). Deletion of the K(V)1.1 potassium channel causes epilepsy in mice. *Neuron* **20**, 809–819.

Specht, C. G., and Schoepfer, R. (2004). Deletion of multimerin-1 in alpha-synuclein-deficient mice. *Genomics* **83**, 1176–1178.

Spires, T. L., Meyer-Luehmann, M., Stern, E. A., McLean, P. J., Skoch, J., Nguyen, P. T., Bacskai, B. J., and Hyman, B. T. (2005). Dendritic spine abnormalities in amyloid precursor protein transgenic mice demonstrated by gene transfer and intravital multiphoton microscopy. *J. Neurosci.* **25**, 7278–7287.

Stagi, M., Gorlovoy, P., Larionov, S., Takahashi, K., and Neumann, H. (2006). Unloading kinesin transported cargoes from the tubulin track via the inflammatory c-Jun N-terminal kinase pathway. *FASEB J.* **20**, 2573–2575.

Stamer, K., Vogel, R., Thies, E., Mandelkow, E., and Mandelkow, E. M. (2002). Tau blocks traffic of organelles, neurofilaments, and APP vesicles in neurons and enhances oxidative stress. *J. Cell Biol.* **156**, 1051–1063.

Stokin, G. B., Lillo, C., Falzone, T. L., Brusch, R. G., Rockenstein, E., Mount, S. L., Raman, R., Davies, P., Masliah, E., Williams, D. S., and Goldstein, L. S. (2005). Axonopathy and transport deficits early in the pathogenesis of Alzheimer's disease. *Science* **307**, 1282–1288.

Stone, J. R., Okonkwo, D. O., Dialo, A. O., Rubin, D. G., Mutlu, L. K., Povlishock, J. T., and Helm, G. A. (2004). Impaired axonal transport and altered axolemmal permeability occur in distinct populations of damaged axons following traumatic brain injury. *Exp. Neurol.* **190**, 59–69.

Stowers, R. S., Megeath, L. J., Gorska-Andrzejak, J., Meinertzhagen, I. A., and Schwarz, T. L. (2002). Axonal transport of mitochondria to synapses depends on milton, a novel Drosophila protein. *Neuron* **36**, 1063–1077.

Suo, Z., Cox, A. A., Bartelli, N., Rasul, I., Festoff, B. W., Premont, R. T., and Arendash, G. W. (2007). GRK5 deficiency leads to early Alzheimer-like pathology and working memory impairment. Neurobiol. Aging 28(12), pp. 1873–1888.

Takei, Y., Teng, J., Harada, A., and Hirokawa, N. (2000). Defects in axonal elongation and neuronal migration in mice with disrupted tau and map1b genes. J. Cell Biol. 150, 989–1000.

Tarrade, A., Fassier, C., Courageot, S., Charvin, D., Vitte, J., Peris, L., Thorel, A., Mouisel, E., Fonknechten, N., Roblot, N., Seilhean, D., Dierich, A., et al. (2006). A mutation of spastin is responsible for swellings and impairment of transport in a region of axon characterized by changes in microtubule composition. Hum. Mol. Genet. 15, 3544–3558.

Tashiro, T., and Komiya, Y. (1994). Impairment of cytoskeletal protein transport due to aging or beta, beta'-iminodipropionitrile intoxication in the rat sciatic nerve. Gerontology 40(Suppl. 2), 36–45.

Tesco, G., Koh, Y. H., Kang, E. L., Cameron, A. N., Das, S., Sena-Esteves, M., Hiltunen, M., Yang, S. H., Zhong, Z., Shen, Y., Simpkins, J. W., and Tanzi, R. E. (2007). Depletion of GGA3 stabilizes BACE and enhances beta-secretase activity. Neuron 54, 721–737.

Tolkovsky, A. M., and Suidan, H. S. (1987). Adenosine 5'-triphosphate synthesis and metabolism localized in neurites of cultured sympathetic neurons. Neuroscience 23, 1133–1142.

Tsaousidou, M. K., Ouahchi, K., Warner, T. T., Yang, Y., Simpson, M. A., Laing, N. G., Wilkinson, P. A., Madrid, R. E., Patel, H., Hentati, F., Patton, M. A., Hentati, A., et al. (2008). Sequence alterations within CYP7B1 implicate defective cholesterol homeostasis in motor-neuron degeneration. Am. J. Hum. Genet. 82, 510–515.

Viancour, T. A., and Kreiter, N. A. (1993). Vesicular fast axonal transport rates in young and old rat axons. Brain Res. 628, 209–217.

Vickers, J. C., Dickson, T. C., Adlard, P. A., Saunders, H. L., King, C. E., and McCormack, G. (2000). The cause of neuronal degeneration in Alzheimer's disease. Prog. Neurobiol. 60, 139–165.

von Bartheld, C. S., Wang, X., and Butowt, R. (2001). Anterograde axonal transport, transcytosis, and recycling of neurotrophic factors: The concept of trophic currencies in neural networks. Mol. Neurobiol. 24, 1–28.

Wakade, A. R., Prat, J. C., and Wakade, T. D. (1985). Sympathetic neurons extend neurites in a culture medium containing cyanide and dinitrophenol but not iodoacetate. FEBS Lett. 190, 95–98.

Wang, P., Yang, G., Mosier, D. R., Chang, P., Zaidi, T., Gong, Y. D., Zhao, N. M., Dominguez, B., Lee, K. F., Gan, W. B., and Zheng, H. (2005). Defective neuromuscular synapses in mice lacking amyloid precursor protein (APP) and APP-Like protein 2. J. Neurosci. 25, 1219–1225.

Wang, X., and Schwarz, T. L. (2009). The mechanism of Ca2+-dependent regulation of kinesin-mediated mitochondrial motility. Cell 136, 163–174.

Wang, X., and Tournier, C. (2006). Regulation of cellular functions by the ERK5 signalling pathway. Cell. Signal. 18, 753–760.

Watson, F. L., Heerssen, H. M., Bhattacharyya, A., Klesse, L., Lin, M. Z., and Segal, R. A. (2001). Neurotrophins use the Erk5 pathway to mediate a retrograde survival response. Nat. Neurosci. 4, 981–988.

Weiss, P., and Hiscoe, H. B. (1948). Experiments on the mechanism of nerve growth. J. Exp. Zool. 107, 315–395.

Whitmarsh, A. J., Kuan, C. Y., Kennedy, N. J., Kelkar, N., Haydar, T. F., Mordes, J. P., Appel, M., Rossini, A. A., Jones, S. N., Flavell, R. A., Rakic, P., and Davis, R. J. (2001). Requirement of the JIP1 scaffold protein for stress-induced JNK activation. Genes Dev. 15, 2421–2432.

Willem, M., Garratt, A. N., Novak, B., Citron, M., Kaufmann, S., Rittger, A., DeStrooper, B., Saftig, P., Birchmeier, C., and Haass, C. (2006). Control of peripheral nerve myelination by the beta-secretase BACE1. Science 314, 664–666.

Willis, D. E., van Niekerk, E. A., Sasaki, Y., Mesngon, M., Merianda, T. T., Williams, G. G., Kendall, M., Smith, D. S., Bassell, G. J., and Twiss, J. L. (2007). Extracellular stimuli specifically regulate localized levels of individual neuronal mRNAs. *J. Cell Biol.* **178**, 965–980.

Xia, C. H., Roberts, E. A., Her, L. S., Liu, X., Williams, D. S., Cleveland, D. W., and Goldstein, L. S. (2003). Abnormal neurofilament transport caused by targeted disruption of neuronal kinesin heavy chain KIF5A. *J. Cell Biol.* **161**, 55–66.

Yamanaka, K., Miller, T. M., McAlonis-Downes, M., Chun, S. J., and Cleveland, D. W. (2006). Progressive spinal axonal degeneration and slowness in ALS2-deficient mice. *Ann. Neurol.* **60**, 95–104.

Yamashita, T., Wu, Y. P., Sandhoff, R., Werth, N., Mizukami, H., Ellis, J. M., Dupree, J. L., Geyer, R., Sandhoff, K., and Proia, R. L. (2005). Interruption of ganglioside synthesis produces central nervous system degeneration and altered axon–glial interactions. *Proc. Natl. Acad. Sci. USA* **102**, 2725–2730.

Yang, D. D., Kuan, C. Y., Whitmarsh, A. J., Rincon, M., Zheng, T. S., Davis, R. J., Rakic, P., and Flavell, R. A. (1997). Absence of excitotoxicity-induced apoptosis in the hippocampus of mice lacking the Jnk3 gene. *Nature* **389**, 865–870.

Yin, X., Crawford, T. O., Griffin, J. W., Tu, P., Lee, V. M., Li, C., Roder, J., and Trapp, B. D. (1998). Myelin-associated glycoprotein is a myelin signal that modulates the caliber of myelinated axons. *J. Neurosci.* **18**, 1953–1962.

Yuan, A., Kumar, A., Peterhoff, C., Duff, K., and Nixon, R. A. (2008). Axonal transport rates in vivo are unaffected by tau deletion or overexpression in mice. *J. Neurosci.* **28**, 1682–1687.

Yuan, A., Sasaki, T., Rao, M. V., Kumar, A., Kanumuri, V., Dunlop, D. S., Liem, R. K., and Nixon, R. A. (2009). Neurofilaments form a highly stable stationary cytoskeleton after reaching a critical level in axons. *J. Neurosci.* **29**, 11316–11329.

Yue, Z., Wang, Q. J., and Komatsu, M. (2008). Neuronal autophagy: Going the distance to the axon. *Autophagy* **4**, 94–96.

Zhao, C., Fancy, S. P., Kotter, M. R., Li, W. W., and Franklin, R. J. (2005). Mechanisms of CNS remyelination–the key to therapeutic advances. *J. Neurol. Sci.* **233**, 87–91.

Zhao, C., Takita, J., Tanaka, Y., Setou, M., Nakagawa, T., Takeda, S., Yang, H. W., Terada, S., Nakata, T., Takei, Y., Saito, M., Tsuji, S., *et al.* (2001). Charcot-Marie-Tooth disease type 2A caused by mutation in a microtubule motor KIF1Bbeta. *Cell* **105**, 587–597.

Zheng, T. S., Hunot, S., Kuida, K., and Flavell, R. A. (1999). Caspase knockouts: Matters of life and death. *Cell Death Differ.* **6**, 1043–1053.

Zhu, Q., Couillard-Despres, S., and Julien, J. P. (1997). Delayed maturation of regenerating myelinated axons in mice lacking neurofilaments. *Exp. Neurol.* **148**, 299–316.

Zhu, Q., Lindenbaum, M., Levavasseur, F., Jacomy, H., and Julien, J. P. (1998). Disruption of the NF-H gene increases axonal microtubule content and velocity of neurofilament transport: Relief of axonopathy resulting from the toxin beta, beta'-iminodipropionitrile. *J. Cell Biol.* **143**, 183–193.

6 Restless Genomes: Humans as a Model Organism for Understanding Host-Retrotransposable Element Dynamics

Dale J. Hedges* and Victoria P. Belancio†

*Hussman Institute for Human Genomics, Dr. John T. Macdonald Foundation Department of Human Genetics, Miller School of Medicine, University of Miami, Miami, Florida, USA
†Department of Structural and Cellular Biology, Tulane School of Medicine, Tulane Cancer Center, Tulane Center for Aging, New Orleans, Louisiana, USA

I. Introduction
  A. Mobilization and classification
  B. LINEs and Tiggers (the diversity and complexity of TE repertoire)
II. The Human Elements
  A. L1
  B. Alu
  C. SVA
  D. Human endogenous retroviruses
  E. Brief evolutionary history
III. Expression and Regulation
  A. Life cycle of human retroelements
  B. TE expression
  C. Mechanisms controlling TE expression and activity
  D. Genetic variation and polymorphism of TE loci
IV. Genomic Instability
  A. DSBs
  B. Retrotransposition
  C. Issues in repair and recombination
V. Impact on Human Health
  A. The potential for L1 to contribute to cancer and aging
  B. A positive note

Advances in Genetics, Vol. 73
Copyright 2011, Elsevier Inc. All rights reserved.

0065-2660/11 $35.00
DOI: 10.1016/B978-0-12-380860-8.00006-9

ABSTRACT

Since their initial discovery in maize, there have been various attempts to categorize the relationship between transposable elements (TEs) and their host organisms. These have ranged from TEs being selfish parasites to their role as essential, functional components of organismal biology. Research over the past several decades has, in many respects, only served to complicate the issue even further. On the one hand, investigators have amassed substantial evidence concerning the negative effects that TE-mutagenic activity can have on host genomes and organismal fitness. On the other hand, we find an increasing number of examples, across several taxa, of TEs being incorporated into functional biological roles for their host organism. Some 45% of our own genomes are comprised of TE copies. While many of these copies are dormant, having lost their ability to mobilize, several lineages continue to actively proliferate in modern human populations. With its complement of ancestral and active TEs, the human genome exhibits key aspects of the host–TE dynamic that has played out since early on in organismal evolution. In this review, we examine what insights the particularly well-characterized human system can provide regarding the nature of the host–TE interaction. © 2011, Elsevier Inc.

I. INTRODUCTION

Transposable elements (TEs) are, at their essence, stretches of DNA that have the capacity to mobilize themselves to different locations throughout a genome. The wealth of new genomic sequencing data made available over the past decade has allowed for significant advances in our understanding of the distribution and diversity of these elements. In conjunction with important experimental work, the analysis of sequencing data has served to highlight the importance of TE mobilization and proliferation on organismal evolution. From the introduction of alternative splice variants to the reshuffling of exons and entire genes, mobile element activity has constituted a potent force shaping genome architecture (Babushok *et al.*, 2007; Belancio *et al.*, 2008b; Lev-Maor *et al.*, 2003; Moran *et al.*, 1999; Sorek *et al.*, 2002; Xing *et al.*, 2006). TE activity is not relegated to the distant evolutionary past, however. Active mobile lineages continue to propagate within the majority of eukaryotic species surveyed thus far, including humans. Whereas the activity of these elements was once believed to be focused exclusively within the germline, there is now emerging evidence for substantial TE activity within somatic tissues. Much remains to be understood concerning the impact of this ongoing somatic activity on organismal fitness and, ultimately, upon human health. Here, we examine recent progress in human TE research, focusing on advances in our understanding of the relationship between element and host. Concentrating on the human genome and its current set of TE inhabitants, we examine both the positive and negative consequences of historical and ongoing TE activity.

A. Mobilization and classification

TE mobilization mechanisms can be divided into two broad categories, based on whether an RNA or DNA intermediate is used during the transposition process. These two varieties of mobilization have been conveniently described as "copy and paste" and "cut and paste," and they form the basis of a common categorization scheme of TEs introduced by Finnegan (1989), which divides TEs into Class 1 and Class 2 elements. According to the original criteria, Class 2 elements are those that mobilize directly from DNA to DNA, without the use of an RNA intermediate (Van Duyne, 2002). With some exceptions, noted below, DNA transposition occurs through a "cut and paste," strategy, where the original double-stranded DNA source element is excised from its existing location and reintroduced to a novel location in the genome. This excision is accomplished by transposase proteins that are encoded directly by the transposons themselves, or, in some cases, hijacked from other transposons in the host genome that possess protein-coding capacity. These same transposase proteins also facilitate the reintegration of the excised element elsewhere in the genome. With some important exceptions that allow for copy number increase, this type of mobilization is conservative in nature, resulting in no overall increase in the total number of TE elements within a host genome. It was representatives of the Class 2 type of elements that McClintock first observed in maize (McCLINTOCK, 1950). Discoveries of additional DNA transposons varieties that, despite their use of a DNA intermediate, did not include a double-stranded DNA removal step during the transposition process, necessitated the updating of Class 2 DNA transposon class to include three distinct subcategories. These include the original "cut and paste transposons," Helitrons, and Mavericks (reviewed in Feschotte and Pritham, 2007). The second major class of TEs, termed Class 1 elements (also commonly referred to as retrotransposons), mobilize by generating RNA transcripts that are subsequently converted into DNA prior to reintegration at a novel location in the host genome. Again, the process is mediated by proteins that are either encoded directly by the retrotransposon itself or hijacked from other protein-coding TEs. This mobilization strategy, termed retrotransposition, is an inherently expansive force within genomes. The total number of elements increases each time a successful retrotransposition event occurs, and, as a direct consequence, so does the total size of the host genome. Retrotranspositional activity has played a tremendously important role in genome expansion and diversification during evolutionary history, providing, in the process, abundant material for natural selection to carve out new functions. It is frequently noted that TEs from various lineages comprise roughly 45–50% of the human genome (Lander et al., 2001). A considerable fraction of the remaining half of our genome likely has an origin in mobile element activity as well, but these elements have since diverged so far from their initial sequences that they can no longer be readily identified through nucleotide or protein homology to known element classes.

B. LINEs and Tiggers (the diversity and complexity of TE repertoire)

Beyond the simple classification scheme of TEs into Class 1 and Class 2 elements, there exists an ever-growing bestiary of mobile element lineages, with colorful names ranging from "SPACE INVADERS" to "Tigger." Identifying and classifying TEs from diverse species has fueled the development of an array of important software tools as well as the establishment of curated databases. Examples include Censor (Jurka et al., 1996) and RepeatMasker (www.repeatmasker.org) for identifying elements based on known libraries of elements, and several methods, including RepeatScout, Recon, RepeatFinder, PILER, and ReAs for the ab initio discovery of elements (Edgar and Myers, 2005; Li et al., 2005; Price et al., 2005; Quesneville et al., 2005; Volfovsky et al., 2001). Repbase and dbRIP provide information about categorized TEs and the polymorphic status of particular inserts, respectively (Jurka et al., 2005; Wang et al., 2006). As the genome of any given organism may hold from a few to hundreds of TE varieties, each of which is diverging independently from its relatives inhabiting the genomes of other organismal species, TE nomenclature has the potential to be orders of magnitude more extensive and cumbersome than analogous systems for organisms. The discovery of TEs that propogate in a "copy and paste" manner without the need for an RNA intermediate served to undermine the basis of the Type1/Type2 classification scheme (Morgante et al., 2005; Wicker et al., 2007). It has, however, proven challenging to devise a taxonomic system for mobile elements that manages to reflect the ancestral relationships among elements, while, at the same time, avoids being too unwieldy for researchers to practically employ on a regular basis. As a consequence of this challenge, multiple categorization schemes have been proposed and are simultaneously in use (Wicker et al., 2007). The past decade's precipitous increase in sequencing capacity, and the "tsunami" of data that continues to ensue from it, has lead to the identification of a growing list of TEs from a increasingly diverse set of sampled organisms. The rising abundance of mobile element literature from this wealth of data poses a formidable challenge when attempting a reasonably comprehensive overview of progress in the area of TE research. We cannot hope, within the scope of this review, to address the full diversity of elements and their associated biology, or even to do adequate justice to describing what is known about the major TE branches. As our own research largely focuses on human TEs, we will primarily concentrate on progress in our understanding of the human host–element relationship (Fig. 6.1). Unlike many other higher eukaryotes, humans are currently bereft of active Class 2 type TEs, although elements of this variety existed relatively recently (evolutionarily speaking) within our primate lineage (Pace and Feschotte, 2007). Despite the human and primate-centric focus of this review, it is of impossible to avoid noting that critical insights into both the

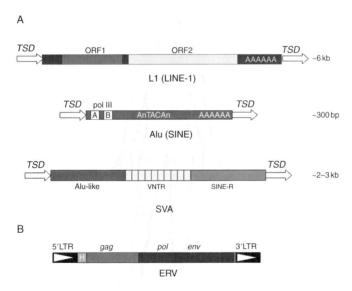

Figure 6.1. The Structure of currently active human TE Lineages. The basic structure of the currently active human TE lineages is represented. The Pol-II transcribed L1 sequence contains an internal bidirectional promoter and encodes a bicistronic transcript that contains two proteins (ORF1 and ORF2). The Alu element consists of two monomeric units united by an A-rich middle region. The SVA element appears to be a chimeric construct, derived from components of both SINE and endogenous retrovirus elements. The structure of HERV elements closely resembles that of the exogenous retroviruses from which they derived from. They are flanked by two long terminal repeats (LTRs) and retain protein-coding regions that are recognizably related to viral *gag*, *pol*, and *env* genes. (See Color Insert.)

molecular biology and evolution of TEs have been obtained from the study of TEs from a wide assortment of nonhuman organisms. We attempt to draw upon these diverse systems throughout our review to illustrate important concepts in the host–element relationship.

Advances in the field of TE research over the past two decades have effectively put to rest lingering questions about their overall relevance within biology. Although once relegated as mere evolutionary curiosities, clear and abundant evidence of their far-reaching impact now extends across diverse disciplines, ranging from evolutionary biology to medical genetics. This new perspective on mobile elements is perhaps best summed up by Lynch (2007), "...they are so ubiquitous, so diverse, and have such a profound effect on eukaryotic chromosomal architecture that one can reasonably argue that an overview of genomic evolution ought to start with them, before moving on to the host genes themselves" (Lynch, 2007). Among other things, the activity of TEs has been an expansionary force in the genome that provided abundant fodder with which natural selection could experiment. TEs have been exapted to perform functional biological roles numerous times during the history of life (reviewed in Bowen and Jordan, 2007; Dooner and Weil, 2007). The fact that such an important role was played by TEs is no longer matter of debate. That this role was cultivated and maintained by natural selection to promote "evolvability" is a stronger claim, and one for which both theoretical and empirical support remains lacking. This last claim is closely related to what has been an enduring question in TE research. Namely, how do we best describe the unusual relationship between host genomes and their TE residents? Is it mutualistic, symbiotic, parasitic, or something that defies our typical classification schemes? Views range from, on the one extreme, TEs as vital components of organismal biology to, at the other extreme, completely parasitic selfish genes undermining organismal fitness. The true nature of the host–TE dynamic no doubt rests somewhere between these polar extremes. A more nuanced view is offered by Kidwell and Lisch (1997), when they write, "the idea that TEs are primarily parasitic is not at all inconsistent with a role for these elements in the evolution of their hosts." Throughout this review, we indicate empirical evidence supporting what is ultimately an ambivalent role for TEs in organismal biology.

II. THE HUMAN ELEMENTS

The TE families remaining active in modern humans include the LINE-1 (L1) element, Alu, SVA, and human endogenous retroviruses (HERVs). All of these lineages belong to the Type 1, retrotransposon class (Fig. 6.1). Their basic structure is outlined below; a more detailed description of their molecular biology is provided in Belancio et al. (2010a).

A. L1

L1 elements are members of the long interspersed element. The L1 sequence encodes a bicistronic polymerase II (Pol-II) transcript that produces two proteins labeled ORF1 and ORF2, both of which are necessary for successful retrotransposition. ORF1's function is poorly understood, although it has a strong affinity for the L1 RNA sequence and is reported to exhibit nucleic acid chaperone activity (Martin and Bushman, 2001; Martin et al., 2005). The larger ORF2 protein encodes a protein harboring both endonuclease (EN) and reverse transcriptase (RT) activities. The protein itself has multiple functions that are conferred by the presence of several distinct domains; these include an N-terminal EN domain (Feng et al., 1996), a central RT domain (Mathias et al., 1991), and a C-terminal cysteine-rich domain of unknown function (Fanning and Singer, 1987). L1 and the nonautonomous retroelements (such as Alu and SVA, described below) replicate via a process known as target-primed reverse transcription (TPRT; Cost et al., 2002; Luan et al., 1993), wherein reverse transcription of the RNA intermediate is primed by an exposed 3′ hydroxyl. The 3′ hydroxyl strand of the host genome is exposed as the result endonucleatic cleavage by ORF2. Based on protein sequence analysis, the L1 lineage descends from a group of non-LTR (long terminal repeat) elements (elements lacking long terminal repeats) that have existed, in some form or another, since early eukaryotic evolution (Xiong and Eickbush, 1990). It remains a matter of debate whether L1 and similar elements were the predecessors or ancestors of modern viruses (Xiong and Eickbush, 1988). There are approximately 500,000 L1 copies in the human genome (haploid), comprising roughly 17% of total genome content (Lander et al., 2001). Only a fraction of these elements are intact, full-length elements that retain the ability to retrotranspose. Most elements are either 5′ truncated during the insertion process, deactivated through postinsertional mutation events, or otherwise removed from the host population due to genetic drift and/or natural selection. There is evidence that the majority of retrotranspositional activity actually results from a relatively small number of particularly active, "hot," L1 elements (Beck et al., 2010; Brouha et al., 2003; Seleme et al., 2006). As is the case with the Alu, SVA, and—to a lesser extent—HERV elements described below, moderns humans are polymorphic with regard to the insertion status of numerous L1 inserts, making L1 an integral component of human genetic diversity (Beck et al., 2010; Ewing and Kazazian, 2010; Huang et al., 2010; Iskow et al., 2010).

B. Alu

Alu are classified as Short INterspersed Elements (SINEs). These 300 bp Pol-III transcribed sequences are comprised of two separate domains tethered by an A-rich linker region. Originally derived from 7SLRNA, which is a member of the signal recognition complex, the current dimeric Alu found in humans formed from the merger of two monomeric units during early primate evolution. There are approximately 1 million copies of Alu in the human genome (haploid), constituting roughly 10% of the total genomic content (Lander et al., 2001). Alu elements do not encode the protein machinery required for their retrotransposition. Instead, it has been experimentally demonstrated that they can effectively commandeer L1 protein products to facilitate their proliferation (Dewannieux et al., 2003). While the ORF2 protein is sufficient for their retrotransposition (Dewannieux et al., 2003), the efficiency of Alu transposition can be improved by the ORF1 expression (Wallace et al., 2008a).

C. SVA

SVA elements arose more recently in the evolutionary timeline than the other active elements, appearing later in primate evolution than Alu (Ostertag et al., 2003; Wang et al., 2005; classified as nonautonomous, they defy easy categorization into existing schemas due to their intermediate size and chimeric nature). These odd elements appear to be cobbled together from several distinct genomic sources, including Alu-like sequence as well as sequence from viral envelope genes and LTR regions (Ostertag et al., 2003). Lacking protein-coding capacity, they are nonautonomous elements that parasitize L1 to obtain the necessary protein machinery for retrotransposition (Ostertag et al., 2003). The presence of several potential Pol-III termination sequences within their consensus sequences argues for a Pol-II transcription and the involvement of the serendipitous adjacent host promoters in driving SVA transcription has been reported (Damert et al., 2009).

D. Human endogenous retroviruses

HERVs are thought to be the ancestors of ancient retroviruses that infected the germline and, subsequently, lost their ability to encode active envelope proteins and move from host to host. Collectively, ERV copies make up ~8% of the human genome with human-specific ERVs (i.e., HERVs) constituting a smaller fraction (Lander et al., 2001). Due to their ability to be brought in through the infection of the germline by exogenous retroviruses, HERVs have been introduced and reintroduced a multitude of times during human evolution. While their rate of proliferation in modern human germlines appears to be limited, with only a modest number of polymorphic locations being identified (Turner et al., 2001), the activity of their remnant promoters has significant biological consequences, having been often adopted in several instances for host genome regulation [reviewed in Maksakova *et al.*, 2006; Sverdlov, 2000]. Also, see Cohen *et al.* (2009) for a more critical examination of the use of LTRs from endogenous retroviruses as functional human promoters (Cohen *et al.*, 2009).

E. Brief evolutionary history

Despite the fact that a spectacularly diverse set of TE lineages proliferated during the course of metazoan evolution, only a handful of families remain active within modern humans (Furano *et al.*, 2004; Pritham and Feschotte, 2007). It has been noted, for example, that mammals in general retain far fewer ancient L1 lineages compared to the fish species that have been examined thus far, and, in primates, this number has been whittled down to a single L1 family during the course of human evolution (Furano *et al.*, 2004). The bulk of ancient transposon diversity appears to have been lost during the evolution of tetrapods, possibly during synapsid evolution (Kordis *et al.*, 2006). The forces leading to this shift in TE genomic ecology remain unclear. Both demographic processes (e.g., population bottlenecks) and molecular changes are likely to have played their part in the transition. Evidence that only a handful of members of the L1 elements—so called "hot" elements—exhibit the highest levels of activity (Brouha *et al.*, 2003; Seleme *et al.*, 2006), suggest that random demographic fluctuations could have profound impacts on the persistence of TE diversity. As ecological sustainable population sizes generally decrease within higher trophic levels, loss of TE diversity could result from the fact that the loss of active lineages due to genetic drift occurs more rapidly than new lineages are emerging. If drift were the predominant force driving the loss of TE diversity, however, we would arguably find more examples of lineages that have lost their entire L1 complement (Cantrell *et al.*, 2008; Casavant *et al.*, 2000). The adaptive evolution of TE nucleotide and protein sequences, possibly to evade host repression mechanisms, likely play a substantial role in determining levels of TE diversity. Boissinot and Furano (2001) provided convincing evidence, based on the ratio of nonsynonymous to synonymous amino acid changes, that adaptive evolution occurred in the L1 lineage during the last 25 myrs of human evolution (Boissinot and Furano, 2001), although the evolutionary pressures driving these changes have not been definitively established.

Alu elements arrived considerably later on the evolutionary scene than L1 and have enjoyed spectacular success, reaching approximately 1 million copies in the haploid human genome. Thought to have evolved from the human *7SLRNA* gene during early primate radiation, Alus have far outperformed their L1, SVA, and HERV counterparts. Examination of sequence data does suggest, however, that the overall amplification rate of both Alu and L1 elements has attenuated over time from a peak rate approximately 40–50 mya (Ohshima *et al.*, 2003). The cause of the amplification burst and subsequent attenuation has not been established, although changes in host molecular biology and/or demographic effects, such as population bottlenecks, have been suggested. Both the evidence of changing levels of TE diversity, as well as for large fluctuation in insertion rates over evolutionary time, demonstrate that the relationship between TE and host genome is not a static one. Both host and TE are concurrently evolving and external perturbations from the environmental can cause the landscape of insertion rates and element diversity to shift, sometimes dramatically. Across taxa, this can be observed in the existence of a diverse genomic ecologies of TEs and host [discussed in Brookfield, 2005]. These range from systems containing a large diversity of elements, each of which have but a few genomic instances, to those systems, like the human, where a select few families achieve very high copy numbers. While many important questions remain to be answered regarding the impact of host demographic history on TE expression, it has become clear that factoring heavily into the resulting TE ecology is the extent to which host molecular biology augments or intervenes with TE proliferation. To understand how the molecular biology of the host cell can help or hinder the mobilization process, we will now examine in more detail the life cycle of the human retrotransposon.

III. EXPRESSION AND REGULATION

In order to survive and impose lasting genetic alterations on future generations, TE-associated modification of the host genome must ultimately occur in the germline—at least within organisms that have a sequestered germline. From the selfish gene perspective, activity in somatic cells would appear to only reduce genetic fitness of the host without resulting in any increase in TE copy number. This relatively simple—yet ultimately incomplete—evolutionary logic resulted in decades of near-exclusive focus on the germline as the principal site of mammalian TE expression. While TE expression and proliferation within the germline remains the only means of ensuring long-term TE survival, as we describe below, there is now emerging evidence that somatic TE activity is considerably higher than once suspected. A more complete characterization of this somatic activity is a vital component of understanding the full impact of TEs on organismal fitness and, ultimately, human health, and well being. On account of the ongoing tension in the TE–host coexistence, a network of defense mechanisms has been erected by host cells to shield the genome from unchecked TE activity. An understanding of the TE life cycle is essential for appreciating the breadth and complexity of this regulatory network.

A. Life cycle of human retroelements

Among the human TEs that have retained their activity throughout the course of evolution are three main groups of non-LTR retrotransposons (L1, Alu, and SVA) that stand apart from the LTR group of retroelements such as HERVs.

Although all TEs are frequently considered to have a parasitic aspect, the nonautonomous group of non-LTR transposons, such as SINEs (represented by Alu elements in the primate lineage) and SVA (hominoid-specific retroelements) take this parasitism to a another level. Because SINEs and SVA elements do not encode any proteins for orchestrating their own mobilization (hence the "nonautonomous") they have evolved to parasitize the retrotransposition machinery of the currently active human autonomous non-LTR retroelement, LINE-1. The retrotransposition process of non-LTR retrotransposons begins with transcription of a retrotranspositionally competent locus within the host genome. L1, Alu, and SVA elements accumulated in the human genome to 500,000, 1,000,000, and 1700 copies, respectively, the majority of which are inactive (Lander et al., 2001). Out of the structurally intact TE copies only a fraction of loci retained functional promoters. Finally, a large proportion of the potentially expressed loci have lost the ability to mobilize due to mutations inactivating functional protein domains or sequences within their RNAs critical for amplification. The former applies only to L1 elements (because only they encode proteins), while the latter is true for all human retrotransposons. As a result there are about 100 predicted active L1 loci per haploid human genome (Penzkofer et al., 2005). L1 elements are transcribed by an atypical bidirectional internal RNA Pol-II promoter located within the L1 5' untranslated region, 5'UTR (Speek, 2001; Swergold, 1990). At least some of the produced L1 mRNAs are capped and RNA capping is reported to stimulate translation of L1 proteins (Dmitriev et al., 2007; Kulpa and Moran, 2005; McMillan and Singer, 1993). As a result of Pol-II transcription, L1 mRNAs are polyadenylated. The presence of this polyA tail at the 3' end is critical for the L1 integration process (Symer et al., 2002).

Very little is known about transcription of SVA elements. They are most likely transcribed by RNA Pol-II because of the presence of multiple RNA Pol-III terminators within their sequence and evidence for posttranscriptional processing of SVA RNA characteristic of the Pol-II generated RNAs (Damert et al., 2009; Hancks et al., 2009; Ostertag et al., 2003; Wang et al., 2005). Bioinformatic analysis of SVA elements identified in the human genome, followed by empirical testing, suggested that SVA retrotransposons do not contain an internal promoter that can be carried over into the new genomic location to ensure efficient transcription of the de novo insertions (Damert et al., 2009). SVA loci appear to rely on the existence of the functional promoters at the sites of their integration (Damert et al., 2009). Along with their relatively recent appearance on the evolutionary scene, the lack of a mobile promoter may explain their relatively low copy numbers.

In contrast to L1 and SVA elements, Alu transcription is carried out by the RNA Pol-III complex as is the case with its 7SLRNA ancestor. Alu RNA is not capped and the artificial generation of the capped Alu transcripts significantly changes requirements for Alu mobilization (Kroutter *et al.*, 2009). Similarly to the L1 and SVA elements, Alu RNAs contain a polyA tail at their 3′ end, but in contrast to the polyA tails of L1 and SVA, it is encoded within genomic Alu sequences rather than being generated through polyadenylation by cellular proteins.

Mammalian L1 elements encode two proteins, open reading frame 1 (ORF1) and open read frame 2 (ORF2), that are absolutely essential for L1 retrotransposition (Fig. 6.1). L1 ORF1 and ORF2 proteins are presumed to be made in the cytoplasm of the cells that permit transcription of a full length, bicistronic L1 mRNA. Very little is known about L1 mRNA translation and even less about the mechanisms that may regulate L1 protein production (Alisch *et al.*, 2006; Dmitriev *et al.*, 2007; Leibold *et al.*, 1990; Li *et al.*, 2006; McMillan and Singer, 1993). Extensive empirical evidence points out that L1 ORF1 plays a crucial role in the formation of ribonucleoprotein (RNP) complexes with L1 mRNA that are believed to represent important retrotransposition intermediates (Kolosha and Martin, 1997, 2003; Kulpa and Moran, 2005; Martin, 1991; Martin *et al.*, 2000). L1 ORF2 is also associated with the L1 RNPs (Kulpa and Moran, 2005; Wei *et al.*, 2001) and these L1 RNA/protein complexes gain access by an undefined passive or active mechanism to genomic DNA. L1 ORF2 contains apurinic-like EN and RT domains that are responsible for a series of sequential enzymatic steps in the LINE, SINE, and SVA integration process (Clements and Singer, 1998; Cost and Boeke, 1998; Cost *et al.*, 2002; Feng *et al.*, 1996; Martin *et al.*, 1998). While the requirement for the functional L1 ORF2 is shared by all three non-LTR retroelements (LINEs, SINEs, and SVA), L1 ORF1 protein has been reported to be indispensable only in the case of L1 mobilization (Dewannieux *et al.*, 2003; Moran *et al.*, 1996). Currently there is very limited understanding of how, when, and where Alu and SVA RNAs hijack the L1 ORF2 protein. The means of gaining access to L1 ORF2 by SINEs and SVAs are complicated by the existence of a very strong cis-preference of L1 proteins for the specific L1 mRNA instance that was the source of their translation (Wei *et al.*, 2001). This cis-preference may be one of the mechanisms that serve to prevent retroelements from wreaking havoc in the host genome through mobilization of otherwise retrotranspositionally incompetent TE RNAs and normal cellular transcripts resulting in the bombardment of the human genome with retro-pseudogene copies. Nevertheless, despite the apparent barriers to commandeering the L1 mobilization apparatus, both SINEs and SVA elements have been extremely successful at doing so (Lander *et al.*, 2001; Wang *et al.*, 2005), suggesting a potential active mechanism of gaining access to the L1 ORF2. Alu elements have been particularly adept at this process, which is likely one of the reasons that allowed them to flourish to over 1,000,000 copies in the human genome surpassing L1 copy number by more than a half (Lander *et al.*, 2001).

The initiation of retrotransposition in the nucleus involves recognition of an EN target site within the host genomic DNA by the L1 ORF2 followed by the first-strand DNA cleavage by the L1 ORF2 EN that produces a DNA nick at the EN site (Cost and Boeke, 1998; Feng *et al.*, 1996; Jurka, 1997). Even though the consensus L1 EN site is deduced to be 5'-TTTTAA-3', some of the sites with a single nucleotide substitution within this sequence are known to be efficiently used by the L1 EN (Jurka, 1997; Szak *et al.*, 2002). The cleavage of genomic DNA at the T-A junction is proposed to generate a single strand (ss), 5'-TTTT-3', DNA available for a noncovalent interaction with the polyA tail at the 3' end of the L1 mRNA (or SINE and SVA RNAs; Feng *et al.*, 1996; Symer *et al.*, 2002). Free 3' end of the ss genomic DNA serves as a primer for the L1 RT to synthesize first-strand L1 cDNA (Martin *et al.*, 1998; Piskareva and Schmatchenko, 2006). This process is known as TPRT and is common among retroelements from species as diverse as the silk worm, *Bombyx mori*, and human, *Homo sapiens* (Christensen and Eickbush, 2005; Cost *et al.*, 2002). The remaining steps of the retrotransposition process that include second strand L1 DNA synthesis (Piskareva and Schmatchenko, 2006), removal of the RNA template, and covalent connection of the *de novo* L1 (or Alu and SVA) insert with genomic DNA at the site of integration continue to be very poorly characterized. It is speculated that cellular DNA repair proteins or other cellular factors are likely to assist in the completion of some or all of these steps. This assumption is based on the apparent lack of certain enzymatic functions within L1 proteins, such as ligase activity, that would be required to carry out these steps. It is further based on the recent discoveries of the opposing effects of cellular proteins from various DNA repair pathways on the efficiency of L1 retrotransposition (Gasior *et al.*, 2007, 2008; Suzuki *et al.*, 2009). The final product of retrotransposition contains distinct structural characteristics such as L1, SINE, SVA, or cellular transcript sequences flanked by the EN recognition site known as the target site duplication (TSD). These signature features unequivocally identify L1 machinery involvement in the generation of integration events.

Even though L1, SINEs, and SVA elements rely completely on the L1 retrotransposition apparatus for their mobilization, different elements exhibit distinct differences between the retrotransposition requirements. One of these prerequisites is the dependence or the lack of it on the L1 ORF1 protein (Dewannieux *et al.*, 2003; Moran *et al.*, 1996). As mentioned above, in contrast to L1, which absolutely requires participation of the functional ORF1 protein in addition to the ORF2 protein for successful retrotransposition, Alu elements can be efficiently mobilized by the L1 ORF2 alone (Dewannieux *et al.*, 2003; Moran *et al.*, 1996). While the presence of the L1 ORF1 protein slightly stimulates Alu retrotransposition in tissue culture (Wallace *et al.*, 2008a), this modest effect supports the observation that Alu retrotransposition is largely independent of the L1 ORF1. The proposed explanations for this discrepancy include the difference in the origin of transcripts through RNA Pol-II in the case of L1 mRNA and Pol-III in the case of Alu, as well as the association with the cellular translation machinery (neither Alu nor SVA transcripts are believed to contain any ORFs; Kroutter *et al.*, 2009). Interestingly, an artificial switch from Pol-III to Pol-II generated Alu transcripts results in the requirement of ORF1 for Alu mobilization (Kroutter *et al.*, 2009).

The multistep life cycle of human TEs provides ample opportunities for the host to erect barriers that would allow prevention of efficient amplification of TEs. Understanding of the TE life cycle makes it very clear that suppression of TE expression can be a very effective way of minimizing their negative impact on the genome stability.

B. TE expression

TEs are families of elements interspersed throughout the human genome, as a result, each of human TEs is represented by hundreds or thousands of nonidentical loci that can be expressed in any given cell (Lander *et al.*, 2001). Thus, endogenous expression detected by conventional techniques is usually a combination of transcripts produced by a subpopulation of loci expressed in the assessed cell type. Due to the sequence variation among the majority of TE loci found in the human genome, each method used to analyze TE expression is often accompanied by its own set of specific detection biases, discussion of which is beyond the scope of this review. Technical difficulties associated with detection of human TEs combined with the presumed relatively low expression levels and incomplete knowledge of rules governing TE expression have accounted for the impediment of progress in this area of TE research. The expression of endogenous L1 elements (mRNA and ORF1 protein) was reported to be restricted to the mouse germline (Branciforte and Martin, 1994; Ostertag *et al.*, 2002). The expression of the L1 proteins was also detected in specific somatic cell types (such as Leydig, Sertoli, and vascular endothelial cells; Ergun *et al.*, 2004) and in normal breast tissues (Asch *et al.*, 1996). Analysis of endogenous L1 expression in various types of human tumors and cancer cell lines established that L1 expression, as a rule, is significantly upregulated in human malignancies (Bratthauer and Fanning, 1992, 1993; Bratthauer *et al.*, 1994). As is obvious from the TE life cycle (described above), ongoing L1 expression is necessary for the activity of Alu and SVA elements. Even though it is deduced from the Alu and SVA retrotransposition events occurred in the germline that both of these elements are expressed at minimum in the germline, the endogenous expression of Alu and SVA elements is very poorly characterized (Fuhrman *et al.*, 1981; Shaikh *et al.*, 1997). Few studies have attempted to assess Alu transcripts in a manner that distinguishes Pol-II- and Pol-III- promoted transcripts (Shaikh *et al.*, 1997).

Several lessons have been derived concerning L1 expression from trans-genic mouse models. Transgenic animals containing L1 expressed from its native promoter supported L1 expression in testis and ovaries (Ostertag *et al.*, 2002). Pol-II driven L1 was also expressed in kidney, lung, intestine, liver, and brain (Ostertag *et al.*, 2002). RT-PCR analysis of germ cell fractions such as pachytene spermatocytes, round spermatids, and condensing spermatids demonstrated L1 expressed with varying efficiency within these three cell types (Ostertag *et al.*, 2002). Another transgenic mouse model of human L1 driven by the mouse pHsp70-2 promoter demonstrated very strong positional effect of the transgene integration site on retrotransposition (Babushok *et al.*, 2006). This observation suggested that L1 transgene expression in the transgenic mouse model can be significantly affected by its specific location, consistent with the reports of positional effects on the expression of endogenous L1 loci (Lavie *et al.*, 2004). Transgenic mouse and rat models of human and mouse L1 retrotransposition driven by their respective endogenous promoters demonstrated L1 mRNA pres-ence in both germ cells and embryo. In these models retrotransposition events occurred in embryogenesis creating nonheritable somatic mosaicism (Kano *et al.*, 2009). In contrast to the previous models, a transgenic animal expressing a single copy of the synthetic mouse L1 element demonstrated significantly higher retro-transposition compared to retrotransposition frequencies detected in animals containing multiple integrated copies of the same L1 transgene (An *et al.*, 2006, 2008). Retrotransposition in a transgenic mouse model was also reported to take place in the mouse brain strongly supporting L1 expression in normal brain cells (Muotri *et al.*, 2005). Analysis of the methylation status of endoge-nous L1 promoters coupled with detection of *de novo* L1 retrotransposition events (Coufal *et al.*, 2009) in normal human brain produced results consistent with ongoing expression of endogenous L1 elements in human brain cells. More comprehensive analysis of endogenous L1 mRNA expression in normal human tissues and adult stem cells demonstrated that the majority of the examined human tissues support endogenous L1 mRNA expression (Belancio *et al.*, 2010b). However, the abundance of the total L1-related transcripts and the amount of the full-length L1 mRNA varies significantly among tissues. This variation appears to be at least in part influenced by the tissue-specific differences in the efficiency and pattern of L1 mRNA processing (Belancio *et al.*, 2010b). The same is true for human cancer cell lines where the amount of the full-length L1 mRNA is dictated by the efficiency of L1 mRNA splicing and premature polyadenylation (Belancio *et al.*, 2010b). Somatic endogenous L1 expression combined with the lack of obvious restrictions for somatic retrotransposition (An *et al.*, 2006, 2008; Kano *et al.*, 2009; Kubo *et al.*, 2006) and L1-induced double-strand DNA break (DSB) formation in normal cells (Belancio *et al.*) suggests that ongoing L1-induced DNA damage is likely to take place in somatic tissues. Because of the reported variation in the L1 promoter strength in different cell types (Swergold, 1990; Yang et al., 2003) and the reported variation in the extent of L1 mRNA processing (Belancio et al., 2010b) the exposure to the L1-associated DNA damage is expected to vary in a tissue-specific manner. Along with the significant broadening of the potential spectrum of L1-induced DNA damage comes the importance of understanding of the time, location, and levels of TE expression in human tissues making unraveling of the somatic L1 expression an important aspect of L1 biology.

C. Mechanisms controlling TE expression and activity

Even though many critical elements controlling TE expression and activity remain unknown, a plethora of cellular mechanisms that control human TE expression and activity has been unveiled to date. Based on the fact that TEs can significantly perturb the normal function of the genome it is not surprising that host organisms employed a multifaceted network of road blocks designed to suppress TE expression and mobilization. Of the three currently active human TEs, L1 expression and activity is the best studied to date, mostly due to the fact that L1s are the driving force of any TE-related unrest in the human genome. Among the well-characterized mechanisms influencing human TE expression and activity are transcriptional regulation, promoter methylation, RNA processing, and cellular pathways involved in controlling TE-induced damage. The number and diversity of the barriers erected by mammalian cells to suppress TE expression speaks to the importance of keeping the production of these elements in check.

One of the first levels of defense controlling TE expression is promoter strength and availability of the transcription factors necessary to drive RNA production. Members of the SOX and RUNX family of transcription factors as well as steroid hormones can regulate L1 expression dictating tissue specificity of L1 transcription (Morales et al., 2002; Tchenio et al., 2000; Yang et al., 2003). YY1 transcription factor does not change transcription efficiency, but it is important for the proper site of transcription initiation at the L1 promoter (Athanikar et al., 2004). Genomic sequences immediately upstream of the L1 integration site can stimulate L1 promoter strength potentially rendering some L1 loci the ability to contribute more molecules than others to the combined pool of L1 transcripts (Lavie et al., 2004). A similar situation is reported for Alu elements; while transcription of Alu promoter alone by Pol-III is relatively weak but upstream genomic sequences can stimulate it by several fold (Roy et al., 2000). Expression of SVA elements is reported to be driven by cellular promoters present upstream of the SVA integration sites suggesting that these elements are likely to exhibit significant variation in their total expression, with different SVA loci differentially contributing to the combined pool of SVA transcripts (Damert et al., 2009).

Promoter function can be significantly regulated through the promoter methylation and epigenetic modifications. Utilization of promoter methylation to repress TE expression is thought to be a common cellular defense mechanism against TE-associated damage [reviewed in Yoder et al., 1997]. The promoters of the majority of the L1 and Alu loci are hypermethylated in normal cells (Alves et al., 1996; Florl et al., 1999; Hata and Sakaki, 1997; Lees-Murdock et al., 2003; Takai et al., 2000; Thayer et al., 1993; Tsutsumi, 2000). The importance of the maintenance of their methylation status is confirmed in the transgenic mice deficient for DnmtL3 protein required for upholding of DNA methylation (Bourc'his and Bestor, 2004). Significant upregulation of the expression of endogenous retroviruses and L1 elements accompanied the lack of proper DNA methylation in these transgenic animals. The effect of deficient methylation of genomic DNA on the expression of SINEs has not been tested in this experimental model. Retrotransposon methylation is also regulated by Piwi family members MILI and MIWI2 (Kuramochi-Miyagawa et al., 2008). Despite hypermethylation, there is a small portion of loci, at least in the case of L1 elements that escape methylation (Coufal et al., 2009). This is consistent with the ongoing endogenous L1 expression in human somatic tissues (Belancio et al., 2010b). Methylation control of TE transcription is likely released during the course of embryogenesis when methyl groups are removed in the global manner to reestablish genome methylation (Hellmann-Blumberg et al., 1993). The loss of L1 and Alu promoter methylation is also reported in the majority of human cancers (Takai et al., 2000; Tsutsumi, 2000) and it is routinely used as a marker of transformation.

Once transcription occurs, mRNA processing, specifically as it relates to L1 expression, plays an important role in dictating how much of the functional L1 mRNA is made. This processing involves both RNA splicing and premature polyadenylation at the polyadenylation (pA) sites abundant within L1 ORFs (Belancio et al., 2006; Han et al., 2004; Perepelitsa-Belancio and Deininger, 2003). Both processes exhibit tissue specificity and can lead to the production of L1-related transcripts that are retrotranspositionally incompetent at the expense of generation of full-length L1 (Belancio et al., 2010b). This level of control exists in both normal and cancer cells (Belancio et al., 2010b). It is currently unclear whether L1 RNA processing in cancer cells changes or remains the same as it was in the normal cells that they originated from. Nevertheless, mRNA processing can limit L1 expression by as much as 10-fold among normal cells and it is responsible for at least fivefold difference in the full-length L1 mRNA expression between cancer cells that otherwise support the same steady-state total L1-related mRNA (Belancio et al., 2010b). Similar to L1, SVA elements also contain functional splice sites that are efficiently used during SVA transcription (Damert et al., 2009). Alu elements that are transcribed by the Pol-III machinery are observed to be immune to the effects of RNA processing observed for L1 and SVA elements.

Once functional TE RNAs are produced, the next defense mechanism targets transcripts stability. The implementation of the cellular machinery to reduce steady-state L1 mRNA levels has been shown through several experimental systems. One such system implicated siRNA production from the antisense L1 transcripts generated by the reverse L1 promoter itself. Endogenously expressed L1 siRNAs were detected in human cells. This regulation through RNA interference was reported to exhibit a twofold effect on L1 expression. Another example of L1 mRNA regulation was shown in transgenic animals lacking PIWI protein. The lack of functional piRNA (or rasiRNA) pathway resulted in multifold upregulation of endogenous retroviruses and L1 elements through their effect on methylation (Kuramochi-Miyagawa *et al.*, 2008). In later studies, siRNA against PIWI in cultured human cells was shown to lead to the upregulation of endogenously expressed L1 transcripts (Lin *et al.*, 2009).

Because very little is known about production of human L1 proteins, particularly the ORF2 protein, some conflicting mechanisms controlling L1 translation have been put forward (Alisch *et al.*, 2006; Dmitriev *et al.*, 2007; Leibold *et al.*, 1990; Li *et al.*, 2006; McMillan and Singer, 1993). It is, however, known that wild-type L1 sequence composition favors suboptimal codon usage (Han and Boeke 2004; Wallace *et al.*, 2008b). Codon optimization of the mouse L1 element that also eliminated the majority of polyadenylation sites present in L1 sequence resulted in a significant increase in L1 mRNA, presumably L1 proteins, and L1 retrotransposition rate (Han and Boeke 2004). Codon optimization of the human L1 ORF2 also lead to a measurable increase in protein production (Wallace *et al.*, unpublished data). The L1 life cycle presumably involves steps that take place in different cellular compartments (nucleus and cytoplasm). Almost nothing is known about the trafficking of L1 mRNA, proteins, and/or RNPs and as a result little is known about targeted intracellular trafficking (Goodier *et al.*, 2004, 2007). While the intermediate steps in L1-mediated integration still remain poorly understood, the involvement of several cellular proteins associated with antiviral defense and DNA repair has been reported. The APOBEC family of proteins that function as RNA editing enzymes exhibit an effect on L1 and Alu retrotransposition that seems to be independent of their enzymatic activity through which they reduce viral RNA (Bogerd *et al.*, 2006; Chiu *et al.*, 2006; Hulme *et al.*, 2007; Muckenfuss *et al.*, 2006; Stenglein and Harris, 2006).

An emerging role of cellular DNA repair machinery in L1 retrotransposition is exciting because it introduces an interesting dilemma into the TE–host relationship. On one hand, DNA repair proteins (ATM) are required for L1 mobilization (Gasior *et al.*, 2006), that is, the host is assisting parasite invasion. On the other hand, ERCC1 protein of the cellular nucleotide excision repair (NER) pathway negatively regulates L1 retrotransposition (Gasior *et al.*, 2008) serving a protective role against L1 assault. Hence the dependence on host proteins for the completion of retrotransposition may serve to provide an important avenue through which the host cell can regulate TE proliferation levels.

Very little is known about posttranscriptional regulation of Alu and SVA elements. In the case of Alu it is mainly due to the technical difficulties associated with the detection of the endogenous Alu transcripts produced through Pol-III vs. Pol-II transcription. Alu inserts are enriched within human genes (Lander et al., 2001), and, as a result, they are transcribed as a part of numerous cellular mRNAs. This makes it extremely challenging to delineate authentic Pol-III Alu-promoted transcripts from Pol-II generated transcripts. It is known that, following transcription, many Alu transcripts appear to be processed into small cytoplasmic (scAlu) transcripts, which, based on genomic sequence data, exhibit greatly reduced retrotransposition efficiency (Shaikh et al., 1997). Although a subset of these scAlus can be attributable to the introduction of cryptic Pol-III termination sites, it is unclear whether the remainder are generated by simple degradation or a more active process (Shaikh et al., 1997). Another reported mechanism that acts downstream of transcription but at the unidentified step in the TE life cycle is down regulation of TE retrotransposition by the APOBEC family of proteins (Chiu et al., 2006; Hulme et al., 2007; Muckenfuss et al., 2006; Stenglein and Harris, 2006). Members of this family of proteins exhibit differential effects on L1 and Alu retrotransposition, highlighting the differences in some yet unknown requirements between mobilizations of the two human retroelements.

The complicated and dynamic nature of the TE–host coexistence is further supported by the observation that artificial upregulation of L1 expression through stable L1 transfection in HeLa cells significantly decreases retrotransposition of both L1 and Alu elements in these cells (Wallace et al., 2010). The phenomenon appears to correlate with the levels of stably expressed L1 ORF2 protein and the presence of the functional L1 EN domain (Wallace et al., 2010). Adaptation of the cellular DNA repair machinery to handle the L1-induced DNA double-strand breaks is implicated as a primary cause of the reduced L1 and Alu retrotransposition in cancer cells overexpressing functional L1. Because of the usage of artificial levels of L1 expression in this experimental system it is not immediately clear what thresh-hold levels of endogenous L1 expression need to be bypassed to institute a similar phenomenon in vivo. Nevertheless, these data strongly suggest that the interplay between TEs and their host is likely an ever-changing process and should be viewed as such, particularly when considering TE expression and activity in cells with genetic defects (Morrish et al., 2002) or cells exposed to environmental stimuli known to affect TE activity (Kale et al., 2005, 2006).

It is obvious from the above examples that keeping TE expression at bay is an important priority for the normal cellular existence. This significance is further supported by the data collected in cultured cells that show that ectopic L1 expression causes significant toxicity in both normal and cancer cells (Belancio et al., 2010b; Gasior et al., 2006; Wallace et al., 2008b), indicating that each cell type may have a threshhold for tolerable L1 expression. This limit may reflect the ability of cells to deal with the TE-induced genomic insult. As a result, unplanned upregulation of endogenous L1 expression may result in elimination of normal—and even some cancerous—cells through apoptosis (Belgnaoui et al., 2006; Bourc'his and Bestor, 2004) or senescence (Belancio et al., 2010b; Wallace et al., 2008b). Thus, the collective TE expression and activity in any given cell type is likely determined by the combination of the factors established to limit TE expression and by the amount of the TE-associated toxicity that can be endured by these cells.

D. Genetic variation and polymorphism of TE loci

Significant progress has been made in understanding the complexity associated with the fact that TEs are families of elements individual members of which are distributed throughout the genome and often harbor significant sequence variations relative to the rest of the group. Several causes are currently known to contribute to this variation. One of them is infidelity of the L1 RT which is estimated to introduce at least one mutation per every 6000 incorporated nucleotides. This corresponds to the size of the human L1. The biological implication of this finding is that every *de novo* full-length L1 insertion generated by the same active L1 locus has a high likelihood of containing at least one point mutation. Another reason for diversity among family members is the natural accumulation of mutations within any given TE locus due to genomic drift.

Other frequent on-arrival or time-inflicted changes include significant alterations in the length of the polyA tail present at the end of each element. Longer A-tails are associated with younger and more active Alu and L1 elements (Lander et al., 2001; Roy-Engel et al., 2002). However, because of the difference in the generation of this A-tail between L1 and Alu elements, the encoded length of the polyA stretch has a particularly dramatic effect on the rate of Alu retrotransposition (Comeaux et al., 2009; Roy-Engel et al., 2002). An A-tail of an individual element can shrink or grow by 10s of nucleotides during, or shortly following retrotransposition. This translates in the several fold difference in its retrotransposition efficiency among individual insertions (Comeaux et al., 2009; Dewannieux et al., 2003; Roy-Engel et al., 2002). The same consequences for Alu retrotransposition can also arise from mutations interrupting the continuity of the A-tail (Comeaux et al., 2009). The variations generated during transposition along with accumulated random mutations influence the relative activity of individual TE loci (Aleman et al., 2000; Bennett et al., 2008). The disparity in the efficiency of retrotransposition is observed not only among the loci of the same TE family but also between the same TE locus isolated from different individuals (Brouha et al., 2003; Lutz et al., 2003; Seleme et al., 2006). Depending on the presence of modifying mutations, the same insertion locus could be highly active in one individual but quiescent in another. This phenomenon was first reported for L1 elements and is likely to be the case for Alu and SVA elements as well.

Regardless of their origin, mutations within TE sequences can have a dual effect on their activity, as already alluded to above regarding A-tail length. On one hand, these mutations can inactivate or reduce the activity of one of the important L1 protein functions required for retrotransposition of L1 or all three human non-LTR retrotransposons. They can also disrupt other critical elements, such as promoter activity, RNA folding, RNA/protein, protein/protein interactions, etc., each one of which may eliminate or diminish retrotransposition. On rare occasions, however, these mutations may potentially lead to a beneficial change in any of the above mentioned activities or interactions leading to an improvement in retrotransposition. L1 elements with highest retrotransposition potential are referred to as "hot" elements (Brouha et al., 2003). "Hot" Alu and SVA elements are likely present in the human genome as well, but are yet to be characterized.

TE loci in any given human genome can be either fixed or polymorphic in the population. Fixed loci are evolutionary older and are present in all members of the population (excluding rare genomic rearrangement events that can remove a TE locus after fixation took place). Because of their evolutionary age they are more likely to harbor inactivating mutations and, as a result, have lost their activity (Aleman et al., 2000; Lander et al., 2001; Moran et al., 1996). The vast majority of TE loci found in the human genome is represented by the fixed-present elements. Polymorphic TE loci are found only in the subset of individuals in the population and their allele frequencies may vary significantly among the populations of different origin (Cordaux et al., 2007; Wang et al., 2005; Watkins et al., 2003). Polymorphic TE loci are likely to be the most active and they are expected to account for the bulk of TE activity in any given genome (Beck et al., 2010; Brouha et al., 2002, 2003; Ewing and Kazazian, 2010; Huang et al., 2010; Iskow et al., 2010). Initial screening suggested that polymorphic L1s are rare and a very small number of L1 loci are "hot" (Brouha et al., 2002, 2003). Advances in the whole genome analysis significantly broadened the estimation of the polymorphic L1 loci per genome. In fact, the most recent estimation is that any two human genomes differ on average by 285 sites with respect to presence or absence of L1 insertion (this includes truncated and full-length L1s; Ewing and Kazazian, 2010). Additionally, over half of the identified polymorphic L1s exhibit levels of activity that can be characterized as "hot" (Beck et al., 2010). The observation that there is most likely a group of polymorphic TE loci that are private, that is unique to individual genomes, further expands the TE-associated genetic diversity between individual human genomes (Beck et al., 2010; Ewing and Kazazian, 2010). There is also a significant increase in the L1-associated variation between normal and some cancer genomes (Iskow et al., 2010). These findings strongly support that there is a considerable interindividual genetic variation due to TE presence or absence and as a consequence, considerable interindividual variation in the combined TE activity. These observations combined with the significant tissue-specific variation in the endogenous L1 expression in somatic human tissues (Belancio *et al.*, 2010b) suggest that TE-associated genetic variation is likely to exist among somatic tissues of the same individual.

IV. GENOMIC INSTABILITY

One of the main conflicts between TEs and their host genomes arises from the fact that multiple aspects of TE activity pose a challenge to the structural integrity of the genome. These include L1-iduced DSBs, L1, Alu, and SVA retrotransposition and retrotransposition-mediated rearrangements (deletions, insertions, inversions; Gilbert *et al.*, 2002, 2005; Moran *et al.*, 1999; Ostertag and Kazazian, 2001), and TE-associated recombination [reviewed in Hedges and Deininger, 2007] (Fig. 6.2).

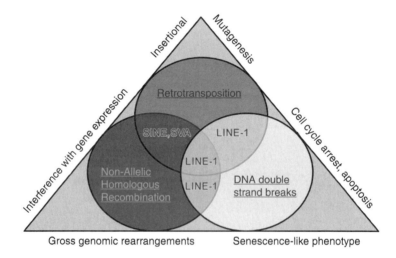

Figure 6.2. L1-associated genomic instability. Human non-LTR transposable elements can trigger a number of events leading to mutagenic alterations in the host genome. Among them are retrotransposition, DNA double-strand breaks (DSBs), and nonallelic homologous recombination (NAHR). Out of the three human non-LTR TEs only L1 elements can induce all of these events. L1, Alu, and SVA retrotransposition causes insertional mutagenesis when the *de novo* integrations reside within or near human genes or functional regulatory elements. Both retrotransposition and NAHR lead to interference with normal gene expression through alterations in the gene regulation and/or processing. NAHR and L1-associated DSBs have a potential to produce gross genomic rearrangements. In addition, L1-induced DSBs are reported to induce cell-cycle arrest, apoptosis, and senescence-like phenotype in normal and cancer cells. (See Color Insert.)

A. DSBs

The discovery of DSB generation by the L1 EN (Belgnaoui *et al.*, 2006; Farkash *et al.*, 2006; Gasior *et al.*, 2006) introduced a possibility of the existence of genomic damage that is likely to impose additional strain on the TE/host relationship. The L1 EN that allows for L1, Alu, and SVA mobilization within the human genome introduces DNA strand breakage which, when not properly resolved, can have deleterious consequences on the host genome (Gasior *et al.*, 2006; Wallace *et al.*, 2008b). These breakages can involve one or both strands of the DNA backbone. While single stranded DNA breaks can typically be repaired by the cell without catastrophic consequences, they can, when in combination with a replication fork or some secondary chemical insult, lead to the generation of DSBs. DSBs are known to have detrimental effects on cells [reviewed in van Gent *et al.*, 2001; Vilenchik and Knudson, 2003]. Consequently, organisms have involved an impressive protein-based surveillance system to detect and mark such breaks for repair (Abraham, 2003; Bakkenist and Kastan, 2003; Lieber, 2010). In humans, both H2AX foci staining and comet assays have been used to demonstrate that L1 EN activity results in the formation of DSBs in both normal and cancer cells (Belancio *et al.*, 2010b; Belgnaoui *et al.*, 2006; Farkash *et al.*, 2006; Gasior *et al.*, 2006). Such breaks were observed to occur in excess of the number of successful insertions (Gasior *et al.*, 2006). The accumulation of DSBs can initiate an arrest of the cell cycle while repair is attempted (Gasior *et al.*, 2006). If the damage is too extensive, apoptosis or senescence will be triggered (Belancio *et al.*, 2010b; Belgnaoui *et al.*, 2006; Gasior *et al.*, 2006; Wallace *et al.*, 2008b). Even though the mutagenic potential of the L1-induced DSBs is not known, they are expected to contribute to genomic instability at the sites of L1 expression that, in addition to germline and tumors (Branciforte and Martin, 1994; Bratthauer and Fanning, 1993; Bratthauer *et al.*, 1994; Martin and Branciforte, 1993; Ostertag *et al.*, 2002), also includes normal human somatic tissues (Belancio *et al.*, 2010b). This assumption is based on the reports that even when dsDNA repair is successful, genomic rearrangements resulting from the repair process can lead to the loss or gain of genomic information and/or the disruption of regulatory signals [reviewed in van Gent *et al.*, 2001; Vilenchik and Knudson, 2003]. Thus, it is likely that L1 EN is responsible for "hidden" L1 damage previously not associated with the activity of these elements. As discussed in more detail in the *TEs and human health* section below, when these effects occur in the germline, they have implications for heritable genetic diseases. When they occur in the soma, there are potential consequences for both cancer and aging.

B. Retrotransposition

While the mutagenic potential of the L1-associated DSBs has not been characterized, the contribution of human TEs to genomic instability through retrotransposition is well established [reviewed in Belancio *et al.*, 2009, 2010a; Cordaux and

Batzer, 2009]. All active human retroelements (L1,Alu, and SVA) contributed with different frequencies to the variety of human disease ranging from hemophilia to cancer [reported human diseases are summarized in Belancio et al., 2008a, 2010a]. Despite the continuously growing list of the TE-induced human diseases, the rate of TE-associated mutagenesis, including retrotransposition and recombination, in humans is likely significantly underestimated due to the difficulty in detection of TE-caused mutations by conventionally used diagnostic methods of screening [reviewed in Belancio et al., 2009]. New studies, however, leveraging 2nd generation sequencing are improving our estimates of baseline transposition rates (Beck et al., 2010; Ewing and Kazazian, 2010; Huang et al., 2010; Iskow et al., 2010).

In addition to insertional mutagenesis resulting in disease, retrotransposition of L1, Alu, and most likely SVA elements can have less drastic phenotypic effects on cell survival or function. Integration of these elements within human intronic regions most of the time does not prevent gene expression, but may reduce cell fitness or function under certain stress conditions or in combination with other genomic alterations [reviewed in Kines and Belancio, 2011]. Because L1 and SVA elements contain functional splice and polyadenylation sites (Belancio et al., 2006, 2008b; Hancks et al., 2009; Perepelitsa-Belancio and Deininger, 2003) and Alu elements are prone to acquiring functional splice sites through random mutations long after the integration process is completed. Intronic integration events are known to interfere with expression of genes in which they integrate (Belancio et al., 2006; Han et al., 2004; Lin et al., 2008; Perepelitsa-Belancio and Deininger, 2003; Sorek et al., 2004; Ustyugova et al., 2006; Wheelan et al., 2005). Full-length L1 insertions in forward orientation have the most pronounced effect on gene expression most likely due to the presence of the functional promoter in addition to the highest content of the splice and polyA sites (Boissinot et al., 2001, 2006; Chen et al., 2006; Ustyugova et al., 2006). Intronic Alu integration events can influence alternative splicing (Sorek et al., 2002) and the presence of inverted Alu sequences within 3'UTRs of genes can alter nuclear retention of mRNAs and strongly represses gene expression (Chen and Carmichael, 2008, 2009; Chen et al., 2008). As discussed below, occasionally, retrotransposition provides genomic rearrangements or cis-acting signals that convey positive benefits to the host organism (Babushok et al., 2007; Xing et al., 2006)

C. Issues in repair and recombination

Another significant contribution of human TEs to genomic instability is through nonallelic homologous recombination (NAHR). Due to their very nature, TEs typically exist in multiple interspersed copies within a given host genome. Depending on their age, the various genomic copies will diverge at the level of nucleotide identity from one another. The longer elements reside in the genome, the more likely that random genetic mutation will level the level of nucleotide identity that they share with their family consensus sequence. The presence of interspersed sequence homology poses a serious challenge for the maintenance of genomic integrity [reviewed in Hedges and Deininger, 2007]; this is particularly the case for higher eukaryotes, such as mammals, that have larger genomes containing higher repetitive element content (Lander *et al.*, 2001). The cellular machinery involved in both recombination and DNA repair can be led astray by homologous sequences present at nonallelic positions (Jasin, 2000). In the case of recombination, this is most often observed in the context of NAHR. The cellular protein machinery that initiates and monitors the strand invasion process cannot always discern interspersed TE sequence homology from truly allelic sequences. As a consequence of NAHR, genetic sequence can be duplicated on one chromosome and lost on the other. In terms of TE/host interactions this result can often have negative fitness consequences due to the loss or alteration of critical genetic information. A significant number of human genetic disorders have resulted from the nonallelic recombination of nearby Alu elements (Callinan and Batzer, 2006; Deininger and Batzer, 1999).

In addition to meiotic recombination, diploid (and higher ploidy) organisms typically rely on the set of homologous chromosomes as templates for the homology-driven (HR) DNA repair. The HR DNA repair system is related to the recombination system and subject to the same limitations in discerning true allelic sequence from interspersed homology. Products of misaligned DNA repair—or repair events in which the only homologous templates available are nearby repetitive sequences—can also generate rearrangements. These rearrangements are, more often than not, intrachromosomal. Nevertheless, this same process can instigate interchromosomal rearrangements (Elliott and Jasin, 2002; Elliott *et al.*, 2005).

While the focus of the deleterious effects of TEs has often centered on the detrimental effects of unchecked insertional mutagenesis, there is evidence that their ability to instigate nonallelic recombination may have an even great impact on organismal short- and long-term fitness. To date, the larger fraction of Alu-related genetic disease has been observed to arise through mutagenic recombination (Callinan and Batzer, 2006; Deininger and Batzer, 1999; Elliott *et al.*, 2005). Song and Boissinot (2006) demonstrated evidence that negative selection in the human genome principally acts upon TEs as a function of TE length. This suggests that mutagenic NAHR, as opposed to insertional mutagenesis, may have the more substantial negative consequences that retrotranspositional activity. Similar evidence for the role of ectopic, nonallelic recombination in determining the fitness consequences of individual TE insertions was found in Drosophila (Petrov *et al.*, 2003). In addition to causing disease, TE-associated reshuffling of the host genetic material has had a significant impact on the host genome architecture during the course of evolution (Han *et al.*, 2005, 2007, 2008; Konkel and Batzer, 2010; Lee *et al.*, 2008; Xing *et al.*, 2009).

The variety of mutations induced by TEs in the human genome argues for the importance of the multitude of defense mechanisms present within human cells that target almost every step of the TE life cycle. It also underscores the importance of the balanced long-term evolutionary existence between the host genome and TEs.

V. IMPACT ON HUMAN HEALTH

L1 has long been regarded as one of the intrinsic factors contributing to genomic instability; however, its effects, such as insertional mutagenesis and NAHR between interspersed L1 or Alu repeats (L1-, Alu-, and SVA-induced diseases are summarized in Belancio et al., 2008a, 2010a), were considered to be restricted to cancer and mendelian genetic disorders, such as hemophilia A & B. While TE-associated risk to human health is accepted, it is extremely challenging to estimate TE contribution to human disease because of the randomness of the TE-associated mutagenesis, the diversity of mutations arising from TE activity, and the lack of adequate high throughput methods for the identification and analysis of these mutations [reviewed in Belancio et al., 2009]. The discovery of the ongoing endogenous L1 expression in human tissues (Asch et al., 1996; Belancio et al., 2010b; Ergun et al., 2004), L1's ability to damage human DNA not only through retrotransposition and recombination but also via induction of DNA DSBs (Belgnaoui et al., 2006; Farkash et al., 2006; Gasior et al., 2006), and L1-induced toxicity (Gasior et al., 2006; Wallace et al., 2008b) provides the basis for reevaluation of the L1 role in human somatic and germline diseases.

Based on the significant estimated variation in the combined L1 activity in any given human genome (Seleme et al., 2006), there is likely a gradient of endogenous L1 activity in the population and as a result, a spectrum of the L1-associated burden imposed on the host genome (Fig. 6.3). This estimated discrepancy in the L1 activity is associated with the presence of polymorphic L1 loci and a significant variation in the activity of the same L1 locus among individuals on account of mutations that modulate their activity (Beck et al., 2010; Brouha et al., 2003; Huang et al., 2010; Iskow et al., 2010; Seleme et al., 2006). Persons lacking polymorphic L1 elements (that are most likely to be the most active L1 loci in the genome) and containing inactive or the least active fixed L1 loci would likely experience the least damage from the endogenous L1 activity (Fig. 6.3). On the other hand, genomes expressing very active and polymorphic L1 loci would probably be exposed to the highest L1 damage. Even though this assumption is relatively straight forward, and it is supported by experimental evidence, the transition from knowing the spectrum of functional L1 loci in any given genome to predicting their impact on human health is not trivial. The difficulty predominantly arises from the existence of numerous defense

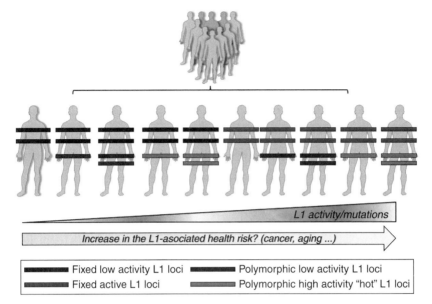

Figure 6.3. Variation in the endogenous L1 activity in the human population. Due to the significant number of the functional L1 loci in any human genome, the variation in the activity of the same L1 locus among individuals, and the presence of polymorphic L1 loci that are likely to be the most active L1s, there as an estimated 300-fold variation in the L1 activity among individuals in the human population. In a simplified model where there is only one fixed and one polymorphic locus, the lowest predicted endogenous L1 activity would be expected in individuals whose genomes harbor fixed L1 loci with low activity (black bars). Individuals heterozygous (one blue/dark gray bar) or homozygous (two blue/dark gray bars) for polymorphic low or high "hot" (green/light gray bars) activity L1 loci are likely to experience a gradual increase in their total endogenous L1 activity. A further augmentation in the endogenous L1 activity is expected in persons that contain fixed active L1 loci (red/white bars) and a combination of these fixed L1s and polymorphic low or high activity L1 loci. This gradient of L1 activity and most likely L1-associated mutagenesis may be one of the contributing factors to the discrepancy in the time of the onset and severity of human diseases associated with genomic instability observed within human population. (See Color Insert.)

mechanisms implemented by the host to combat the insult from TEs (see above sections). A whole new chapter in human TE research has been opened by the discovery of the intimate connection between the cellular DNA repair machinery and L1-associated DNA damage (Gasior et al., 2006; Gasior et al., 2008; Suzuki et al., 2009). Thus, adequate DNA repair in individual genomes is expected to control L1-associated damage. On the other hand, any decline in the efficacy of the relevant DNA repair pathways, whether due to the loss of function or to the decreased activity associated with certain genotypes, would likely lead to an increased rate of accumulation of the L1-induced DNA damage.

The same assumption is true for a long list of cellular processes, ranging from DNA methylation, transcription, and RNA processing to antiviral defense, reported to control various steps of the L1 life cycle. The existence of these additional dimensions influencing potential L1-associated health risk suggests a scenario under which individuals with the same cumulative endogenous L1 activity may endure drastically different burden from the L1-induced DNA damage. On the other hand, persons with relatively low endogenous L1 activity may have comparable load of the L1-associated mutagenesis to individuals with significantly higher endogenous L1 activity depending on the strength of their anti-TE shield. Thus, the consideration of the status of all the players of the TE-controlling network is necessary for the proper assessment of the TE-associated individual health risk and further thorough evaluation of the complex relationship between TEs and their hosts need to take place.

Inherited defects in any of the pathways controlling TE expression and/or activity would result in systemic or tissue- or development-specific increase in the TE-associated damage, while acquired somatic mutations are likely to be limited to the cells in which they occurred. Even though little is known about the combined effect of L1 activity with either somatic or germline mutations, it can potentially be a significant contributor to human diseases. For example, high risk of breast cancer associated with inactivating BRCA1 mutations is proposed to be linked the mutagenic repair of DSBs induced by response to estrogen (Fu et al., 2003). Despite the fact that L1 expression and the role of the L1-induced DSBs in mammary gland tumorigenesis remain unexplored, it may be one of the factors contributing to genomic instability during breast cancer development.

A. The potential for L1 to contribute to cancer and aging

The finding of extensive endogenous L1 expression in normal human somatic tissues combined with the L1's ability to induce DSBs not only in human cancer but also normal cells brings a possibility that ongoing endogenous L1 expression can be a continuous source of DNA damage in somatic tissues (Belancio et al., 2010b). DNA damage is known to promote genomic instability, which is one of the contributing factors of the normal aging process and age-associated diseases particularly cancer (Campisi and Vijg, 2009; Coppede and Migliore, 2010; Erol, 2010). One of the accepted theories of mammalian aging is destabilization of the genome through accumulation of DNA damage over the course of life span of an organism (Alexander, 1967). Low levels of endogenous L1 expression in the majority of human somatic tissues and adult stem cells strongly suggests that ongoing L1 expression can contribute to genomic instability over the life span of an organism through retrotransposition of L1, SINEs, and SVA elements, recombination between interspersed copies of L1 and SINEs, and DNA DSBs induced by L1 EN. Endogenous L1 expression varies significantly among different human tissues and adult stem cells (Belancio *et al.*, 2010b) suggesting tissue-specific variation in the L1-induced DNA damage. Our findings suggest that adult stem cells as well as some human tissues may be protected from the endogenous L1-induced DNA damage relative to other tissues due to the extensive L1 RNA processing that leads to the production of retrotranspositionally defective L1-related mRNAs (Belancio *et al.*, 2010b). Accumulation of the L1-related DNA damage in somatic and adult stem cells with time may result in reduced performance, or, in the case of adult stem cells, in the decreased tissue-renewal capacity and production of differentiated progeny cells that inherit genomic defects accumulated in stem cells (Fig. 6.4). L1-induced DNA damage may also increase with time due to age-related alterations in DNA repair and other cellular pathways that control L1 activity (Barbot *et al.*, 2002; Richardson, 2003; Seluanov *et al.*, 2004; Singhal *et al.*, 1987). Additionally, the significant variation in the endogenous L1 activity estimated to exist in the human population (Seleme *et al.*, 2006; Fig. 6.3) correlates with the remarkable variation in the rate of individual aging, the length of life span, and the onset and severity of the age-associated diseases.

With the exception of an increasing number of evolutionary advantageous examples of acquisition of diverse beneficial functions by TEs within their host genomes, the largely supported view of the TE influence on the genome stability is that of a negative nature. While this view is applicable most of the time to the genome of a single cell, it may only represent one side of the coin when considering the fate of the cells that support TE expression in terms of the overall health of the tissue and organism in which they reside. The discovery of the L1-associated toxicity (Gasior *et al.*, 2006; Wallace *et al.*, 2008b) that manifests itself in the form of apoptosis and cellular senescence in both cancer and normal (Belancio *et al.*, 2010b) human cells suggests that the outcome of TE expression needs to be considered in the context of the specific genomic environment. While TE activity promotes genomic instability through a plethora of mechanisms, TE-associated genomic rearrangements that can lead to the onset or progression of disease may also trigger cell-cycle arrest, followed by elimination of aberrant cell(s) from the population. Based on the reported experimental evidence, deregulated L1 expression, which often happens in response to some external or internal stimuli, in cells with fully or partially functional cell cycle check points can trigger apoptosis or senescence leading to the removal of these cells from further propagation. Interestingly, accumulation of senescent cells in normal tissues with age has been reported (Herbig *et al.*, 2006; Jeyapalan *et al.*, 2007). On the contrary, cells defective in the DNA damage surveillance would escape elimination and continue to exist with accumulated genetic defects. The double-edged-sword hypothesis of L1 expression (Belancio *et al.*, 2010a) implies that L1, through precisely the same types of DNA damage, may play a role in human diseases such as cancer or normal biological processes such as aging.

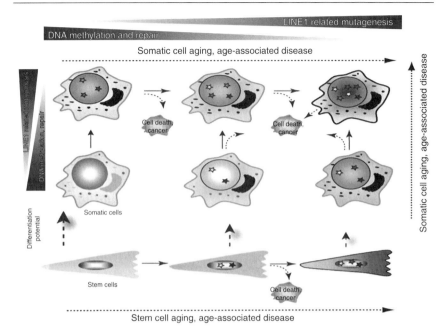

Figure 6.4. Model of L1 involvement in aging and age-associated diseases. A stem cell in the bottom left corner of the diagram represents a young adult stem cell that gives rise to healthy differentiated progeny cells. With time endogenous L1 activity most likely in the form of L1 ORF2 expression from SpORF2 mRNA (Belancio *et al.*, 2010a,b) leads to accumulation of somatic mutations (different color stars in the nucleus) that may result in generation of differentiated progeny cells that carry those mutations, decrease in differentiation potential (vertical dashed arrows), or cell death or malignant transformation. Gradual loss of fitness or ability to perform preprogrammed functions due to accumulation of mutations is reflected by the increased intensity of the gray color in the stem cells with age. At the level of differentiated somatic cells, a healthy replicatively young cell (the very left cell in the middle horizontal row) with time will progress into a cell (top right corner of the diagram) that has accumulated a number of mutations in the cellular DNA due to the endogenous L1 activity via insertional mutagenesis of full-length L1, SINEs, and SVA elements and via integration independent mutagenic activity of L1 and SpORF2 expression through unfaithful repair of the L1-induced DSBs (different color stars in the nucleus). L1 activity in somatic cells may also result in cell death, cellular senescence, or malignant transformation. An acceleration of L1-induced mutagenesis with time may result from either increased L1 expression due to the age-associated hypomethylation of the cellular DNA, decrease in the efficiency of the DNA repair machinery with age, and/or age-associated alterations in the expression and/or function of other cellular factors that may play a role in L1 expression. (See Color Insert.)

B. A positive note

The previous section highlighted several avenues by which TE activity can potentially compromise genomic integrity and, as a consequence, human health. At the same time, it is also safe to say that evolution of humans would have taken a dramatically different course—if it could even have been possible at all—were it not for the periodic "intervention" of TE activity. There are now numerous established examples of TEs being coopted for functional roles across many different taxa. Perhaps most famously, the vertebrate adaptive immune system was made possible, in part, by the invasion of immunoglobulin-coding genes by a RAG transposon (reviewed in Flajnik and Kasahara, 2010). The recombination signal sequences (RSS), introduced into these genes by the transposon conferred the ability for directed somatic recombination, increasing the ability to rapidly generate antibody diversity. A second example is that of the syncytin gene, which is derived from the envelop gene of the human endogenous retrovirus (HERV-W; Mi et al., 2000). Yet another important example of the contribution of TEs to human evolution was the discovery that the SETMAR gene was derived from the fusion of the SET histone methyltransferase gene with a mariner-like Hsmar1 transposon (Cordaux et al., 2006). SETMAR was observed to have inherited several biochemical properties from its transposon source, although its molecular function remains poorly understood (Liu et al., 2007). In addition to the creation of novel genes by the direct contribution of genetic sequence, TEs activity can result in the rearrangement, duplication, and/or reshuffling of existing exons or entire genes (Goodier et al., 2000; Moran et al., 1999; Pickeral et al., 2000). A demonstration of this process in humans was provided by Xing et al. (2006), where the authors observed that three copies of the AMAC gene in primates had been generated by L1 transduction (Xing et al., 2006). Several lines of evidence, including intact ORFs and active expression, suggest these newly emerged copies play a functional role in the genome. These are but a few examples from an increasing number of examples of TE domestication events. Across the entire genome, it has been reported that TE sequences are incorporated in as much as 4% of human protein-coding genes (Kim et al., 2010a,b). As our knowledge of the genome increases, demonstrations of TE contributions to organismal biology will continue to mount. There nevertheless remains a large theoretical and empirical leap to be made between the observation of extensive evidence of TE domestication, to the conclusion that TE lineage activity is maintained by natural selection fulfill such an evolutionary role. Once again, it is important to be reminded of Kidwell and Lisch's observation that the role of TEs in the evolution of its host is not at odds with their having an essentially parasitic relationship to the genome. As discussed below, how precisely to characterize that relationship depends, in part, on empirical data that remains to be collected.

VI. CONCLUSIONS

A. The human model

The human organism, in many respects, serves as an exemplary system within which to consider the interplay between host genomes and their TE inhabitants. As detailed above, we find in our own species clear examples of both the positive and negative attributes of TE activity. We have discussed at length the extensive systems the host organism has evolved to curtail TE activity. This serves a strong indicator that TE activity, left unchecked, presents a significant danger to the host organism. We must, however, consider that the unchecked activity of many "functional" biological processes would also be detrimental to the organism. When such processes are themselves curtailed, we are apt to say they are "kept in balance" or "regulated," as opposed to "defended against." The presence of host repression mechanisms, in and of themselves, is not sufficient to exclude a possible functional role for persistant TE activity, at least among some taxa.

Perhaps what distinguishes TEs from these more essential systems is the ambiguity concerning the extent to which they are components of the "organismal machine," as opposed to separate entities pursuing their own agenda. As Dawkins illustrated in his *The Selfish Gene* (1976), any gene can be conceived of as pursuing its own agenda of propagation within subsequent generation, potentially at the expense of what might appear to be the wellbeing of its host organism. When the gene happens to be fundamental to a species' survival, however, we tend to have a more tolerant view towards this sort of scheming. The ability of some TEs to be horizontally transferred across species boundaries strongly suggests that their presence can be nonessential for organismal function. Presumably the host species receiving the horizontally transferred TE was getting along well enough before the TE arrived on the scene. At the same time, we perceive TEs to be diligently pursuing their own agenda of proliferation, occasionally generating negative consequences for the organism. From this perspective, their behavior is more like a parasite, one that just so happens to reside in our own germline. But, as discussed above, TEs have occasionally made positive functional contributions to organismal biology during evolution. The overall frequency of such contributions, however, remains to be determined. Is it sufficiently high enough to pay the rent, so to speak. For further insight, we can look to the study of the human microbiome. The increasing realization that mammals and other multicellular eukaryotes function as super-organisms, comprised of a number of interacting entities exhibiting various degrees of interdependence. In any such system, the line between self and nonself is not so easily delineated. In the case of the human organism, a spectrum emerges wherein, on the one extreme, there is the mitochondria organelle, an entity fully integrated and essential to organismal survival. On the other side are members of the microbial and viral community that give and take to various extents from the entity as a whole. Where along this gradient TEs will ultimately be placed has yet to be determined, as the relative degree of positive and negative contributions to the human organism remain to be elucidated.

B. The road ahead

The direction of TE research in the next decades, as with most scientific endeavors, will be driven largely by technological and methodological improvements. Figuring heavily in these changes will be the availability of relatively inexpensive whole genome sequencing. Low cost, ubiquitous genomic sequencing will allow for better evaluation of "natural", unmodified, TE activity, decreasing reliance on artificially tagged retrotransposition constructs. While such constructs have been of tremendous importance for mobile element research and will continue to be critical for evaluating certain molecular hypotheses, sequencing-based experiments will be able to address questions surrounding how closely modified TE constructs mirror the biological activity their natural counterparts. We further expect that the sheer volume of sequencing data anticipated over the coming few years will also bring changes to the *focus* of TE research. While novel TE biology no doubt awaits discovery amidst the multitude of genomes that have yet to be sampled, saturation with TE diversity information will likely lead to a deemphasis of TE discovery and annotation as an end-in-itself, and the increasing relegation of identification and taxonomical assignment to automated computational pipelines. Emerging technologies will also allow us to peer more deeply into somatic activity of TEs within different organisms, including the mechanisms of their genetic and epigenetic regulation. As discussed above, new evidence suggesting that TE activity in the soma is higher than once believed opens several avenues for exploration in the arena of impacts on organismal fitness and on human health in particular. The ultimate consequences of somatic TE activity on both cancer and aging is currently unknown, and the elucidation of TE impact in these areas will require detailed analysis across diverse tissues. Arguably, the most important findings of the past several years have revolved around the discovery novel cellular regulatory systems for controlling TE activity. These include the regulation of retrotransposition by posttranscriptional processing of TE expression and by members of the APOBEC protein family, the evidence of small RNA pathways playing a role in repressing mammalian retrotransposons, and the involvement of DNA repair in regulation of TE activity. The presence of these systems across diverse taxa speaks to the threat posed by unchecked TE proliferation. Further understanding of these regulatory mechanisms, and how they may interact with gene regulatory networks, will continue to be fertile areas of investigation. In addition, the ability to survey epigenetic markers at a genomic scale, in multiple tissues and developmental stages, will likely provide novel insights into the relationship between interspersed TEs sequences, chromatin structure, and genetic regulation. We anticipate that, despite the scientific advances that await us, no simplistic, pithy description of the host–element relationship will emerge. At best, we only be able to more deeply appreciate the complex relationship dynamic, where cooperation, antagonism, and serendipity each have their roles.

Acknowledgments

VPB is supported by P20RR020152, NIH/NIA 5K01AG030074-02 and The Ellison Medical Foundation New Scholar in Aging award.

References

Abraham, R. T. (2003). Checkpoint signaling: Epigenetic events sound the DNA strand-breaks alarm to the ATM protein kinase. *Bioessays* **25,** 627–630.

Aleman, C., Roy-Engel, A. M., Shaikh, T. H., and Deininger, P. L. (2000). Cis-acting influences on Alu RNA levels. *Nucleic Acids Res.* **28,** 4755–4761.

Alexander, P. (1967). The role of DNA lesions in the processes leading to aging in mice. *Symp. Soc. Exp. Biol.* **21,** 29–50.

Alisch, R. S., Garcia-Perez, J. L., Muotri, A. R., Gage, F. H., and Moran, J. V. (2006). Unconventional translation of mammalian LINE-1 retrotransposons. *Genes Dev.* **20,** 210–224.

Alves, G., Tatro, A., and Fanning, T. (1996). Differential methylation of human LINE-1 retrotransposons in malignant cells. *Gene* **176,** 39–44.

An, W., Han, J. S., Wheelan, S. J., Davis, E. S., Coombes, C. E., Ye, P., Triplett, C., and Boeke, J. D. (2006). Active retrotransposition by a synthetic L1 element in mice. *Proc. Natl. Acad. Sci. USA* **103,** 18662–18667.

An, W., Han, J. S., Schrum, C. M., Maitra, A., Koentgen, F., and Boeke, J. D. (2008). Conditional activation of a single-copy L1 transgene in mice by Cre. *Genesis* **46,** 373–383.

Athanikar, J. N., Badge, R. M., and Moran, J. V. (2004). A YY1-binding site is required for accurate human LINE-1 transcription initiation. *Nucleic Acids Res.* **32,** 3846–3855.

Asch, A. L., Eliacin, E., Fanning, T. G., Connolly, J. L., Bratthauer, G., and Asch, B. B. (1996). Comparative expression of the LINE-1 p40 protein in human breast carcinomas and normal breast tissues. *Oncol. Res.* **8**(6), 239–247.

Babushok, D. V., Ostertag, E. M., Courtney, C. E., Choi, J. M., and Kazazian, H. H., Jr. (2006). L1 integration in a transgenic mouse model. *Genome Res.* **16,** 240–250.

Babushok, D. V., Ohshima, K., Ostertag, E. M., Chen, X., Wang, Y., Mandal, P. K., Okada, N., Abrams, C. S., and Kazazian, H. H., Jr. (2007). A novel testis ubiquitin-binding protein gene arose by exon shuffling in hominoids. *Genome Res.* **17,** 1129–1138.

Bakkenist, C. J., and Kastan, M. B. (2003). DNA damage activates ATM through intermolecular autophosphorylation and dimer dissociation. *Nature* **421,** 499–506.

Barbot, W., Dupressoir, A., Lazar, V., and Heidmann, T. (2002). Epigenetic regulation of an IAP retrotransposon in the aging mouse: Progressive demethylation and de-silencing of the element by its repetitive induction. *Nucleic Acids Res.* **30,** 2365–2373.

Beck, C. R., Collier, P., Macfarlane, C., Malig, M., Kidd, J. M., Eichler, E. E., Badge, R. M., and Moran, J. V. (2010). LINE-1 retrotransposition activity in human genomes. *Cell* **141,** 1159–1170.

Belancio, V. P., Hedges, D. J., and Deininger, P. (2006). LINE-1 RNA splicing and influences on mammalian gene expression. *Nucleic Acids Res.* **34,** 1512–1521.

Belancio, V. P., Hedges, D. J., and Deininger, P. (2008a). Mammalian non-LTR retrotransposons: For better or worse, in sickness and in health. *Genome Res.* **18,** 343–358.

Belancio, V. P., Roy-Engel, A. M., and Deininger, P. (2008b). The impact of multiple splice sites in human L1 elements. *Gene* **411,** 38–45.

Belancio, V. P., Deininger, P. L., and Roy-Engel, A. M. (2009). LINE dancing in the human genome: Transposable elements and disease. *Genome Med.* **1,** 97.

Belancio, V. P., Roy-Engel, A. M., and Deininger, P. L. (2010a). All y'all need to know 'bout retroelements in cancer. *Semin. Cancer Biol* **20**(4), 200–210.

Belancio, V. P., Roy-Engel, A. M., Pochampally, R. R., and Deininger, P. (2010b). Somatic expression of LINE-1 elements in human tissues. *Nucleic Acids Res.* **38**(12), 3909–3922.

Belgnaoui, S. M., Gosden, R. G., Semmes, O. J., and Haoudi, A. (2006). Human LINE-1 retrotransposon induces DNA damage and apoptosis in cancer cells. *Cancer Cell Int.* **6,** 13.

Bennett, E. A., Keller, H., Mills, R. E., Schmidt, S., Moran, J. V., Weichenrieder, O., and Devine, S. E. (2008). Active Alu retrotransposons in the human genome. *Genome Res.* **18,** 1875–1883.

Bogerd, H. P., Wiegand, H. L., Hulme, A. E., Garcia-Perez, J. L., O'Shea, K. S., Moran, J. V., and Cullen, B. R. (2006). Cellular inhibitors of long interspersed element 1 and Alu retrotransposition. *Proc. Natl. Acad. Sci. USA* **103,** 8780–8785.

Boissinot, S., and Furano, A. V. (2001). Adaptive evolution in LINE-1 retrotransposons. *Mol. Biol. Evol.* **18,** 2186–2194.

Boissinot, S., Entezam, A., and Furano, A. V. (2001). Selection against deleterious LINE-1-containing loci in the human lineage. *Mol. Biol. Evol.* **18,** 926–935.

Boissinot, S., Davis, J., Entezam, A., Petrov, D., and Furano, A. V. (2006). Fitness cost of LINE-1 (L1) activity in humans. *Proc. Natl. Acad. Sci. USA* **103,** 9590–9594.

Bourc'his, D., and Bestor, T. H. (2004). Meiotic catastrophe and retrotransposon reactivation in male germ cells lacking Dnmt3L. *Nature* **431,** 96–99.

Bowen, N. J., and Jordan, I. K. (2007). Exaptation of protein coding sequences from transposable elements. *Genome Dyn.* **3,** 147–162.

Branciforte, D., and Martin, S. L. (1994). Developmental and cell type specificity of LINE-1 expression in mouse testis: Implications for transposition. *Mol. Cell. Biol.* **14,** 2584–2592.

Bratthauer, G. L., and Fanning, T. G. (1992). Active line-1 retrotransposons in human testicular cancer. *Oncogene* **7,** 507–510.

Bratthauer, G. L., and Fanning, T. G. (1993). Line-1 retrotransposon expression in pediatric germ-cell tumors. *Cancer* **71,** 2383–2386.

Bratthauer, G. L., Cardiff, R. D., and Fanning, T. G. (1994). Expression of LINE-1 retrotransposons in human breast cancer. *Cancer* **73,** 2333–2336.

Brookfield, J. F. Y. (2005). The ecology of the genome—Mobile DNA elements and their hosts. *Nat. Rev. Genet.* **6,** 128–136.

Brouha, B., Badge, R., Schustak, J., Moran, J., and Kazazian, H. (2002). Active L1 retrotransposons in the human genome. *Am. J. Hum. Genet.* **71,** 410.

Brouha, B., Schustak, J., Badge, R. M., Lutz-Prigge, S., Farley, A. H., Moran, J. V., and Kazazian, H. H., Jr. (2003). Hot L1s account for the bulk of retrotransposition in the human population. *Proc. Natl. Acad. Sci. USA* **100,** 5280–5285.

Callinan, P. A., and Batzer, M. A. (2006). Retrotransposable elements and human disease. *Genome Dyn.* **1,** 104–115.

Campisi, J., and Vijg, J. (2009). Does damage to DNA and other macromolecules play a role in aging? If so, how? *J. Gerontol. A Biol Sci. Med. Sci.* **64,** 175–178.

Cantrell, M. A., Scott, L., Brown, C. J., Martinez, A. R., and Wichman, H. A. (2008). Loss of LINE-1 activity in the megabats. *Genetics* **178,** 393–404.

Casavant, N. C., Scott, L., Cantrell, M. A., Wiggins, L. E., Baker, R. J., and Wichman, H. A. (2000). The end of the LINE?: Lack of recent L1 activity in a group of South American rodents. *Genetics* **154,** 1809–1817.

Chen, L. L., and Carmichael, G. G. (2008). Gene regulation by SINES and inosines: Biological consequences of A-to-I editing of Alu element inverted repeats. *Cell Cycle* **7,** 3294–3301.

Chen, L. L., and Carmichael, G. G. (2009). Altered nuclear retention of mRNAs containing inverted repeats in human embryonic stem cells: Functional role of a nuclear noncoding RNA. *Mol. Cell* **35,** 467–478.

Chen, J., Rattner, A., and Nathans, J. (2006). Effects of L1 retrotransposon insertion on transcript processing, localization and accumulation: Lessons from the retinal degeneration 7 mouse and implications for the genomic ecology of L1 elements. *Hum. Mol. Genet.* **15,** 2146–2156.

Chen, L. L., DeCerbo, J. N., and Carmichael, G. G. (2008). Alu element-mediated gene silencing. *EMBO J.* **27,** 1694–1705.

Chiu, Y. L., Witkowska, H. E., Hall, S. C., Santiago, M., Soros, V. B., Esnault, C., Heidmann, T., and Greene, W. C. (2006). High-molecular-mass APOBEC3G complexes restrict Alu retrotransposition. *Proc. Natl. Acad. Sci. USA* **103,** 15588–15593.

Christensen, S. M., and Eickbush, T. H. (2005). R2 target-primed reverse transcription: Ordered cleavage and polymerization steps by protein subunits asymmetrically bound to the target DNA. *Mol. Cell. Biol.* **25,** 6617–6628.

Clements, A. P., and Singer, M. F. (1998). The human LINE-1 reverse transcriptase: Effect of deletions outside the common reverse transcriptase domain. *Nucleic Acids Res.* **26,** 3528–3535.

Cohen, C. J., Lock, W. M., and Mager, D. L. (2009). Endogenous retroviral LTRs as promoters for human genes: A critical assessment. *Gene* **448,** 105–114.

Comeaux, M. S., Roy-Engel, A. M., Hedges, D. J., and Deininger, P. L. (2009). Diverse cis factors controlling Alu retrotransposition: What causes Alu elements to die? *Genome Res.* **19,** 545–555.

Coppede, F., and Migliore, L. (2010). DNA repair in premature aging disorders and neurodegeneration. *Curr. Aging Sci.* **3,** 3–19.

Cordaux, R., and Batzer, M. A. (2009). The impact of retrotransposons on human genome evolution. *Nat. Rev. Genet.* **10,** 691–703.

Cordaux, R., Udit, S., Batzer, M. A., and Feschotte, C. (2006). Birth of a chimeric primate gene by capture of the transposase gene from a mobile element. *Proc. Natl. Acad. Sci. USA* **103,** 8101–8106.

Cordaux, R., Srikanta, D., Lee, J., Stoneking, M., and Batzer, M. A. (2007). In search of polymorphic Alu insertions with restricted geographic distributions. *Genomics* **90,** 154–158.

Cost, G. J., and Boeke, J. D. (1998). Targeting of human retrotransposon integration is directed by the specificity of the L1 endonuclease for regions of unusual DNA structure. *Biochemistry* **37,** 18081–18093.

Cost, G. J., Feng, Q., Jacquier, A., and Boeke, J. D. (2002). Human L1 element target-primed reverse transcription in vitro. *EMBO J.* **21,** 5899–5910.

Coufal, N. G., Garcia-Perez, J. L., Peng, G. E., Yeo, G. W., Mu, Y., Lovci, M. T., Morell, M., O'Shea, K. S., Moran, J. V., and Gage, F. H. (2009). L1 retrotransposition in human neural progenitor cells. *Nature* **460,** 1127–1131.

Damert, A., Raiz, J., Horn, A. V., Lower, J., Wang, H., Xing, J., Batzer, M. A., Lower, R., and Schumann, G. G. (2009). 5′-Transducing SVA retrotransposon groups spread efficiently throughout the human genome. *Genome Res.* **19,** 1992–2008.

Deininger, P. L., and Batzer, M. A. (1999). Alu repeats and human disease. *Mol. Genet. Metab.* **67,** 183–193.

Dewannieux, M., Esnault, C., and Heidmann, T. (2003). LINE-mediated retrotransposition of marked Alu sequences. *Nat. Genet.* **35**, 41–48.

Dmitriev, S. E., Andreev, D. E., Terenin, I. M., Olovnikov, I. A., Prassolov, V. S., Merrick, W. C., and Shatsky, I. N. (2007). Efficient translation initiation directed by the 900-nucleotide-long and GC-rich 5′ untranslated region of the human retrotransposon LINE-1 mRNA is strictly cap dependent rather than internal ribosome entry site mediated. *Mol. Cell. Biol.* **27**, 4685–4697.

Dooner, H. K., and Weil, C. F. (2007). Give-and-take: Interactions between DNA transposons and their host plant genomes. *Curr. Opin. Genet. Dev.* **17**, 486–492.

Edgar, R. C., and Myers, E. W. (2005). PILER: Identification and classification of genomic repeats. *Bioinformatics* **21**(Suppl. 1), i152–i158.

Elliott, B., and Jasin, M. (2002). Double-strand breaks and translocations in cancer. *Cell. Mol. Life Sci.* **59**, 373–385.

Elliott, B., Richardson, C., and Jasin, M. (2005). Chromosomal translocation mechanisms at intronic alu elements in mammalian cells. *Mol. Cell* **17**, 885–894.

Ergun, S., Buschmann, C., Heukeshoven, J., Dammann, K., Schnieders, F., Lauke, H., Chalajour, F., Kilic, N., Stratling, W. H., and Schumann, G. G. (2004). Cell type-specific expression of LINE-1 open reading frames 1 and 2 in fetal and adult human tissues. *J. Biol. Chem.* **279**, 27753–27763.

Erol, A. (2010). Systemic DNA damage response and metabolic syndrome as a premalignant state. *Curr. Mol. Med.* **10**, 321–334.

Ewing, A. D., and Kazazian, H. H., Jr. (2010). High-throughput sequencing reveals extensive variation in human-specific L1 content in individual human genomes. *Genome Res.*

Farkash, E. A., Kao, G. D., Horman, S. R., and Prak, E. T. (2006). Gamma radiation increases endonuclease-dependent L1 retrotransposition in a cultured cell assay. *Nucleic Acids Res.* **34**, 1196–1204.

Feng, Q., Moran, J. V., Kazazian, H. H., Jr., and Boeke, J. D. (1996). Human L1 retrotransposon encodes a conserved endonuclease required for retrotransposition. *Cell* **87**, 905–916.

Feschotte, C., and Pritham, E. J. (2007). DNA transposons and the evolution of eukaryotic genomes. *Annu. Rev. Genet.* **41**, 331–368.

Finnegan, D. J. (1989). Eukaryotic transposable elements and genome evolution. *Trends Genet.* **5**, 103–107.

Flajnik, M. F., and Kasahara, M. (2010). Origin and evolution of the adaptive immune system: Genetic events and selective pressures. *Nat. Rev. Genet.* **11**, 47–59.

Florl, A. R., Lower, R., Schmitz-Drager, B. J., and Schulz, W. A. (1999). DNA methylation and expression of LINE-1 and HERV-K provirus sequences in urothelial and renal cell carcinomas. *Br. J. Cancer* **80**, 1312–1321.

Fu, Y. P., Yu, J. C., Cheng, T. C., Lou, M. A., Hsu, G. C., Wu, C. Y., Chen, S. T., Wu, H. S., Wu, P. E., and Shen, C. Y. (2003). Breast cancer risk associated with genotypic polymorphism of the nonhomologous end-joining genes: A multigenic study on cancer susceptibility. *Cancer Res.* **63**, 2440–2446.

Fuhrman, S. A., Deininger, P. L., LaPorte, P., Friedmann, T., and Geiduschek, E. P. (1981). Analysis of transcription of the human Alu family ubiquitous repeating element by eukaryotic RNA polymerase III. *Nucleic Acids Res.* **9**, 6439–6456.

Furano, A. V., Duvernell, D. D., and Boissinot, S. (2004). L1 (LINE-1) retrotransposon diversity differs dramatically between mammals and fish. *Trends Genet.* **20**, 9–14.

Gasior, S. L., Wakeman, T. P., Xu, B., and Deininger, P. L. (2006). The human LINE-1 retrotransposon creates DNA double-strand breaks. *J. Mol. Biol.* **357**, 1383–1393.

Gasior, S. L., Preston, G., Hedges, D. J., Gilbert, N., Moran, J. V., and Deininger, P. L. (2007). Characterization of pre-insertion loci of de novo L1 insertions. *Gene* **390**, 190–198.

Gasior, S. L., Roy-Engel, A. M., and Deininger, P. L. (2008). ERCC1/XPF limits L1 retrotransposition. *DNA Repair (Amst.)* **7**, 983–989.

Gilbert, N., Lutz-Prigge, S., and Moran, J. V. (2002). Genomic deletions created upon LINE-1 retrotransposition. *Cell* **110**, 315–325.

Gilbert, N., Lutz, S., Morrish, T. A., and Moran, J. V. (2005). Multiple fates of L1 retrotransposition intermediates in cultured human cells. *Mol. Cell. Biol.* **25**, 7780–7795.

Goodier, J. L., Ostertag, E. M., and Kazazian, H. H., Jr. (2000). Transduction of 3′-flanking sequences is common in L1 retrotransposition. *Hum. Mol. Genet.* **9**, 653–657.

Goodier, J. L., Ostertag, E. M., Engleka, K. A., Seleme, M. C., and Kazazian, H. H., Jr. (2004). A potential role for the nucleolus in L1 retrotransposition. *Hum. Mol. Genet.* **13**, 1041–1048.

Goodier, J. L., Zhang, L., Vetter, M. R., and Kazazian, H. H., Jr. (2007). LINE-1 ORF1 protein localizes in stress granules with other RNA-binding proteins, including components of RNA interference RNA-induced silencing complex. *Mol. Cell. Biol.* **27**, 6469–6483.

Han, J. S., and Boeke, J. D. (2004). A highly active synthetic mammalian retrotransposon. *Nature* **429**, 314–318.

Han, J. S., Szak, S. T., and Boeke, J. D. (2004). Transcriptional disruption by the L1 retrotransposon and implications for mammalian transcriptomes. *Nature* **429**, 268–274.

Han, K., Sen, S. K., Wang, J., Callinan, P. A., Lee, J., Cordaux, R., Liang, P., and Batzer, M. A. (2005). Genomic rearrangements by LINE-1 insertion-mediated deletion in the human and chimpanzee lineages. *Nucleic Acids Res.* **33**, 4040–4052.

Han, K., Lee, J., Meyer, T. J., Wang, J., Sen, S. K., Srikanta, D., Liang, P., and Batzer, M. A. (2007). Alu recombination-mediated structural deletions in the chimpanzee genome. *PLoS Genet.* **3**, 1939–1949.

Han, K., Lee, J., Meyer, T. J., Remedios, P., Goodwin, L., and Batzer, M. A. (2008). L1 recombination-associated deletions generate human genomic variation. *Proc. Natl. Acad. Sci. USA* **105**, 19366–19371.

Hancks, D. C., Ewing, A. D., Chen, J. E., Tokunaga, K., and Kazazian, H. H., Jr. (2009). Exon-trapping mediated by the human retrotransposon SVA. *Genome Res.* **19**, 1983–1991.

Hata, K., and Sakaki, Y. (1997). Identification of critical CpG sites for repression of L1 transcription by DNA methylation. *Gene* **189**, 227–234.

Hedges, D. J., and Deininger, P. L. (2007). Inviting instability: Transposable elements, double-strand breaks, and the maintenance of genome integrity. *Mutat. Res.* **616**, 46–59.

Hellmann-Blumberg, U., Hintz, M. F., Gatewood, J. M., and Schmid, C. W. (1993). Developmental differences in methylation of human Alu repeats. *Mol. Cell. Biol.* **13**, 4523–4530.

Herbig, U., Ferreira, M., Condel, L., Carey, D., and Sedivy, J. M. (2006). Cellular senescence in aging primates. *Science* **311**, 1257.

Huang, C. R., Schneider, A. M., Lu, Y., Niranjan, T., Shen, P., Robinson, M. A., Steranka, J. P., Valle, D., Civin, C. I., Wang, T., et al. (2010). Mobile interspersed repeats are major structural variants in the human genome. *Cell* **141**, 1171–1182.

Hulme, A. E., Bogerd, H. P., Cullen, B. R., and Moran, J. V. (2007). Selective inhibition of Alu retrotransposition by APOBEC3G. *Gene* **390**, 199–205.

Iskow, R. C., McCabe, M. T., Mills, R. E., Torene, S., Pittard, W. S., Neuwald, A. F., Van Meir, E. G., Vertino, P. M., and Devine, S. E. (2010). Natural mutagenesis of human genomes by endogenous retrotransposons. *Cell* **141**, 1253–1261.

Jasin, M. (2000). Chromosome breaks and genomic instability. *Cancer Invest.* **18**, 78–86.

Jeyapalan, J. C., Ferreira, M., Sedivy, J. M., and Herbig, U. (2007). Accumulation of senescent cells in mitotic tissue of aging primates. *Mech. Ageing Dev.* **128**, 36–44.

Jurka, J. (1997). Sequence patterns indicate an enzymatic involvement in integration of mammalian retroposons. *Proc. Natl. Acad. Sci. USA* **94**, 1872–1877.

Jurka, J., Klonowski, P., Dagman, V., and Pelton, P. (1996). CENSOR–a program for identification and elimination of repetitive elements from DNA sequences. *Comput. Chem.* **20**, 119–121.

Jurka, J., Kapitonov, V. V., Pavlicek, A., Klonowski, P., Kohany, O., and Walichiewicz, J. (2005). Repbase Update, a database of eukaryotic repetitive elements. *Cytogenet. Genome Res.* **110,** 462–467.

Kale, S. P., Moore, L., Deininger, P. L., and Roy-Engel, A. M. (2005). Heavy metals stimulate human LINE-1 retrotransposition. *Int. J. Environ. Res. Public Health* **2,** 14–23.

Kale, S. P., Carmichael, M. C., Harris, K., and Roy-Engel, A. M. (2006). The L1 retrotranspositional stimulation by particulate and soluble cadmium exposure is independent of the generation of DNA breaks. *Int. J. Environ. Res. Public Health* **3,** 121–128.

Kano, H., Godoy, I., Courtney, C., Vetter, M. R., Gerton, G. L., Ostertag, E. M., and Kazazian, H. H., Jr. (2009). L1 retrotransposition occurs mainly in embryogenesis and creates somatic mosaicism. *Genes Dev.* **23,** 1303–1312.

Kidwell, M. G., and Lisch, D. (1997). Transposable elements as sources of variation in animals and plants. *Proc. Natl. Acad. Sci. USA* **94,** 7704–7711.

Kim, D. S., Huh, J. W., Kim, Y. H., Park, S. J., and Chang, K. T. (2010a). Functional impact of transposable elements using bioinformatic analysis and a comparative genomic approach. *Mol. Cells* **30,** 77–87.

Kim, D. S., Huh, J. W., Kim, Y. H., Park, S. J., Kim, H. S., and Chang, K. T. (2010b). Bioinformatic analysis of TE-spliced new exons within human, mouse and zebrafish genomes. *Genomics* **96**(5), 266–271.

Kines, K. J., and Belancio, V. P. (2011). Expressing genes do not forget their LINEs; interference of transposable elements with mammalian gene expression. *Front. Biosci.* (in press).

Kolosha, V. O., and Martin, S. L. (1997). In vitro properties of the first ORF protein from mouse LINE-1 support its role in ribonucleoprotein particle formation during retrotransposition. *Proc. Natl. Acad. Sci. USA* **94,** 10155–10160.

Kolosha, V. O., and Martin, S. L. (2003). High-affinity, non-sequence-specific RNA binding by the open reading frame 1 (ORF1) protein from long interspersed nuclear element 1 (LINE-1). *J. Biol. Chem.* **278,** 8112–8117.

Konkel, M. K., and Batzer, M. A. (2010). A mobile threat to genome stability: The impact of non-LTR retrotransposons upon the human genome. *Semin. Cancer Biol.* **20**(4), 211–221.

Kordis, D., Lovsin, N., and Gubensek, F. (2006). Phylogenomic analysis of the L1 retrotransposons in Deuterostomia. *Syst. Biol.* **55,** 886–901.

Kroutter, E. N., Belancio, V. P., Wagstaff, B. J., and Roy-Engel, A. M. (2009). The RNA polymerase dictates ORF1 requirement and timing of LINE and SINE retrotransposition. *PLoS Genet.* **5,** e1000458.

Kubo, S., Seleme, M. C., Soifer, H. S., Perez, J. L., Moran, J. V., Kazazian, H. H., Jr., and Kasahara, N. (2006). L1 retrotransposition in nondividing and primary human somatic cells. *Proc. Natl. Acad. Sci. USA* **103,** 8036–8041.

Kulpa, D. A., and Moran, J. V. (2005). Ribonucleoprotein particle formation is necessary but not sufficient for LINE-1 retrotransposition. *Hum. Mol. Genet.* **14,** 3237–3248.

Kuramochi-Miyagawa, S., Watanabe, T., Gotoh, K., Totoki, Y., Toyoda, A., Ikawa, M., Asada, N., Kojima, K., Yamaguchi, Y., Ijiri, T. W., *et al.* (2008). DNA methylation of retrotransposon genes is regulated by Piwi family members MILI and MIWI2 in murine fetal testes. *Genes Dev.* **22,** 908–917.

Lander, E. S., Linton, L. M., Birren, B., Nusbaum, C., Zody, M. C., Baldwin, J., Devon, K., Dewar, K., Doyle, M., FitzHugh, W., *et al.* (2001). Initial sequencing and analysis of the human genome. *Nature* **409,** 860–921.

Lavie, L., Maldener, E., Brouha, B., Meese, E. U., and Mayer, J. (2004). The human L1 promoter: Variable transcription initiation sites and a major impact of upstream flanking sequence on promoter activity. *Genome Res.* **14,** 2253–2260.

Lee, J., Han, K., Meyer, T. J., Kim, H. S., and Batzer, M. A. (2008). Chromosomal inversions between human and chimpanzee lineages caused by retrotransposons. *PLoS ONE* **3**, e4047.

Lees-Murdock, D. J., De Felici, M., and Walsh, C. P. (2003). Methylation dynamics of repetitive DNA elements in the mouse germ cell lineage. *Genomics* **82**, 230–237.

Leibold, D. M., Swergold, G. D., Singer, M. F., Thayer, R. E., Dombroski, B. A., and Fanning, T. G. (1990). Translation of LINE-1 DNA elements in vitro and in human cells. *Proc. Natl. Acad. Sci. USA* **87**, 6990–6994.

Lev-Maor, G., Sorek, R., Shomron, N., and Ast, G. (2003). The birth of an alternatively spliced exon: 3′ splice-site selection in Alu exons. *Science* **300**, 1288–1291.

Li, R., Ye, J., Li, S., Wang, J., Han, Y., Ye, C., Wang, J., Yang, H., Yu, J., Wong, G. K., *et al.* (2005). ReAS: Recovery of ancestral sequences for transposable elements from the unassembled reads of a whole genome shotgun. *PLoS Comput. Biol.* **1**, e43.

Li, P. W., Li, J., Timmerman, S. L., Krushel, L. A., and Martin, S. L. (2006). The dicistronic RNA from the mouse LINE-1 retrotransposon contains an internal ribosome entry site upstream of each ORF: Implications for retrotransposition. *Nucleic Acids Res.* **34**, 853–864.

Lieber, M. R. (2010). The mechanism of double-strand DNA break repair by the nonhomologous DNA end-joining pathway. *Annu. Rev. Biochem.* **79**, 181–211.

Lin, L., Shen, S., Tye, A., Cai, J. J., Jiang, P., Davidson, B. L., and Xing, Y. (2008). Diverse splicing patterns of exonized Alu elements in human tissues. *PLoS Genet.* **4**, e1000225.

Lin, C., Yang, L., Tanasa, B., Hutt, K., Ju, B. G., Ohgi, K., Zhang, J., Rose, D. W., Fu, X. D., Glass, C. K., *et al.* (2009). Nuclear receptor-induced chromosomal proximity and DNA breaks underlie specific translocations in cancer. *Cell* **139**, 1069–1083.

Liu, D., Bischerour, J., Siddique, A., Buisine, N., Bigot, Y., and Chalmers, R. (2007). The human SETMAR protein preserves most of the activities of the ancestral Hsmar1 transposase. *Mol. Cell. Biol.* **27**, 1125–1132.

Luan, D. D., Korman, M. H., Jakubczak, J. L., and Eickbush, T. H. (1993). Reverse transcription of R2Bm RNA is primed by a nick at the chromosomal target site: A mechanism for non-LTR retrotransposition. *Cell* **72**, 595–605.

Lutz, S. M., Vincent, B. J., Kazazian, H. H., Jr., Batzer, M. A., and Moran, J. V. (2003). Allelic heterogeneity in LINE-1 retrotransposition activity. *Am. J. Hum. Genet.* **73**, 1431–1437.

Lynch, M. (2007). The Origins of Genome Architecture. Sinauer Associates, Inc., Sunderland, MA.

Maksakova, I. A., Romanish, M. T., Gagnier, L., Dunn, C. A., van de Lagemaat, L. N., and Mager, D. L. (2006). Retroviral elements and their hosts: Insertional mutagenesis in the mouse germ line. *PLoS Genet.* **2**, e2.

Martin, S. L. (1991). Ribonucleoprotein particles with LINE-1 RNA in mouse embryonal carcinoma cells. *Mol. Cell. Biol.* **11**, 4804–4807.

Martin, S. L., and Branciforte, D. (1993). Synchronous expression of LINE-1 RNA and protein in mouse embryonal carcinoma cells. *Mol. Cell. Biol.* **13**, 5383–5392.

Martin, S. L., and Bushman, F. D. (2001). Nucleic acid chaperone activity of the ORF1 protein from the mouse LINE-1 retrotransposon. *Mol. Cell. Biol.* **21**, 467–475.

Martin, S. L., Li, J., Epperson, L. E., and Lieberman, B. (1998). Functional reverse transcriptases encoded by A-type mouse LINE-1: Defining the minimal domain by deletion analysis. *Gene* **215**, 69–75.

Martin, S. L., Li, J., and Weisz, J. A. (2000). Deletion analysis defines distinct functional domains for protein-protein and nucleic acid interactions in the ORF1 protein of mouse LINE-1. *J. Mol. Biol.* **304**, 11–20.

Martin, S. L., Cruceanu, M., Branciforte, D., Wai-Lun, L. P., Kwok, S. C., Hodges, R. S., and Williams, M. C. (2005). LINE-1 retrotransposition requires the nucleic acid chaperone activity of the ORF1 protein. *J. Mol. Biol.* **348**, 549–561.

Mathias, S. L., Scott, A. F., Kazazian, H. H., Jr., Boeke, J. D., and Gabriel, A. (1991). Reverse transcriptase encoded by a human transposable element. *Science* **254,** 1808–1810.

McCLINTOCK, B. (1950). The origin and behavior of mutable loci in maize. *Proc. Natl. Acad. Sci. USA* **36,** 344–355.

McMillan, J. P., and Singer, M. F. (1993). Translation of the human LINE-1 element, L1Hs. *Proc. Natl. Acad. Sci. USA* **90,** 11533–11537.

Mi, S., Lee, X., Li, X., Veldman, G. M., Finnerty, H., Racie, L., LaVallie, E., Tang, X. Y., Edouard, P., Howes, S., *et al.* (2000). Syncytin is a captive retroviral envelope protein involved in human placental morphogenesis. *Nature* **403,** 785–789.

Morales, J. F., Snow, E. T., and Murnane, J. P. (2002). Environmental factors affecting transcription of the human L1 retrotransposon. I. Steroid hormone-like agents. *Mutagenesis* **17,** 193–200.

Moran, J. V., Holmes, S. E., Naas, T. P., DeBerardinis, R. J., Boeke, J. D., and Kazazian, H. H., Jr. (1996). High frequency retrotransposition in cultured mammalian cells. *Cell* **87,** 917–927.

Moran, J. V., DeBerardinis, R. J., and Kazazian, H. H., Jr. (1999). Exon shuffling by L1 retrotransposition. *Science* **283,** 1530–1534.

Morgante, M., Brunner, S., Pea, G., Fengler, K., Zuccolo, A., and Rafalski, A. (2005). Gene duplication and exon shuffling by helitron-like transposons generate intraspecies diversity in maize. *Nat. Genet.* **37,** 997–1002.

Morrish, T. A., Gilbert, N., Myers, J. S., Vincent, B. J., Stamato, T. D., Taccioli, G. E., Batzer, M. A., and Moran, J. V. (2002). DNA repair mediated by endonuclease-independent LINE-1 retrotransposition. *Nat. Genet.* **31,** 159–165.

Muckenfuss, H., Hamdorf, M., Held, U., Perkovic, M., Lower, J., Cichutek, K., Flory, E., Schumann, G. G., and Munk, C. (2006). APOBEC3 Proteins Inhibit Human LINE-1 Retrotransposition. *J. Biol. Chem.* **281,** 22161–22172.

Muotri, A. R., Chu, V. T., Marchetto, M. C., Deng, W., Moran, J. V., and Gage, F. H. (2005). Somatic mosaicism in neuronal precursor cells mediated by L1 retrotransposition. *Nature* **435,** 903–910.

Ohshima, K., Hattori, M., Yada, T., Gojobori, T., Sakaki, Y., and Okada, N. (2003). Whole-genome screening indicates a possible burst of formation of processed pseudogenes and Alu repeats by particular L1 subfamilies in ancestral primates. *Genome Biol.* **4,** R74.

Ostertag, E. M., and Kazazian, H. H. (2001). Twin priming: A proposed mechanism for the creation of inversions in L1 retrotransposition. *Genome Res.* **11,** 2059–2065.

Ostertag, E. M., DeBerardinis, R. J., Goodier, J. L., Zhang, Y., Yang, N., Gerton, G. L., and Kazazian, H. H. (2002). A mouse model of human L1 retrotransposition. *Nat. Genet.* **32,** 655–660.

Ostertag, E. M., Goodier, J. L., Zhang, Y., and Kazazian, H. H., Jr. (2003). SVA elements are nonautonomous retrotransposons that cause disease in humans. *Am. J. Hum. Genet.* **73,** 1444–1451.

Pace, J. K., and Feschotte, C. (2007). The evolutionary history of human DNA transposons: Evidence for intense activity in the primate lineage. *Genome Res.* **17,** 422–432.

Penzkofer, T., Dandekar, T., and Zemojtel, T. (2005). L1Base: From functional annotation to prediction of active LINE-1 elements. *Nucleic Acids Res.* **33,** D498–D500.

Perepelitsa-Belancio, V., and Deininger, P. (2003). RNA truncation by premature polyadenylation attenuates human mobile element activity. *Nat. Genet.* **35,** 363–366.

Petrov, D. A., Aminetzach, Y. T., Davis, J. C., Bensasson, D., and Hirsh, A. E. (2003). Size matters: Non-LTR retrotransposable elements and ectopic recombination in Drosophila. *Mol. Biol. Evol.* **20,** 880–892.

Pickeral, O. K., Makalowski, W., Boguski, M. S., and Boeke, J. D. (2000). Frequent human genomic DNA transduction driven by LINE-1 retrotransposition. *Genome Res.* **10,** 411–415.

Piskareva, O., and Schmatchenko, V. (2006). DNA polymerization by the reverse transcriptase of the human L1 retrotransposon on its own template in vitro. *FEBS Lett.* **580,** 661–668.

Price, A. L., Jones, N. C., and Pevzner, P. A. (2005). De novo identification of repeat families in large genomes. *Bioinformatics* **21**(Suppl. 1), i351–i358.

Pritham, E. J., and Feschotte, C. (2007). Massive amplification of rolling-circle transposons in the lineage of the bat Myotis lucifugus. *Proc. Natl. Acad. Sci. USA* **104,** 1895–1900.

Quesneville, H., Bergman, C. M., Andrieu, O., Autard, D., Nouaud, D., Ashburner, M., and Anxolabehere, D. (2005). Combined evidence annotation of transposable elements in genome sequences. *PLoS Comput. Biol.* **1,** 166–175.

Richardson, B. (2003). Impact of aging on DNA methylation. *Ageing Res. Rev.* **2,** 245–261.

Roy, A. M., West, N. C., Rao, A., Adhikari, P., Aleman, C., Barnes, A. P., and Deininger, P. L. (2000). Upstream flanking sequences and transcription of SINEs. *J. Mol. Biol.* **302,** 17–25.

Roy-Engel, A. M., Salem, A. H., Oyeniran, O. O., Deininger, L., Hedges, D. J., Kilroy, G. E., Batzer, M. A., and Deininger, P. L. (2002). Active Alu element "A-tails": Size does matter. *Genome Res.* **12,** 1333–1344.

Seleme, M. C., Vetter, M. R., Cordaux, R., Bastone, L., Batzer, M. A., and Kazazian, H. H., Jr. (2006). Extensive individual variation in L1 retrotransposition capability contributes to human genetic diversity. *Proc. Natl. Acad. Sci. USA* **103,** 6611–6616.

Seluanov, A., Mittelman, D., Pereira-Smith, O. M., Wilson, J. H., and Gorbunova, V. (2004). DNA end joining becomes less efficient and more error-prone during cellular senescence. *Proc. Natl. Acad. Sci. USA* **101,** 7624–7629.

Shaikh, T. H., Roy, A. M., Kim, J., Batzer, M. A., and Deininger, P. L. (1997). cDNAs derived from primary and small cytoplasmic Alu (scAlu) transcripts. *J. Mol. Biol.* **271,** 222–234.

Singhal, R. P., Mays-Hoopes, L. L., and Eichhorn, G. L. (1987). DNA methylation in aging of mice. *Mech. Ageing Dev.* **41,** 199–210.

Song, M., and Boissinot, S. (2006). Selection against LINE-1 retrotransposons results principally from their ability to mediate ectopic recombination. *Gene.* **390**(1–2), 206–213.

Sorek, R., Ast, G., and Graur, D. (2002). Alu-containing exons are alternatively spliced. *Genome Res.* **12,** 1060–1067.

Sorek, R., Lev-Maor, G., Reznik, M., Dagan, T., Belinky, F., Graur, D., and Ast, G. (2004). Minimal conditions for exonization of intronic sequences: 5′ splice site formation in alu exons. *Mol. Cell* **14,** 221–231.

Speek, M. (2001). Antisense promoter of human L1 retrotransposon drives transcription of adjacent cellular genes. *Mol. Cell. Biol.* **21,** 1973–1985.

Stenglein, M. D., and Harris, R. S. (2006). APOBEC3B and APOBEC3F inhibit L1 retrotransposition by a DNA deamination-independent mechanism. *J. Biol. Chem.* **281,** 16837–16841.

Suzuki, J., Yamaguchi, K., Kajikawa, M., Ichiyanagi, K., Adachi, N., Koyama, H., Takeda, S., and Okada, N. (2009). Genetic evidence that the non-homologous end-joining repair pathway is involved in LINE retrotransposition. *PLoS Genet.* **5,** e1000461.

Sverdlov, E. D. (2000). Retroviruses and primate evolution. *Bioessays* **22,** 161–171.

Swergold, G. D. (1990). Identification, characterization, and cell specificity of a human LINE-1 promoter. *Mol. Cell. Biol.* **10,** 6718–6729.

Symer, D. E., Connelly, C., Szak, S. T., Caputo, E. M., Cost, G. J., Parmigiani, G., and Boeke, J. D. (2002). Human l1 retrotransposition is associated with genetic instability in vivo. *Cell* **110,** 327–338.

Szak, S. T., Pickeral, O. K., Makalowski, W., Boguski, M. S., Landsman, D., and Boeke, J. D. (2002). Molecular archeology of L1 insertions in the human genome. *Genome Biol.* **3,** (research0052).

Takai, D., Yagi, Y., Habib, N., Sugimura, T., and Ushijima, T. (2000). Hypomethylation of LINE1 retrotransposon in human hepatocellular carcinomas, but not in surrounding liver cirrhosis. *Jpn. J. Clin. Oncol.* **30,** 306–309.

Tchenio, T., Casella, J. F., and Heidmann, T. (2000). Members of the SRY family regulate the human LINE retrotransposons. *Nucleic Acids Res.* **28,** 411–415.

Thayer, R. E., Singer, M. F., and Fanning, T. (1993). Undermethylation of specific LINE-1 sequences in human cells producing a LINE-1-encoded protein. *Gene* **133,** 273–277.

Tsutsumi, Y. (2000). Hypomethylation of the retrotransposon LINE-1 in malignancy. *Jpn. J. Clin. Oncol.* **30,** 289–290.

Turner, G., Barbulescu, M., Su, M., Jensen-Seaman, M. I., Kidd, K. K., and Lenz, J. (2001). Insertional polymorphisms of full-length endogenous retroviruses in humans. *Curr. Biol.* **11,** 1531–1535.

Ustyugova, S. V., Lebedev, Y. B., and Sverdlov, E. D. (2006). Long L1 insertions in human gene introns specifically reduce the content of corresponding primary transcripts. *Genetica* **128,** 261–272.

Van Duyne, G. (2002). Mobile DNA II. *In* A structural view of tyrosine recombinase site-specific recombination (N. L. Craig, R. Craigie, M. Gellert, and A. M. Lambowitz, eds.). ASM Press, Washington, DC.

van Gent, D. C., Hoeijmakers, J. H., and Kanaar, R. (2001). Chromosomal stability and the DNA double-stranded break connection. *Nat. Rev. Genet.* **2,** 196–206.

Vilenchik, M. M., and Knudson, A. G. (2003). Endogenous DNA double-strand breaks: Production, fidelity of repair, and induction of cancer. *Proc. Natl. Acad. Sci. USA* **100,** 12871–12876.

Volfovsky, N., Haas, B. J., and Salzberg, S. L. (2001). A clustering method for repeat analysis in DNA sequences. *Genome Biol.* **2,** (RESEARCH0027).

Wallace, N., Wagstaff, B. J., Deininger, P. L., and Roy-Engel, A. M. (2008a). LINE-1 ORF1 protein enhances Alu SINE retrotransposition. *Gene* **419,** 1–6.

Wallace, N. A., Belancio, V. P., and Deininger, P. L. (2008b). L1 mobile element expression causes multiple types of toxicity. *Gene* **419,** 75–81.

Wallace, N. A., Belancio, V. P., Faber, Z., and Deininger, P. (2010). Feedback inhibition of L1 and Alu retrotransposition through altered double strand break repair kinetics. *Mob. DNA* **1,** 22.

Wang, H., Xing, J., Grover, D., Hedges, D. J., Han, K., Walker, J. A., and Batzer, M. A. (2005). SVA elements: A hominid-specific retroposon family. *J. Mol. Biol.* **354,** 994–1007.

Wang, J., Song, L., Grover, D., Azrak, S., Batzer, M. A., and Liang, P. (2006). dbRIP: A highly integrated database of retrotransposon insertion polymorphisms in humans. *Hum. Mutat.* **27,** 323–329.

Watkins, W. S., Rogers, A. R., Ostler, C. T., Wooding, S., Bamshad, M. J., Brassington, A. M., Carroll, M. L., Nguyen, S. V., Walker, J. A., Prasad, B. V., *et al.* (2003). Genetic variation among world populations: Inferences from 100 Alu insertion polymorphisms. *Genome Res.* **13,** 1607–1618.

Wei, W., Gilbert, N., Ooi, S. L., Lawler, J. F., Ostertag, E. M., Kazazian, H. H., Boeke, J. D., and Moran, J. V. (2001). Human L1 retrotransposition: cis preference versus trans complementation. *Mol. Cell. Biol.* **21,** 1429–1439.

Wheelan, S. J., Aizawa, Y., Han, J. S., and Boeke, J. D. (2005). Gene-breaking: A new paradigm for human retrotransposon-mediated gene evolution. *Genome Res.* **15,** 1073–1078.

Wicker, T., Sabot, F., Hua-Van, A., Bennetzen, J. L., Capy, P., Chalhoub, B., Flavell, A., Leroy, P., Morgante, M., Panaud, O., *et al.* (2007). A unified classification system for eukaryotic transposable elements. *Nat. Rev. Genet.* **8,** 973–982.

Xing, J., Wang, H., Belancio, V. P., Cordaux, R., Deininger, P. L., and Batzer, M. A. (2006). Emergence of primate genes by retrotransposon-mediated sequence transduction. *Proc. Natl. Acad. Sci. USA* **103,** 17608–17613.

Xing, J., Zhang, Y., Han, K., Salem, A. H., Sen, S. K., Huff, C. D., Zhou, Q., Kirkness, E. F., Levy, S., Batzer, M. A., *et al.* (2009). Mobile elements create structural variation: Analysis of a complete human genome. *Genome Res.* **19**(9), 1516–1526.

Xiong, Y., and Eickbush, T. H. (1988). Similarity of reverse transcriptase-like sequences of viruses, transposable elements, and mitochondrial introns. *Mol. Biol. Evol.* **5,** 675–690.

Xiong, Y., and Eickbush, T. H. (1990). Origin and evolution of retroelements based upon their reverse transcriptase sequences. *EMBO J.* **9,** 3353–3362.

Yang, N., Zhang, L., Zhang, Y., and Kazazian, H. H. (2003). An important role for RUNX3 in human L1 transcription and retrotransposition. *Nucleic Acids Res.* **31,** 4929–4940.

Yoder, J. A., Walsh, C. P., and Bestor, T. H. (1997). Cytosine methylation and the ecology of intragenomic parasites. *Trends Genet.* **13,** 335–340.

Index

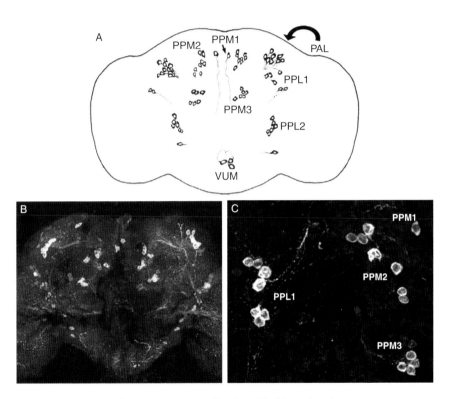

Chapter 1, Figure 1.1 (See Page 10 of this volume).

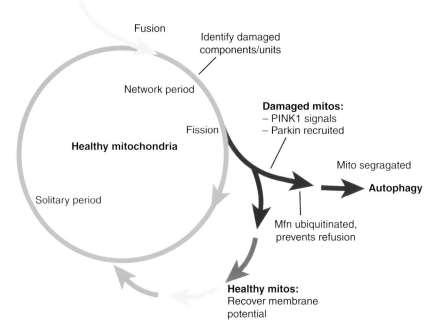

Chapter 1, Figure 1.2 (See Page 34 of this volume).

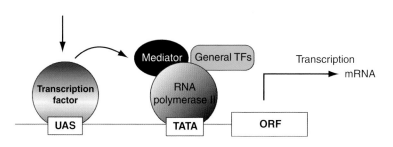

Chapter 2, Figure 2.1 (See Page 54 of this volume).

Budding yeast

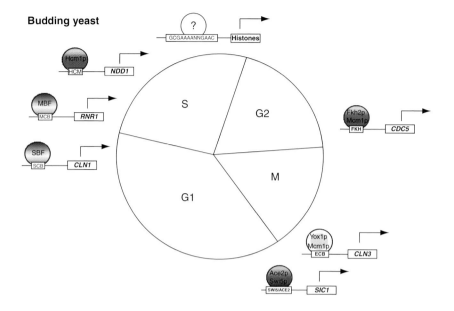

Fission yeast

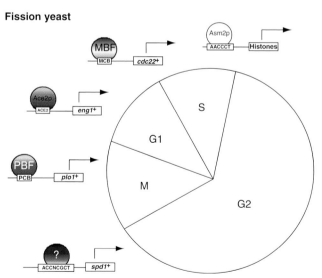

Chapter 2, Figure 2.2 (See Page 57 of this volume).

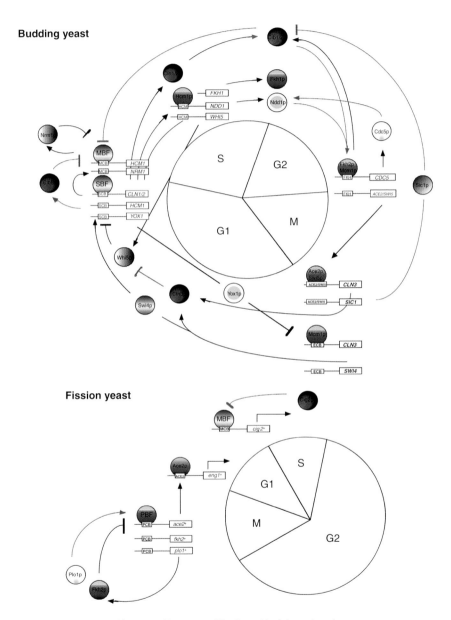

Chapter 2, Figure 2.3 (See Page 66 of this volume).

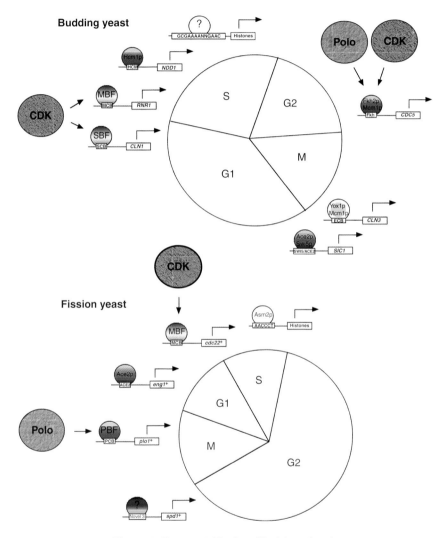

Chapter 2, Figure 2.4 (See Page 73 of this volume).

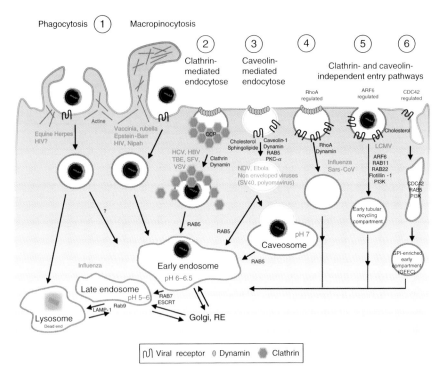

Chapter 4, Figure 4.1 (See Page 128 of this volume).

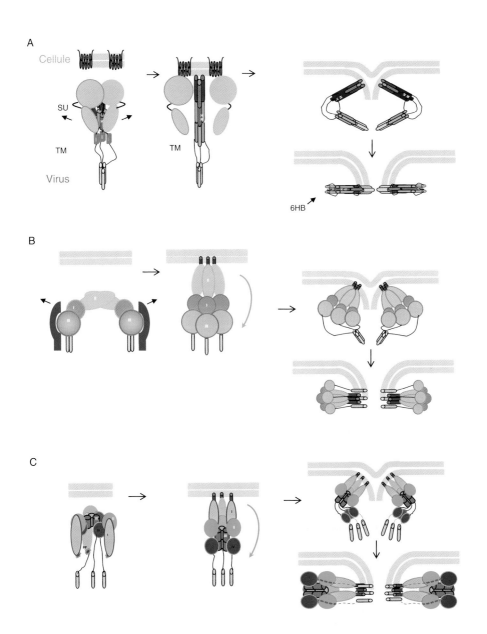

Chapter 4, Figure 4.2 (See Page 136 of this volume).

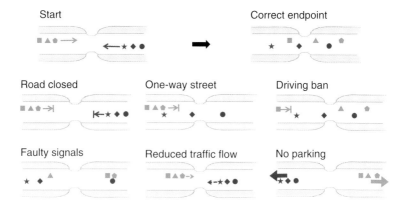

Chapter 5, Figure 5.1 (See Page 189 of this volume).

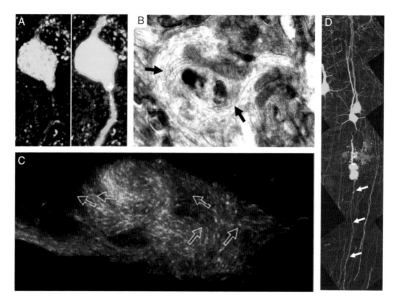

Chapter 5, Figure 5.2 (See Page 192 of this volume).

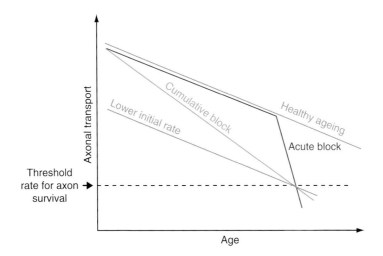

Chapter 5, Figure 5.3 (See Page 193 of this volume).

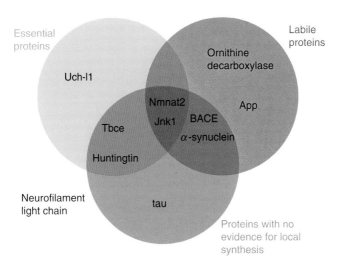

Chapter 5, Figure 5.4 (See Page 203 of this volume).

A

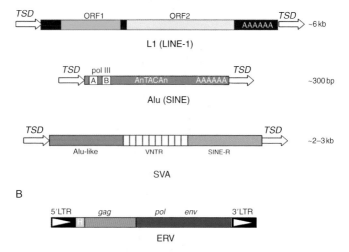

L1 (LINE-1)   ~6 kb

Alu (SINE)   ~300 bp

SVA   ~2–3 kb

B

ERV

Chapter 6, Figure 6.1 (See Page 223 of this volume).

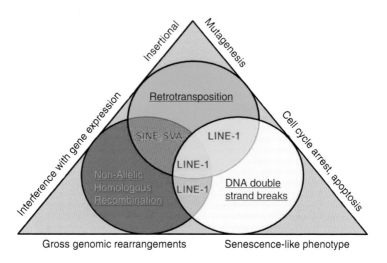

Chapter 6, Figure 6.2 (See Page 240 of this volume).

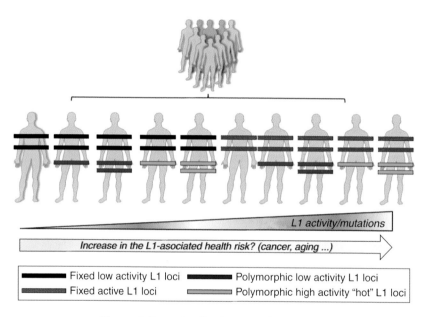

Chapter 6, Figure 6.3 (See Page 245 of this volume).

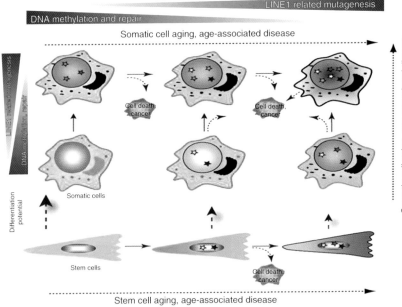

Chapter 6, Figure 6.4 (See Page 248 of this volume).